Prentice-Hall Foundations of Modern Organic Chemistry Series

KENNETH L. RINEHART, JR., Editor

Volumes published or in preparation

N. L. Allinger and J. Allinger	STRUCTURES OF ORGANIC MOLECULES (1965)
Trahanovsky	FUNCTIONAL GROUPS IN ORGANIC COMPOUNDS (1971)
Stewart	THE INVESTIGATION OF ORGANIC REACTIONS (1966)
Saunders	IONIC ALIPHATIC REACTIONS (1965)
Gutsche	THE CHEMISTRY OF CARBONYL COMPOUNDS (1967)
Pryor	INTRODUCTION TO FREE RADICAL CHEMISTRY (1966)
Stock	AROMATIC SUBSTITUTION REACTIONS (1968)
Rinehart	OXIDATION AND REDUCTION OF ORGANIC COMPOUNDS (1966)
DePuy and Chapman	MOLECULAR REACTIONS AND PHOTOCHEMISTRY (1972)
Ireland	ORGANIC SYNTHESIS (1969)
Dyer	APPLICATIONS OF ABSORPTION SPECTROSCOPY OF ORGANIC COMPOUNDS (1965)
Bates and Schaefer	RESEARCH TECHNIQUES IN ORGANIC CHEMISTRY (1971)
Taylor	HETEROCYCLIC COMPOUNDS
Hill	COMPOUNDS OF NATURE
Barker	ORGANIC CHEMISTRY OF BIOLOGICAL COMPOUNDS (1971)
Stille	INDUSTRIAL ORGANIC CHEMISTRY (1968)
Rinehart et al.	X-RAY CRYSTALLOGRAPHY, MASS SPECTROMETRY AND ELECTRON SPIN RESONANCE OF ORGANIC COMPOUNDS
Battiste	NON-BENZENOID AROMATIC COMPOUNDS
Dyer	ORGANIC SPECTRAL PROBLEMS (1972)

ORGANIC SPECTRAL PROBLEMS

John R. Dyer

Department of Chemistry
Georgia Institute of Technology

PRENTICE-HALL, INC., ENGLEWOOD CLIFFS, NEW JERSEY

Printed in the United States of America
ISBN: 0-13-640862-1

PRENTICE-HALL INTERNATIONAL, INC., London
PRENTICE-HALL OF AUSTRALIA PTY. LTD., Sydney
PRENTICE-HALL OF CANADA, LTD., Toronto
PRENTICE-HALL OF INDIA (PRIVATE) LTD., New Delhi
PRENTICE-HALL OF JAPAN, INC., Tokyo

10 9 8 7 6 5 4 3 2 1

Contents

Preface

Within the last decade, the availability of modern instrumentation has redefined the role of the organic chemist involved with any problem of synthetic or natural product structure elucidation. Whether involved with natural or synthetic products, or even many aspects of physical organic chemistry, the advent of easily used nuclear magnetic resonance and mass spectrometers, coupled with better ultraviolet and infrared spectrophotometers, has greatly eased the solution of many problems that arise in organic chemistry. Interestingly, these techniques also provide the methods for studying problems that could not previously even be considered. These methods have become so very powerful that most of the classical methods of organic analysis—determination of molecular weights, elemental composition, functional groups, etc.—have practically become obsolete. It is with this trend in mind that this book of problems was designed. The problems consist of spectral data *only* for 95 organic compounds.

Currently available spectral results are frequently easier to obtain and usually more reliable than classical functional group analyses. Although one recognizes that chemical results are still the foundation of structure elucidation for complicated structures, one must also realize that the use and knowledge of the continually changing modern spectral techniques must not be ignored. These techniques cannot only assist in the elucidation of complicated structures, they may be sufficient for simple structures. In the opinion of the author, the best method to teach interpretation of spectral data is to provide problems of graded difficulty that involve *no* chemistry. This book is designed to promote that philosophy.

This book is intended to be used both by undergraduate students and by beginning graduate students. For each problem, photographic reductions of the infrared, nuclear magnetic resonance, and mass spectra are given; the ultraviolet spectral data are stated, and the mass spectral data are given in tabular form. The problems are arranged approximately in order of increasing difficulty and are divided into three groups.

The problems of the first group should be easily solved by students having a basic knowledge of the four spectral techniques now usually standard in modern undergraduate textbooks. Although the problems of the first group do not require a sophisticated evaluation of the spectral data a unique solution may oblige the reader to evaluate all the spectral data in combination. All of the compounds in the first group show a molecular ion in the mass spectrum of the compound. The first four problems of this group are provided with solutions involving reasonable deductive analysis. These should be examined carefully before proceeding with additional problems.

The problems of the second group require a somewhat better understanding of the interpretation of the spectral data, and may be a challenge to the intellect of the undergraduate student. Again, the first four problems of section two are provided with solutions that involve somewhat more sophisticated interpretation of the spectral data provided.

Some of the problems of group three may be difficult even for graduate students. For groups two and three, a knowledge of interpretation of spectra similar to that found in *Applications of Absorption Spectroscopy of Organic Compounds*, by John R. Dyer, Prentice-Hall, Inc., 1965 and *Interpretation of Mass Spectra*, by Fred W. McLafferty, W. A. Benjamin, Inc., 1966 are sufficient.

The compounds chosen as problems for this book were selected to represent a balanced variety of functional groups and elemental composition. In addition to carbon, hydrogen, or oxygen, the compounds may contain nitrogen, sulfur, or any of the halogens. The compounds all have molecular weights of 350 or less. They were purified by crystallization or distillation and had melting or boiling points consistent with the literature values.

I am very happy to thank those students and faculty of the Georgia Institute of Technology who solved and commented upon the problems in this book. This greatly assisted the arrangement of the problems in the order of increasing difficulty. I also wish to thank Gerald O'Brien, George Turner, and especially Robert M. Calhoun for their expert assistance in purifying the compounds and obtaining the spectral data.

John R. Dyer
Atlanta, Georgia

Recording Conditions

Every effort has been made to obtain spectra of the highest quality possible.

Infrared spectra were recorded using a Perkin-Elmer model 457 grating instrument. The time for scanning the spectrum, 2.5-40 μ (4000-250 cm^{-1}), was six minutes. Sodium chloride plates were used when spectra of liquid films were recorded; equivalent sodium chloride plates were placed in the reference beam. Sodium chloride cells (0.2 mm thickness) were used when spectra of solutions were determined. The figure legend of every spectrum defines the method of determination.

When possible, ultraviolet spectral data were obtained from Vols. I-IV of *Organic Electronic Spectral Data*, Interscience Publishers, 1952-1963. Otherwise, the data were obtained using a Cary model 14 spectrophotometer. In all cases, the solvent used for the determination is stated with the data. The symbol (sh) after a stated wavelength denotes an absorption shoulder rather than a maximum.

Nuclear magnetic resonance spectra were recorded at 60 MHz using a Varian model A-60 D instrument. Tetramethylsilane (TMS) was used as the internal reference when organic solvents were used; the methyl groups of sodium 2,2-dimethyl-2-silapentanesulfonate (DSS) served as the internal reference when deuterium oxide was the solvent. The 500 sec sweep time was used to record the spectrum; the 100 sec sweep time was used to record the integral tracing. Either the 500 Hz or the 1000 Hz sweep width was used to record the normal spectrum. When expansions of certain regions of the spectrum are given, the sweep width (in Hz) and sweep offset (in Hz) are inserted beneath the expansion. The sweep width, solvent, and concentration are stated in the figure legend for each spectrum.

The mass spectra were recorded using a Varian model M-66 spectrometer. The conditions used to record the spectrum were: electron beam, 70 eV; pressure range, 7×10^{-7} to 2×10^{-6} torr. The sweep time for the flat-bed recorder was 2.5 min for 100 amu. Sweep widths of 10-110, 110-210, 210-310, and 310-410 amu were used. Immediately after the flat-bed record of the mass spectrum was obtained, the mass spectrum was recorded using a Varian Statos recorder at intensity ratios of 1, 4,

8, and 32 using a sweep time of 10 amu/sec. Thus, the relative abundances of small ions were determined, relative to the assigned abundance of the base peak as 100.0, to the nearest 0.2, and frequently to 0.1 abundance unit.

For a number of reasons, the abundances of ions in the low mass region (especially under m/e 50) are not exactly reproducible. These ions are more reproducible and more accurately determined when faster scanning speeds are used. For this reason, the mass spectrum recorded using the flat-bed recorder should be used primarily for the initial inspection of the mass spectrum and to note the differences in mass units between the major ions. The tabulated mass spectral data should be used to examine more closely the abundances of the ions and to calculate a molecular formula of a particular ion.

Organic Spectral Problems

Introduction: An Approach to Solving Spectral Problems

Every person who attempts to determine an organic structure that uniquely satisfies the spectral data given in these problems—IR, UV, NMR, and MS—will undoubtedly develop his own approach to working them. This section briefly outlines a suggested approach to solving these problems. It should be anticipated that the approach may qualitatively vary from problem to problem.

It must be remembered that in addition to carbon, hydrogen, and oxygen, the compounds may contain sulfur, nitrogen, or any of the halogens. Also, all of the problems of group one show a molecular ion in the MS, and all of the compounds in this book have molecular weights less than 350.

First, glance briefly at the IR spectrum and attempt to recognize the presence of any of the easily recognizable functional groups in the region 2.5-6.5 μ (4000-1538 cm^{-1}). For all spectral techniques, it should be remembered that the power of *negative* evidence is greater than that of positive evidence; any functional group that is (by correlation tables) supposed to absorb strongly in a particular region will normally do so. The *absence* of such absorption is *powerful evidence* for the absence of such a functional group. Functional groups that are reported to exhibit only weak absorption in a particular region may or may not be observable in that region, and their presence or absence should be treated initially with question. The tentative identification of a functional group in this IR region ($-OH$, $>NH$, $-CHO$, $-CO_2H$, $-\overset{|}{\underset{|}{C}}-H$, $-C{\equiv}C-$, $>C{=}C<$, $-C{\equiv}N$, $>C{=}O$, or others) should be remembered when examining the other spectral data. It will be necessary to give more attention to the IR spectrum later.

Next, examine the UV spectral data. If no absorption is present, a wide variety of functional groups may be tentatively *excluded* from being present in the compound. If the compound does display a UV absorption maximum, its position, intensity, and complexity should be noted. Along with functional groups suspected (present or absent) from the IR spectrum, those suspected from the UV spectral data should be remembered when analyzing *all* of the spectral data. It will occasionally be necessary to return to the UV spectral data and analyze it more carefully.

Then, analyze the NMR spectrum as is best possible to a first approximation. Remember to check whether the sweep width is 500 or 1000 Hz. Here, it is strongly desirable to examine carefully the relative number of hydrogens present in the absorption, as determined from the integration, the positions of the absorptions, and the multiplicities, whether simple (singlet, doublet, triplet, quartet, etc.) or complicated (broad, complex, etc.). The magnitude of J, the coupling constant, should be noted and compared with those from correlation tables. For this analysis, it may be desirable to construct a table such as the following:

τ	*multiplicity*	*J, Hz*	*integration squares*	*relative number of hydrogens*

From such an analysis of the NMR spectrum, it will frequently be possible to write suspected part structures for the compound such as $CH_3-\overset{|}{\underset{|}{C}}-$, CH_3-O-, $CH_3CH_2\overset{|}{\underset{|}{C}}-$, CH_3CH_2O-, $CH_2{=}CH-$, $CH_3-C{\overset{\displaystyle O}{\lessgtr}}$, C_6H_5-, $-C_6H_4-$, $-CHO$, $-CO_2H$, etc. Such part structures that may be suspected should be remembered in the further analysis of the spectral data available. It should be noted that only the *proton* magnetic resonance spectra are reproduced.

Of the other elements that may be present, only fluorine and nitrogen may have an effect on the proton NMR spectrum. Fluorine has a spin of 1/2, and if a compound contains fluorine, nearby hydrogens will couple with that nucleus in the usual manner; H–F coupling constants are significantly larger than H–H coupling constants. Because of the electric quadrupole moment of nitrogen, which has a spin number of 1, the proton NMR absorption of hydrogens attached to that nucleus are generally broadened. Under certain circumstances, a hydrogen bound to nitrogen may couple with that nucleus. Also, in the absence of rapid chemical exchange of >N–H hydrogens, protons adjacent (e.g. –C–H) may be coupled with that hydrogen.

Frequently, a considerable amount of attention should be devoted to the examination of the mass spectrum and the tabulated MS data. As has been indicated in the section "recording conditions," the relative intensities of the ions in the mass spectrum are more accurate in tabulated form than the record of the flatbed recorder. Probably the most important data that can be deduced from the accumulated suspicions derived from examination of the IR, UV, and NMR spectra, combined with a close examination of the MS data, is a proposed molecular formula for the compound. Since the compounds given may contain the elements H, C, N, O, S, F, Cl, Br, and/or I, particular attention should be given to the abundance of M + 1, M + 2, and other M + n ions resulting from the natural isotopic abundance to be anticipated. Tables that present theoretical relative abundances of these common isotopes and ions containing chlorine and/or bromine atoms (halogen clusters) are given inside the back cover of this book.

It is frequently reasonable to assume that an ion in the high mass region may be the molecular ion (molecular weight). Occasionally, a particular compound will not produce a molecular ion, or will produce one of very small intensity. If the molecular ion is fairly intense, however, it may be possible to calculate the molecular formula of the compound (also using data deduced or presumed from the other spectral methods).

A possible molecular formula is calculated by examining the relative abundances of a major ion (M) in the mass spectrum and those ions (M + 1, M + 2, etc.) immediately following. These following ions result mainly from the natural isotopic distributions of the elements in organic compounds (except fluorine and iodine, which are monoisotopic). For example, since ^{13}C has a natural distribution of 1.12% relative to that of ^{12}C at 100.00%, an ion of mass M will be followed by an ion at M + 1 having a theoretical abundance of 1.12% of M *per carbon atom*. Thus, benzene shows a molecular ion at *m/e* 78, and in theory, should show an ion at *m/e* 79 that has a relative abundance of at least 6.72% (6 × 1.12) of the abundance of the ion at *m/e* 78.

The major ions in the mass spectrum should be examined to determine if a formula can be deduced. Thus, ions at M and M + 2 of nearly equal abundance suggest that the ions contain one bromine atom. An M + 2 peak with a relative abundance of about 4.4% suggests that this and the M ion contain one sulfur atom. Examination of the table of isotopic abundances inside the back cover of this book should allow the arithmetic calculation of the theoretical abundances of isotopic ions relative to any ion M. At one glance, one may be able to affirm or exclude the presence of ions containing bromine and/or chlorine atoms (see the table also for the relative abundances of ions in "halogen clusters").

Because of the experimental errors caused by the use of different instruments, and the reactions that occur within the mass spectrometer, the abundances of the ions that follow (M + 1, M + 2, etc.) a particular ion (M^+) are not exactly those values that would be calculated from the natural isotopic abundance tables. An error of several percent should be allowed for the observed abundance of these ions. As an example, for benzene (M^+ = 78), the theoretical abundance of the ion at *m/e* 79 (resulting from the natural abundances of ^{13}C) has been quoted to be 6.72%; if a reasonable error of ± 5% is assumed, the experimentally observed abundance could range from 6.38% to 7.06%. Such a range of observed abundances should be considered in evaluating the importance of these ions.

In addition to an analysis of the fragmentation patterns occurring in the mass spectrum of an organic compound, the estimation of the formula of a given ion is very important. For such an analysis, regardless of the abundance observed for a given ion, the ion (M) must be assigned a relative abundance of 100.0. The ions that follow (M + 1, M + 2, etc.) then have to be normalized to the abundance assigned to M as 100.0. It is important that the reader fully understands the simple arithmetic involved in these situations. Although the following two examples are the *reverse* of the procedure you would follow in an analysis of the formula of a particular ion—these analyses produce theoretical rather than experimental abundances of the ions—they are examples of the methods to be used. Theoretical contributions to M^+ (the molecular ion) and the following ions from the relative natural abundances of the isotopes present are given for the following two compounds. Because of the very small abundances of the isotopes ^2H and ^{17}O, these values are not ordinarily used in the calculations.

2-Nitrothiophene, (thiophene ring, S, NO_2), $C_4H_3NO_2S$. [The molecular ion should occur at *m/e* 129: $^{12}C_4{}^1H_3{}^{14}N^{16}O_2{}^{32}S$.]

m/e	M^+	$^{13}C_4$	^{15}N	$^{16}O_2$	^{32}S	Total
129	100.0					100.0
130		$\frac{4 \times 100.0 \times 1.12}{100} = 4.48$	$\frac{1 \times 100.0 \times 0.37}{100} = 0.37$		$\frac{1 \times 100.0 \times 0.80}{100} = 0.80$	5.65
131				$\frac{2 \times 100.0 \times 0.20}{100} = 0.40$	$\frac{1 \times 100.0 \times 4.44}{100} = 4.44$	4.84

If the formula of this compound were not known, the abundance of the ion at *m/e* 130 (presumably mainly from the natural abundance of ^{13}C) would indicate that the compound (assumed molecular weight 129) could contain a maximum of five $\left(\frac{5.65}{1.12} = 5.04\right)$ carbon atoms. The abundance of the ion at *m/e* 131 should also be carefully examined. If 129 is the molecular weight of the compound, the abundance of the ion at *m/e* 131 strongly suggests the presence of one sulfur atom (in the table of natural abundances, the only isotopic abundance close to the value of 4.84% is ^{34}S, which has a natural abundance of 4.44%). If this assumption is correct, one must also consider the natural abundance of ^{33}S, which is 0.80%. This value should then be subtracted from the abundance of the ion *m/e* 130 so that the suspected contribution to *m/e* 130 from ^{13}C would be 4.85% (5.65 − 0.80). This value then suggests that the compound could contain a maximum of four carbon atoms $\left(\frac{4.85}{1.12} = 4.33\right)$.

From such simple arithmetic, a tentative deduction that the compound contains four carbon atoms and one sulfur atom may be made. The reverse of this procedure may be applied to estimate the formula of any abundant ion that occurs in the mass spectrum. The "extra" abundance (theoretically) of the ions at *m/e* 130 and 131 suggest the presence of a nitrogen atom and two oxygen atoms, but these abundances are too small to be conclusive. More conclusive evidence could be deduced from the assumed molecular weight of the compound and the other spectral data available.

Bromoacetyl chloride, $BrCH_2COCl$, C_2H_2BrClO. [The molecular ion should occur at *m/e* 156: $^{12}C_2{}^1H_2{}^{79}Br^{35}Cl^{16}O$.]

Again, if the formula of this compound is not known, assuming that a molecular ion is present, the complexity of the abundant ions in the region *m/e* 156 to 162 strongly suggests the

m/e	M^+ (Cl + Br)	$^{13}C_2$	^{18}O	Total	Values Normalized to 100.0
156	76.8			76.8	76.7
157		$\frac{2 \times 76.8 \times 1.12}{100}$ = 1.72		1.72	1.72
158	100.0		$\frac{76.8 \times 0.20}{100}$ = 0.15	100.15	100.0
159		$\frac{2 \times 100.0 \times 1.12}{100}$ = 2.24		2.24	2.24
160	24.3		$\frac{100.0 \times 0.20}{100}$ = 0.20	24.50	24.46
161		$\frac{2 \times 24.3 \times 1.12}{100}$ = 0.54		0.54	0.54
162			$\frac{24.3 \times 0.20}{100}$ = 0.05	0.05	0.05

presence of chlorine and/or bromine atoms. These atoms are the only common ones that produce intensely abundant M + 2, M + 4, etc., ions. The other halogens, fluorine and iodine, are monoisotopic and do not produce "halogen clusters." Because of the normalized abundances of the ions at *m/e* 156, 158, and 160, the presence of one chlorine atom and one bromine atom is strongly suggested. The abundance of the ion at *m/e* 159 (2.24%) relative to that at *m/e* 158 (100.0%) dictates that the compound can contain a maximum of two carbon atoms. If these deductions are correct, a partial formula of $^{12}C_2{}^{35}Cl^{79}Br$ can be proposed. The total mass of these atoms is 138, which is only 20 mass units less than the presumed molecular mass (158). Because of the abundance of the ion at *m/e* 162 (0.05%) relative to that ion at *m/e* 160 (24.46%) [when the abundances of these respective ions are normalized to 100, the abundances would be 0.20:100.0, respectively] the ion at *m/e* 160 *could* contain one oxygen atom, which would account for an additional 18 mass units. If this were the case, the remaining mass units (2) could only be accounted for by two hydrogen atoms, and the formula for the compound would be suggested, from these deductions, to be C_2H_2BrClO. For this compound, the main clue for a partial formula is the halogen cluster that dictates one bromine atom and one chlorine atom. The other deductions should be considered somewhat tenuous. Data obtained from the other spectral techniques would be most helpful in the elucidation of the structure of this compound.

At this stage, having examined all of the spectral data available, and hopefully having been able to propose a molecular formula for the compound, it is very useful to calculate the number of rings and/or double bonds that the compound may contain. Regardless of the formula of the compound, this process simply involves "reduction" of the suspected formula to the most saturated noncyclic formula possible containing those elements. A difference in these two formulas of 2n hydrogen atoms, accordingly, indicates the presence of n rings and/or double bonds. Thus, for a hydrocarbon, the suspected formula would be compared with that of C_nH_{2n+2} (a saturated aliphatic hydrocarbon). For example, benzene, C_6H_6, would be compared with C_6H_{14}. The calculated difference of eight hydrogen atoms between these two formulas dictates that benzene contains four rings and/or double bonds, which is exactly what would be anticipated.

Normally, the presence of halogen atoms

in an organic compound merely replaces the corresponding number of hydrogen atoms for this type of calculation. If the presence of halogen atoms is suspected, this numerical relationship should be remembered.

Because of their oxidation states, the presence of oxygen and/or sulfur atoms in a molecule causes no complications in these calculations. Thus cyclohexanone ($C_6H_{10}O$) would be compared with C_6H_{14}, and the difference of four hydrogen atoms between these formulas dictates that cyclohexanone contains two rings and/or double bonds. Also, dimethyl sulfide (C_2H_6S) would be compared with C_2H_6, and the exact correspondence of the number of hydrogen atoms in these formulas dictates that dimethyl sulfide is fully saturated.

The presence of nitrogen in an organic compound results in a modification of these calculations. The formula of the compound that such data might be compared with most simply is methylamine (CH_3NH_2, CH_5N). This formula clearly indicates the presence of one additional hydrogen atom per nitrogen atom in addition to the simple hydrocarbon formula of C_nH_{2n+2}. Thus, for *every* nitrogen atom present in a molecule, *one* additional hydrogen atom must be added to the basic formula. An example of a compound containing nitrogen atoms might be 3-aminopyrrole, ([structure: pyrrole ring with NH_2 at position 3, N–H], $C_4H_6N_2$).

Based on the above reasons, this formula would be reduced to C_4H_6 which differs from the basic formula (C_4H_{12}) by six hydrogen atoms. This analysis would dictate that the compound contains six hydrogen atoms less than a saturated noncyclic hydrocarbon, and this indicates that the compound would contain three double bonds and/or rings. This is of course indicated from the structure of the example chosen.

In connection with the possible presence of nitrogen in an organic compound (especially in the absence of halogen atoms), an important point should be indicated: If the molecule contains an odd number of nitrogen atoms, the molecular mass will be odd, and if the molecule contains an even number of nitrogen atoms, the molecular mass will be even.

The reproduced mass spectrum, as well as the MS data in tabular form, should be closely examined to determine the *m/e* values and the relative intensities of the most abundant ions. In particular, careful attention should be given to the loss of certain mass units that may relate one ion to another. As examples, a difference of 15 mass units, possibly corresponding to a loss of $-CH_3$; of 29, $-C_2H_5$; of 35, Cl; of 79, Br; or others, should be noted.

At this stage, one should be able to combine the suspected chromophoric groups from the IR and UV spectra, the first-order analysis of the NMR spectrum, and the information derived from the MS data, and derive a molecular formula for the compound. For simple compounds, a structural formula that uniquely satisfies *all* the data should follow without difficulty.

In case confusion still exists about a suitable unique structure, as will be the case frequently with compounds in group three and occasionally for compounds in group two, a more detailed examination of the reproduced spectra will be necessary. The IR and UV spectral data may have to be examined in more detail. The NMR spectrum, if not entirely a "first-order" spectrum, may have to be analyzed for accurate coupling constants and absorption positions. This will especially be necessary to determine *cis-* or *trans*-isomerism, aromatic, and heteroaromatic substitution types, etc. Also, particularly for group three compounds, a detailed examination of the MS fragmentation patterns may possibly be necessary.

Finally, after a satisfactory unique structure has been derived, it is instructive to examine *all* of the data in terms of that structure. Are any absorptions anomalous compared with standard reference tables and why? Can you explain the major MS fragmentation patterns and assign structures to the major ions present? Are all the IR, UV, and NMR absorptions expected or explainable? The answers to these questions may be the most valuable result of having worked these problems.

COMPOUND 1-1

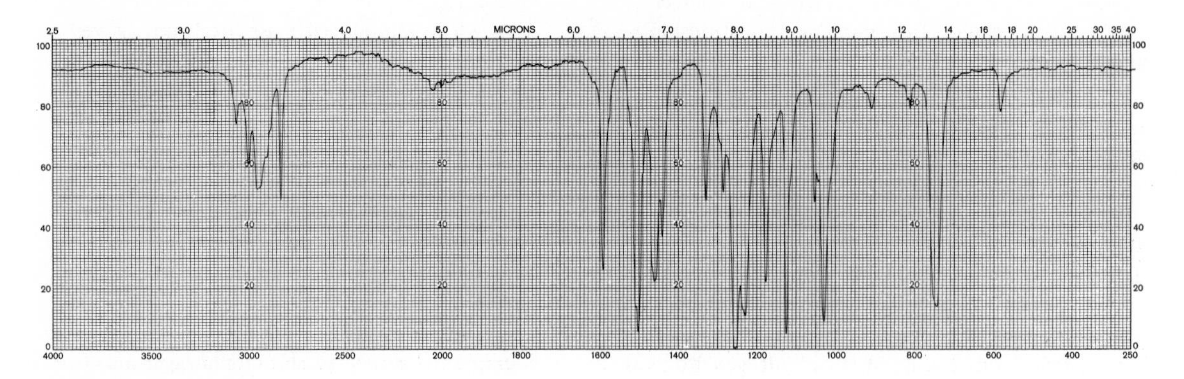

INFRARED SPECTRUM: Liquid film

ULTRAVIOLET DATA: λ_{max}^{Hexane} 225 nm (ϵ 7,100); 271 nm (ϵ 2,500); 275 nm (ϵ 2,600); 280 nm (ϵ 1,900).

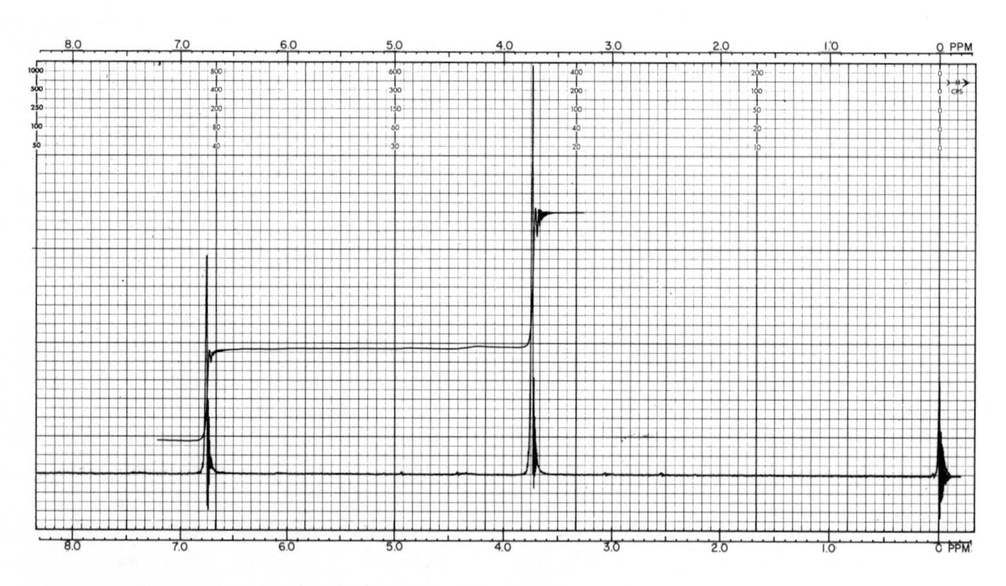

NUCLEAR MAGNETIC RESONANCE SPECTRUM:
15% in carbon tetrachloride, 500 Hz sweep width

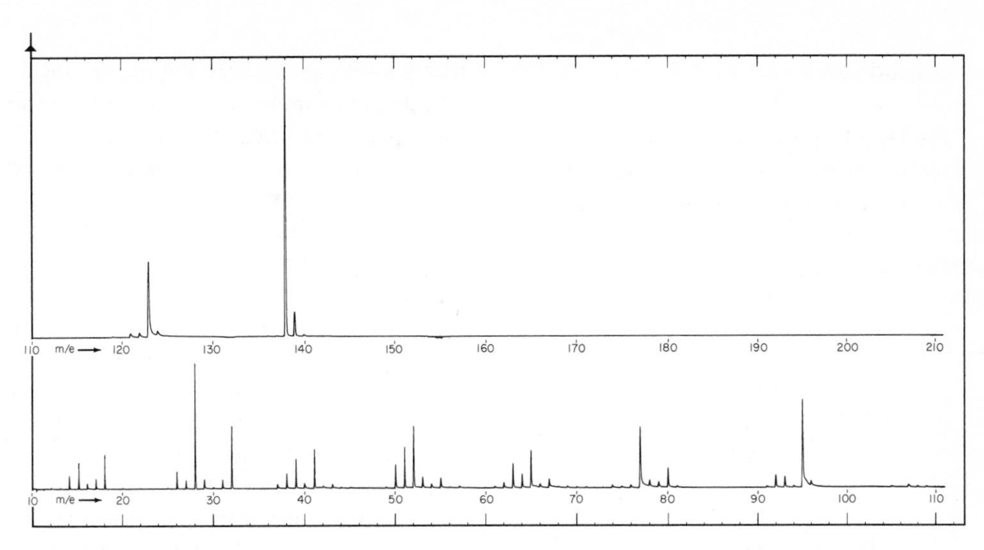

MASS SPECTRUM

MASS SPECTRAL DATA

m/e	*relative intensity*	*m/e*	*relative intensity*	*m/e*	*relative intensity*	*m/e*	*relative intensity*
14	5.7	39	10.2	62	2.4	91	0.5
15	14.2	40	2.0	63	9.2	92	4.6
16	2.0	41	13.4	64	5.8	93	4.0
17	2.3	42	1.1	65	14.2	94	0.6
18	10.5	43	1.6	66	1.6	95	26.0
26	7.3	49	0.7	67	3.8	96	2.6
27	3.8	50	6.6	74	1.3	121	2.0
28	61.0	51	16.0	75	0.9	122	2.0
29	4.2	52	21.8	76	1.5	123	26.3
30	0.7	53	4.6	77	20.7	124	2.6
31	3.7	54	1.7	78	3.0	138	100.0
32	14.7	55	3.8	79	2.4	139	9.0
37	1.7	57	0.7	80	7.8	140	0.9
38	5.9	61	0.7	81	0.9		

Solution to Problem 1-1

The IR spectrum of compound **1-1** shows the absence of hydroxyl, amino (–NH), acetylenic, and carbonyl functional groups. The UV spectrum is complicated, and definitely indicates the presence of a complex chromophoric group. The NMR spectrum of the compound is simple:

τ	*multiplicity*	*integration squares*	*relative number of hydrogens*
3.26	s	9.8	2
6.27	s	14.4	3

This simple NMR spectrum would suggest the presence of hydrogens attached to an aromatic nucleus (τ 3.26) and possibly the presence of an $-OCH_3$ group or groups.

The MS of this compound is most informative. The base peak occurs at m/e 138; since there are no ions present in large abundance at m/e greater than 138, one might assume as an initial hypothesis that the mass of the compound is 138. The abundance of the ion at m/e 139 is 9.0% relative to the abundance of the ion at m/e 138; the abundance of this ion results from the natural abundance of ^{13}C present in the ion at m/e 138. Because of the numerical relationship, the ion at m/e 138 can contain a maximum of eight carbon atoms. The presence of an ion with a significant abundance at m/e 140 suggests the presence of oxygen (which is consistent with the preliminary NMR analysis), but because of the abundance, would exclude the presence of sulfur, chlorine, or bromine.

Occasionally it is possible to derive elemental composition from the MS. If one assumes that the mass of the compound is 138 (the base peak in the MS), the compound can contain a maximum of eight carbon atoms ($8 \times 12 = 96$ mass units). Subtracting 96 from 138 (the presumed molecular weight of the compound) leaves 42 mass units unassigned. Since these 42 mass units must be accounted for only by hydrogen and oxygen (the presence of sulfur, chlorine, or bromine has been excluded by the MS; nitrogen is unlikely because an ion containing one nitrogen usually has an odd mass; the absence of fluorine may be inferred from the simplicity of the NMR spectrum), it would seem reasonable to propose that the compound contains two oxygen atoms. If this were the case, the formula of the compound would be $C_8H_{10}O_2$, dictating the presence of four double bonds and/or rings.

After analyzing these spectral data in a preliminary way, one should re-evaluate all these data more carefully. Relatively little additional information can be obtained from the MS of the compound, except for the presence of the ion at m/e 123, which corresponds to $M^+ - CH_3$. Again, this is in agreement with the preliminary NMR analysis. If the formula of the compound is $C_8H_{10}O_2$, then the ratio of hydrogen atoms would be 4:6 rather than 2:3. From correlation tables it would appear reasonable that the four hydrogens in the NMR spectrum would be attached to an aromatic nucleus, and that the six-hydrogen singlet at τ 6.27 would be satisfied by the presence of two equivalent methoxyl groups.

Based on this reasoning it would seem logical that the compound is a dimethoxy benzene. Such a general structure would satisfy the UV spectrum, the MS, and the NMR spectrum. The best evidence for the substitution pattern on a benzene nucleus is obtained from a careful examination of the IR spectrum. The intense absorption at 13.9 μ (719 cm^{-1}) dictates that the compound is an *o*-disubstituted benzene. Thus the structure of the compound is *o*-dimethoxybenzene (veratrole, catechol dimethyl ether).

Solution to Problem 1-2

The IR spectrum of compound 1-2 shows the absence of hydroxyl, amino, (–NH), acetylenic, and carbonyl functional groups. The UV spectrum indicates the absence of any chromophoric group absorbing above 205 nm. The NMR spectrum of compound **1-2** can be analyzed as follows:

τ	*multiplicity*	*J, Hz*	*integration squares*	*relative number of hydrogens*
6.63	t	7.0	6.9	2
8.23	complex	–	10.7	3
9.06	d	7.0	20.5	6

The NMR spectrum indicates the presence of a two-hydrogen absorption at τ 6.63 (t, J = 7.0 Hz) that is adjacent to two hydrogens; these hydrogens attached to carbon *must* be adjacent to some electronegative group that would shift the absorption downfield to this τ value. Also, the six-hydrogen absorption (d, J = 7.0 Hz) indicates the presence of six equivalent protons adjacent to a single hydrogen. This would strongly indicate an isopropyl group. The complex absorptions at τ 8.23 cannot be analyzed.

The MS of the compound has ions of moderate abundance in the region *m/e* 150; no ions are present at *m/e* greater than 153. Because of the nearly equal abundances of the ions at *m/e* 150 and 152, the presence of one bromine atom is suggested. Another prominent "halogen cluster" appears at *m/e* 107 and 109. This fragmentation corresponds to M^+ - 43 (C_3H_7, probably an isopropyl group, which would be consistent with the preliminary NMR analysis). It is also interesting to note that the base peak in the spectrum occurs at *m/e* 43 (C_3H_7). If the mass of the compound is 150, and if one subtracts the mass of bromine (^{79}Br) from this, one obtains a difference of 71 mass units for the remainder of the molecule. Considering all other spectral data this mass difference would appear to be best satisfied by the hydrocarbon residue C_5H_{11} (mass 71). Abundant ions occur at *m/e* 70 and 71. These correspond to M^+ - HBr and M^+ - Br, respectively. Thus it would appear that the formula of the compound is $C_5H_{11}Br$. This formula dictates that the compound is fully saturated and contains no rings and/or double bonds.

The NMR and MS evidence strongly indicates the presence of an isopropyl group. The NMR absorption at τ 6.63 (a two-hydrogen triplet) suggests the presence of a methylene adjacent to an electronegative substituent. This could best be satisfied by the group $-CH_2Br$. The remaining NMR absorption, a complex three-hydrogen multiplet centered at τ 8.23 would appear to consist of the group $>CH-CH_2-$, where the methylene hydrogens of this group would be coupled with the triplet $-CH_2-$ adjacent to bromine, and the single hydrogen of that group would be coupled with the methyl groups (a six-hydrogen doublet) to form the isopropyl group.

Thus the structure of the compound is isoamyl bromide (3-methyl-l-bromobutane).

COMPOUND 1-2

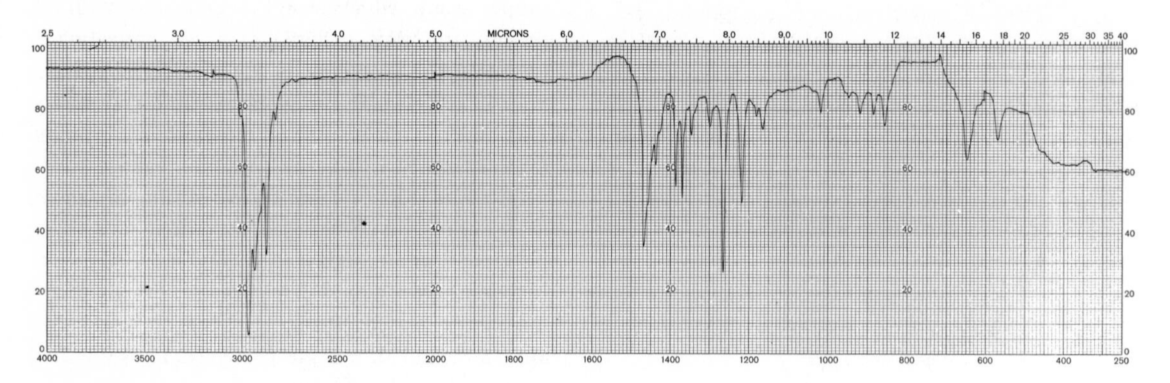

INFRARED SPECTRUM: 5.3% in carbon tetrachloride

ULTRAVIOLET DATA: no λ_{max}^{EtOH} above 205 nm

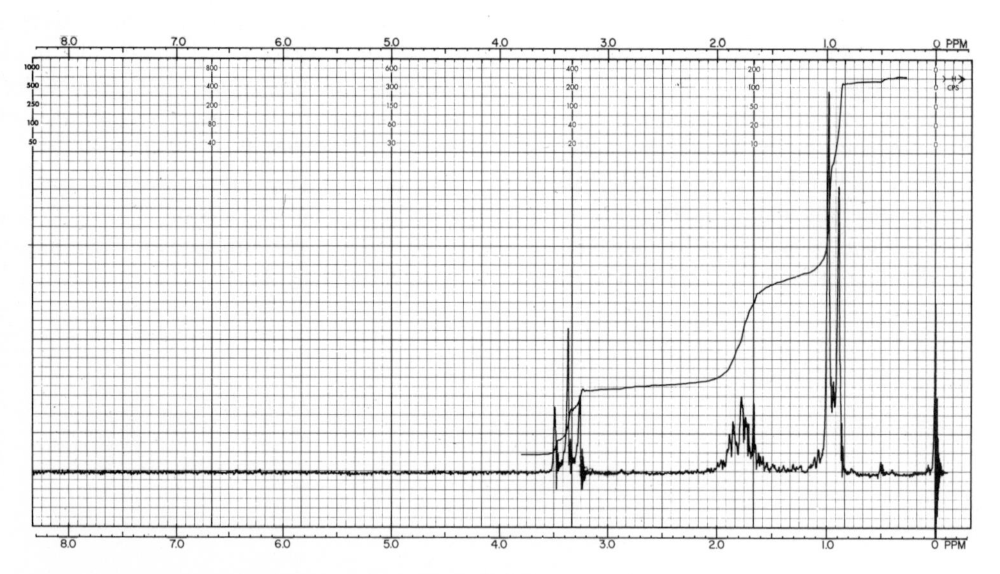

NUCLEAR MAGNETIC RESONANCE SPECTRUM:
15% in carbon tetrachloride, 500 Hz sweep width

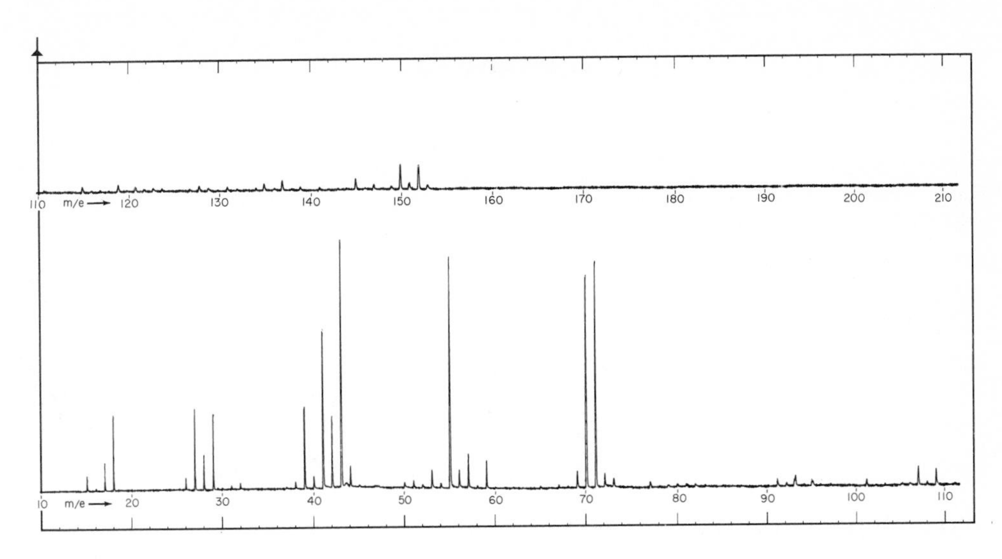

MASS SPECTRUM

MASS SPECTRAL DATA

m/e	relative intensity	m/e	relative intensity	m/e	relative intensity	m/e	relative intensity
14	0.3	54	1.1	93	3.1	129	1.3
15	1.7	55	45.0	94	0.5	130	0.3
16	0.3	56	3.9	95	1.9	131	1.3
17	3.3	57	8.2	101	2.3	132	0.6
18	14.9	58	0.8	102	0.5	134	0.7
26	2.3	59	6.5	103	0.7	135	2.0
27	21.6	65	0.5	105	0.8	136	0.9
28	7.3	66	0.4	106	0.8	137	3.0
29	14.2	67	0.9	107	5.0	138	0.5
30	0.5	68	0.7	108	0.8	139	0.9
31	0.8	69	4.6	109	4.5	140	0.2
32	0.8	70	50.0	110	0.2	141	1.0
37	0.5	71	55.0	115	1.8	142	0.3
38	1.6	72	3.9	116	0.5	143	0.5
39	16.8	73	2.5	117	0.5	144	0.5
40	2.7	77	1.5	118	0.4	145	4.0
41	35.0	78	0.5	119	2.3	146	0.5
42	13.2	79	0.7	120	0.4	147	1.5
43	100.0	80	1.2	121	1.3	148	0.4
44	4.3	81	0.6	122	0.6	149	0.8
45	0.7	82	1.1	123	1.0	150	8.0
50	1.3	83	0.4	124	0.5	151	1.9
51	2.0	91	2.1	127	0.8	152	7.8
52	0.9	92	0.8	128	1.8	153	1.8
53	4.1						

COMPOUND 1-3

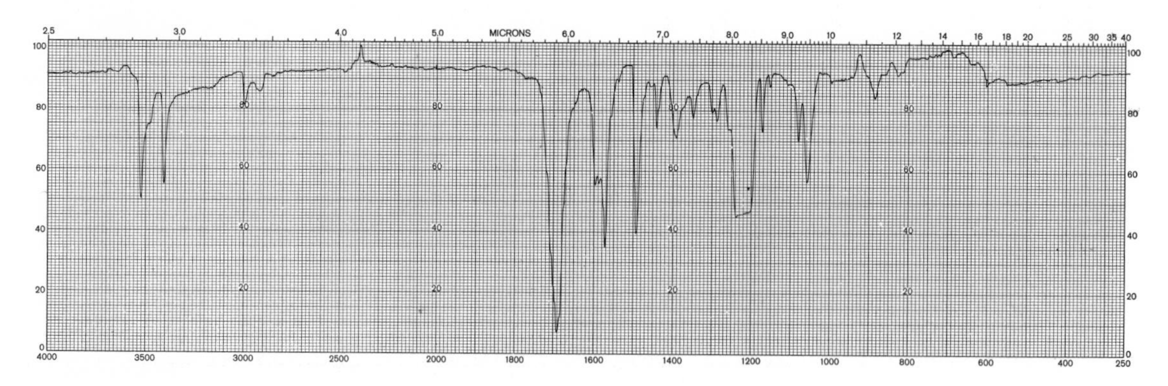

INFRARED SPECTRUM: 1.1% in chloroform

ULTRAVIOLET DATA: λ_{max}^{EtOH} 218 nm (ϵ 6,900); 263 nm (ϵ 970); 269 nm (ϵ 1,400); 275 nm (ϵ 1,170).

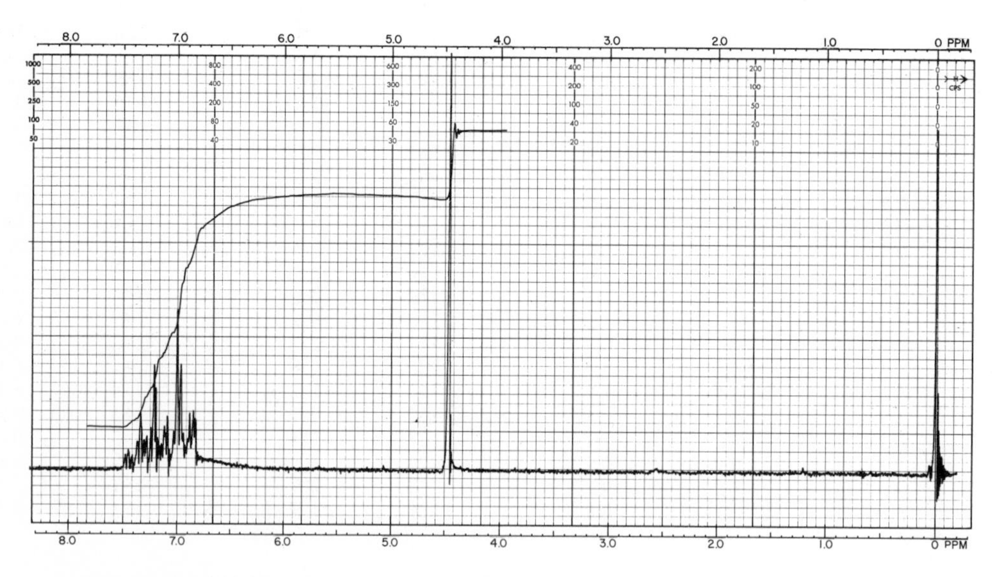

NUCLEAR MAGNETIC RESONANCE SPECTRUM:
15% in deuterochloroform, 500 Hz sweep width

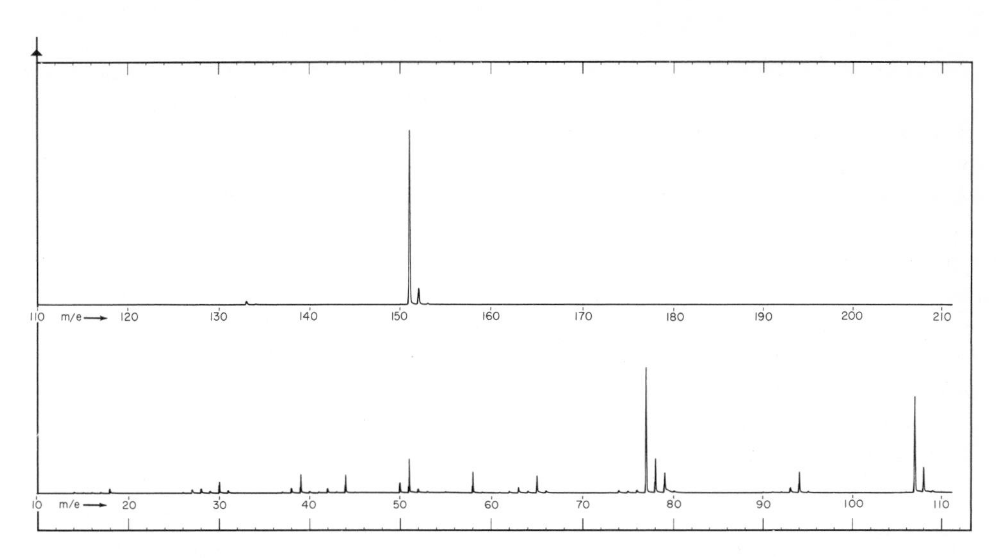

MASS SPECTRUM

MASS SPECTRAL DATA

m/e	*relative intensity*	*m/e*	*relative intensity*	*m/e*	*relative intensity*	*m/e*	*relative intensity*
14	0.4	43	0.7	64	1.3	95	1.3
15	0.3	44	7.8	65	9.0	96	0.2
16	0.4	45	0.3	66	1.5	105	1.7
17	0.4	49	0.2	73	0.3	106	0.4
18	1.3	50	5.1	74	1.8	107	59.0
26	0.4	51	15.1	75	1.4	108	16.5
27	1.4	52	2.1	76	1.9	109	1.5
28	2.1	53	0.6	77	69.0	122	0.4
29	1.1	53.5	0.3	78	19.1	123	0.4
30	4.2	54	0.2	79	11.6	133	2.3
31	1.0	55	0.4	80	1.3	134	0.7
37	0.6	58	10.1	90	0.6	135	0.3
38	2.2	59	0.3	91	0.5	150	0.3
39	8.0	61	0.5	92	0.5	151	100.0
40	0.9	62	1.1	93	3.6	152	9.2
41	0.6	63	3.3	94	12.5	153	0.9
42	2.2						

Solution to Problem 1-3

The IR spectrum of compound **1-3** shows absorption at 2.83 and 2.93 μ (3534 and 3413 cm^{-1}), which would suggest the presence of –NH and/or –OH functional groups. There is also a very strong absorption at 5.91 μ (1692 cm^{-1}), which may suggest the presence of a carbonyl group of some kind. The UV spectral data is complex and cannot be interpreted at this time, however, the data indicate that a significant chromophoric group is present in the molecule. The NMR spectral data may be assigned as follows:

τ	*multiplicity*	*integration squares*	*relative number of hydrogens*
ca 2.9	complex	24.7	7
5.53	s	7.2	2

The presence of complex absorption in the region τ 3 suggests the presence of hydrogens attached to an aromatic nucleus. From NMR correlation tables the presence of hydroxyl or amino hydrogens would appear to be excluded. One reasonable functional group that would satisfy the IR data would be an amide, and since there are two N–H stretching absorptions, it is quite possible that the compound is a primary amide. It should also be noted that this functional group would satisfy the strong carbonyl stretching absorption observed.

The base peak in the MS of the compound occurs at *m/e* 151. There are no abundant ions beyond this mass, and so it may be presumed that 151 is the molecular weight of the compound. It should be recognized that, if this is the case, an odd number for the molecular weight is consistent with the compound containing one nitrogen atom. The abundance of the ion at *m/e* 152 indicates that the ion at *m/e* 151 contains a maximum of eight carbon atoms. Other prominent ions in the MS occur at *m/e* 107 (M^+ - 44; $-CONH_2$ has mass 44) and at *m/e* 77 (this ion frequently corresponds to the benzenium ion, $C_6H_5{}^+$ which would suggest a monosubstituted phenyl group).

It would now be appropriate to determine the molecular formula of the compound. It has been suggested that the compound contains primary amide ($-CONH_2$) and monosubstituted phenyl (C_6H_5-) groups. The hydrogens attached to the aromatic nucleus and those of the primary amide functional group could be accounted for by the complex seven-hydrogen absorption centered near τ 3 in the NMR spectrum. In addition there is a two-hydrogen absorption in the NMR spectrum at τ 5.83 that could best be satisfied by a methylene group, not coupled to any other hydrogens, but in a strongly electronegative environment. If one adds together the masses of these groups (C_6H_5- = 77; $-CH_2-$ = 14; $-CONH_2$ = 44), the sum is mass 135. This mass is different from the mass of the presumed molecular ion (151) by 16 mass units. This difference can best be satisfied by the presence of an oxygen atom. Thus it is reasonable that the molecular formula of the compound is $C_8H_9O_2N$. This formula corresponds to the presence of five rings and/or double bonds.

Using the various correlation tables available, the only reasonable way in which the above functional groups (C_6H_5-, $-CH_2-$, $-CONH_2$, –O–) can be put together is $C_6H_5OCH_2CONH_2$. The compound is phenoxyacetamide.

Solution to Problem 1-4

The IR spectrum of compound **1-4** strongly suggests the presence of hydroxyl and/or amino (N–H) functional groups (strong, broad absorption at 2.99 μ (3344 cm^{-1}). The absence of acetylenic, carbonyl, and olefinic groups is indicated. The UV spectral data indicates the absence of any chromophoric group that would have λ_{max} above 205 nm. The NMR spectrum may be analyzed as follows:

τ	*multiplicity*	*integration squares*	*relative number of hydrogens*
5.29	broad s	3.4	2
5.95	complex	2.0	1
8.33-9.10	complex	19.2	11

A comment is in order with regard to the measurement of the integration of the NMR spectrum of this compound and all others. Ideally, the integration of the least intense absorption present is assigned to be one hydrogen atom, and all of the other absorptions present in the spectrum are related to it. It must be emphasized, however, that the integral of a broad absorption of low intensity is the *least* accurate. For this spectrum, if the absorption at τ 5.95 were assigned 1.0 H, the ratio of the hydrogens present in the groups would be 1.7:1.0:9.6. On the other hand if the absorption at τ 5.29 were assigned the value 2.0 H, the ratio of the hydrogens present in the groups would be 2.0:1.0:11.3. Thus, *it is not always possible to obtain an extremely accurate hydrogen count from the NMR spectrum*. However, the spectrum does indicate the presence of three hydrogens (τ 5.29 and τ 5.95) in a rather electronegative environment and about eleven hydrogens (τ 8.33-9.10) in a "saturated" environment.

A closer examination of the region τ 8.33-9.10 is possible. In this region the NMR absorptions at τ 8.79-9.10 appear to be relatively simple and to contain 9 hydrogen atoms from the integration. This absorption appears to have a six-hydrogen singlet at τ 8.82, and a three-hydrogen doublet at τ 8.84, J = 10 Hz. If this analysis is correct, the functional groups $CH_3-\overset{|}{\underset{|}{C}}-CH_3$ and $CH_3-\overset{|}{\underset{|}{C}H}$ are suggested (total mass 70).

The ions in the MS of the compound that occur at highest *m/e* values are at *m/e* 116 and 118. It should be noted that these ions have nearly equal intensity, which might suggest the presence of one bromine atom. From the previous NMR analysis, however, it is not possible that the compound contains one bromine atom because the sum of those masses (70 + 79 = 149) exceeds the masses of the ions in the high mass region of the mass spectrum.

Examination of the two-hydrogen NMR absorption at τ 8.33 - 8.66 reveals that it is complex and not a simple multiplet. This absorption might best be satisfied by the non-equivalent hydrogens of a methylene group adjacent to an asymmetric center. Thus, tentatively, the fragments $CH_3-\overset{|}{\underset{|}{C}}-CH_3$ and $CH_3-\underset{|}{C}H-CH_2-$ (total mass 84) might be suggested. If the ion at *m/e* 118 is the molecular ion, the remaining mass difference (118 - 84) is 34 mass units. Since the IR spectrum of the compound demonstrated very strong hydroxyl absorption (and since there is no evidence for the presence of >N–H), the 34 mass units might best be satisfied by the presence of two hydroxyl groups.

If the above assumptions are correct, the formula of the compound would be $C_6H_{14}O_2$ (mass 118). The above fragments can be assembled to produce only one logical structure for the compound, 2-methyl-2, 4-dihydroxypentane (diacetone alcohol).

COMPOUND 1-4

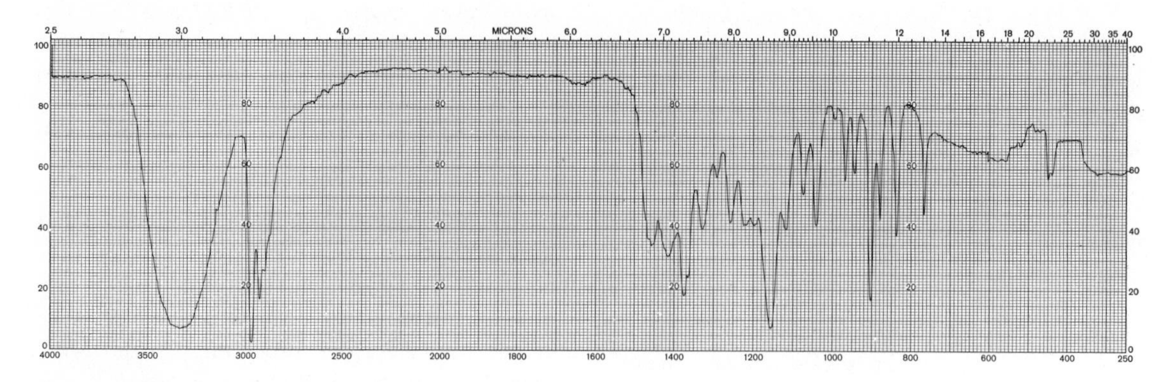

INFRARED SPECTRUM: Liquid film

ULTRAVIOLET DATA: no λ_{max} above 205 nm

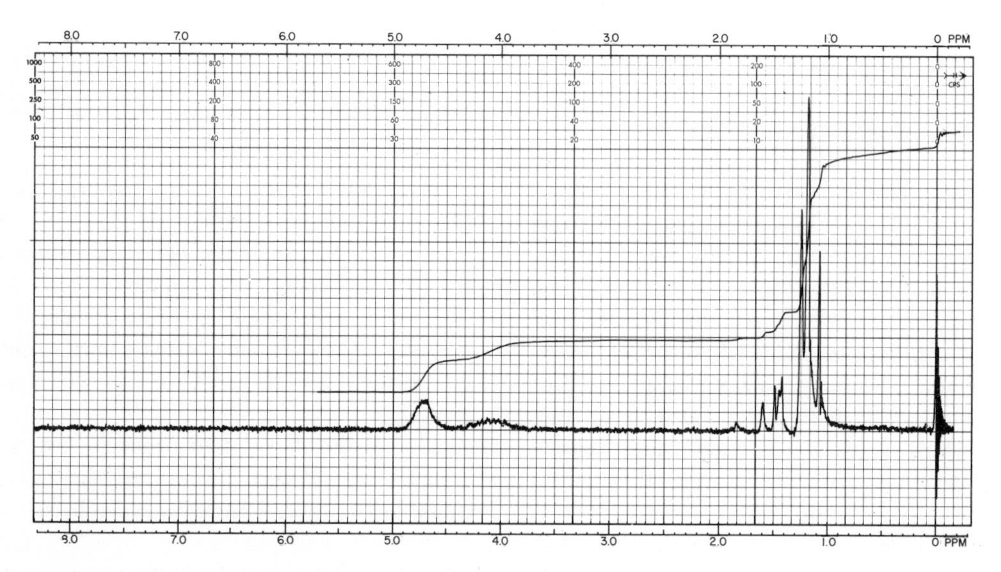

NUCLEAR MAGNETIC RESONANCE SPECTRUM:
15% in carbon tetrachloride, 500 Hz sweep width

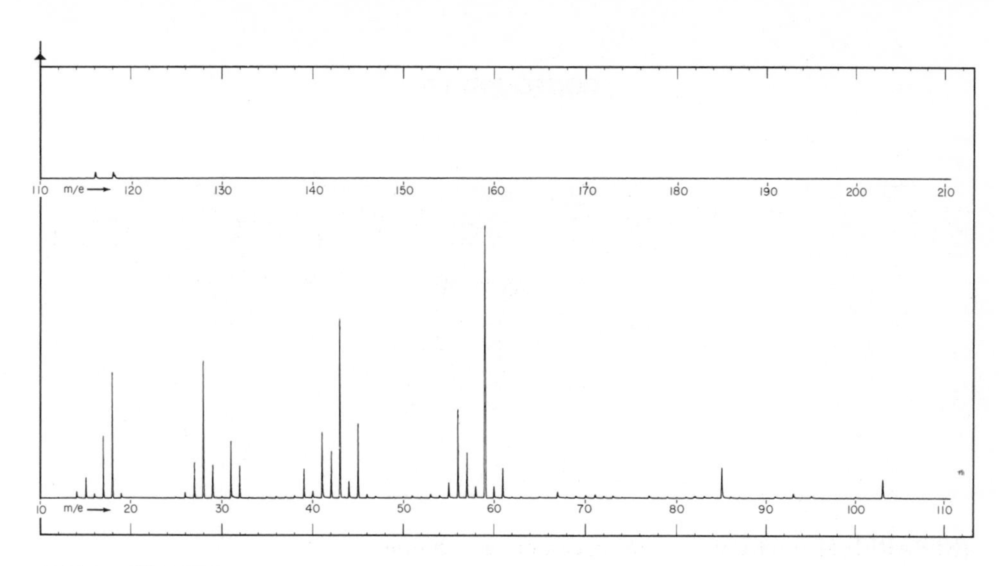

MASS SPECTRUM

MASS SPECTRAL DATA

m/e	*relative intensity*	*m/e*	*relative intensity*	*m/e*	*relative intensity*	*m/e*	*relative intensity*
14	0.9	37	0.3	56	27.3	79	0.5
15	4.1	38	0.7	57	13.7	80	0.3
16	0.8	39	7.1	58	3.4	81	0.5
17	6.3	40	2.1	59	100.0	82	0.8
18	30.0	41	19.7	60	3.7	83	0.7
19	1.2	42	12.7	61	9.2	85	10.4
25	0.3	43	69.0	63	0.5	86	0.7
26	1.2	44	4.6	65	0.4	87	0.5
27	8.6	45	20.8	66	0.3	91	0.5
28	25.1	46	1.1	67	2.1	93	1.2
29	9.4	47	0.5	69	0.8	95	0.8
30	0.6	51	0.7	70	0.8	100	0.6
31	12.3	52	0.4	71	1.1	103	7.0
32	4.5	53	1.1	72	0.6	104	0.5
35	0.3	54	0.7	73	0.8	116	1.7
36	0.6	55	4.5	77	0.8	118	1.8

COMPOUND 1-5

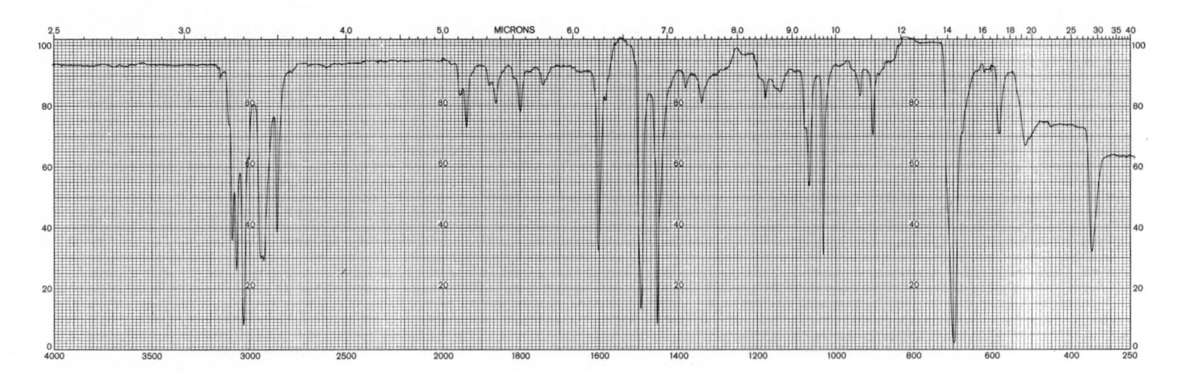

INFRARED SPECTRUM: 9.6% in carbon tetrachloride

ULTRAVIOLET DATA: λ_{max}^{Hexane} 260 nm (ϵ 420)

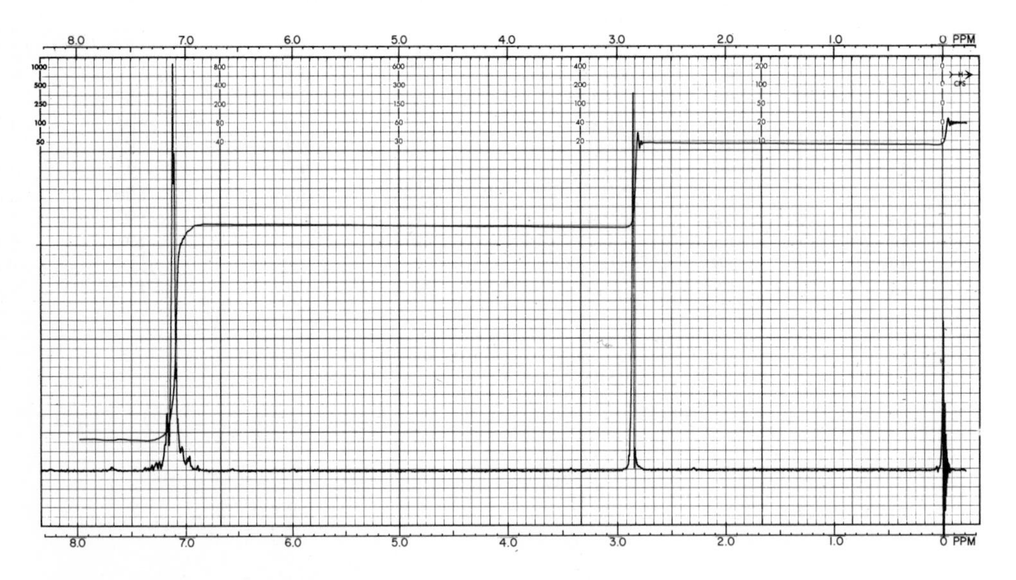

NUCLEAR MAGNETIC RESONANCE SPECTRUM:
15% in carbon tetrachloride, 500 Hz sweep width

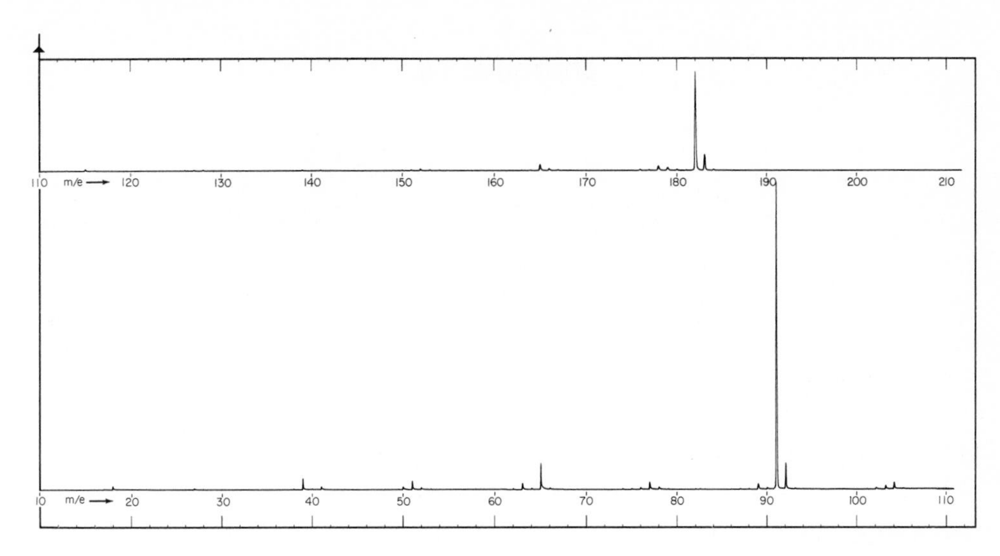

MASS SPECTRUM

MASS SPECTRAL DATA

m/e	*relative intensity*	*m/e*	*relative intensity*	*m/e*	*relative intensity*	*m/e*	*relative intensity*
18	0.9	74	0.9	102	1.7	153	0.7
26	0.3	75	1.2	103	3.7	154	0.3
27	1.7	76	2.3	104	6.0	163	0.2
28	0.3	77	6.0	105	0.9	164	0.2
38	1.0	78	3.1	113	0.3	165	4.8
39	10.2	79	0.6	114	0.2	166	1.8
40	0.8	82	0.5	115	1.8	167	0.5
41	3.1	82.5	0.4	116	0.3	175	0.1
50	3.1	83	0.5	126	0.5	176	1.0
51	8.5	86	0.2	127	0.6	177	0.6
52	2.1	87	0.5	128	0.8	178	3.6
53	0.7	88	0.6	129	0.3	179	2.4
61	0.2	89	3.8	139	0.7	180	1.0
62	1.3	90	1.2	140	0.2	181	0.4
63	5.6	91	100.0	141	0.6	182	69.0
64	1.8	92	15.0	150	0.4	183	11.9
65	19.0	93	1.1	151	0.8	184	1.0
66	1.8	101	0.3	152	1.9		

COMPOUND 1-6

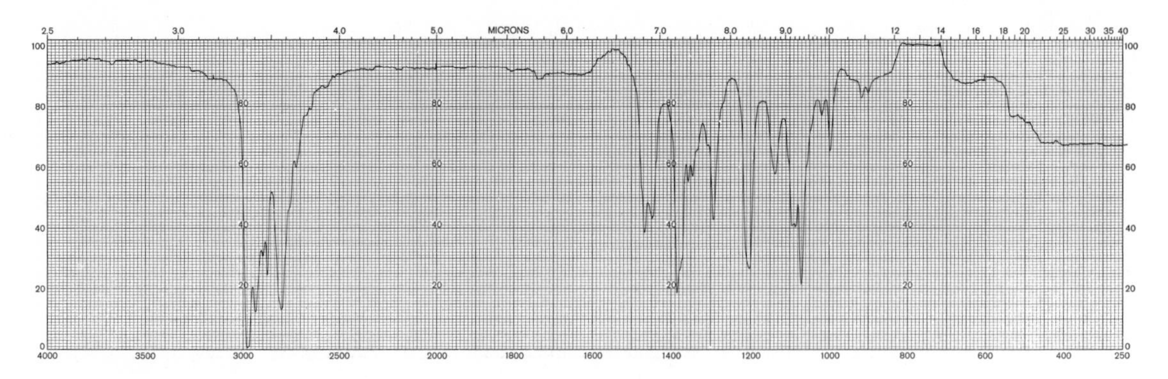

INFRARED SPECTRUM: 4.7% in carbon tetrachloride

ULTRAVIOLET DATA: λ_{max}^{Hexane} 196 nm (ϵ 5,000)

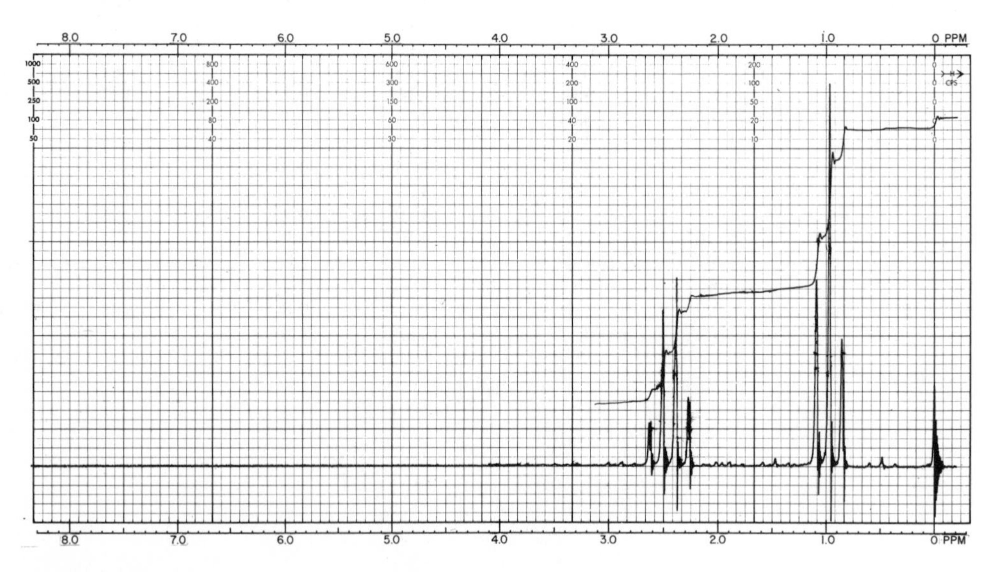

NUCLEAR MAGNETIC RESONANCE SPECTRUM:
15% in carbon tetrachloride, 500 Hz sweep width

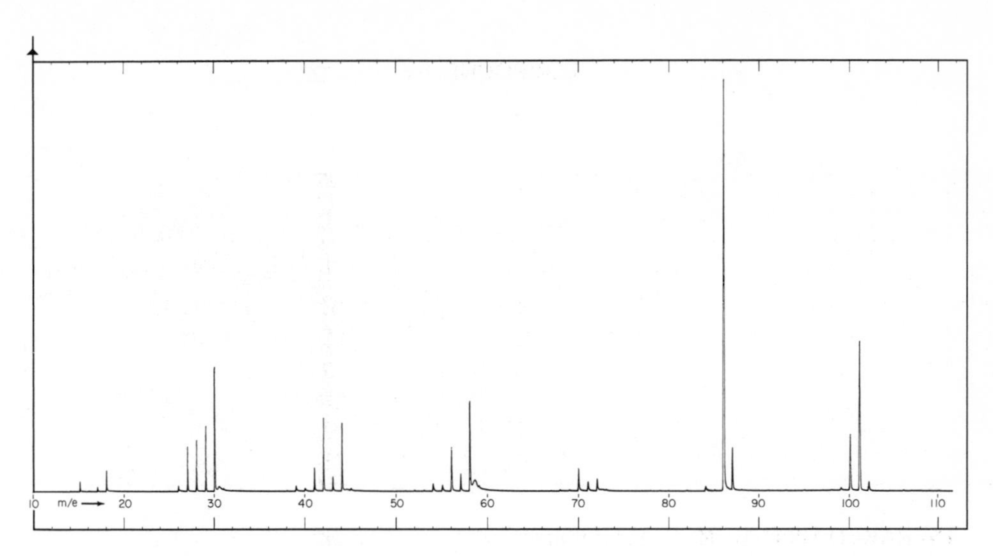

MASS SPECTRUM

MASS SPECTRAL DATA

m/e	*relative intensity*	*m/e*	*relative intensity*	*m/e*	*relative intensity*	*m/e*	*relative intensity*
15	1.9	32	0.3	55	0.9	72	1.6
17	0.3	39	0.6	56	4.8	84	0.7
18	2.3	40	0.4	57	2.3	86	100.0
26	1.0	41	2.6	58	10.9	87	5.9
27	7.8	42	8.2	59	0.7	99	0.5
28	8.3	43	1.7	68	0.2	100	8.1
29	9.5	44	7.4	69	0.2	101	20.1
30	15.9	45	0.4	70	2.9	102	1.7
31	0.4	54	1.0	71	1.2		

COMPOUND 1-7

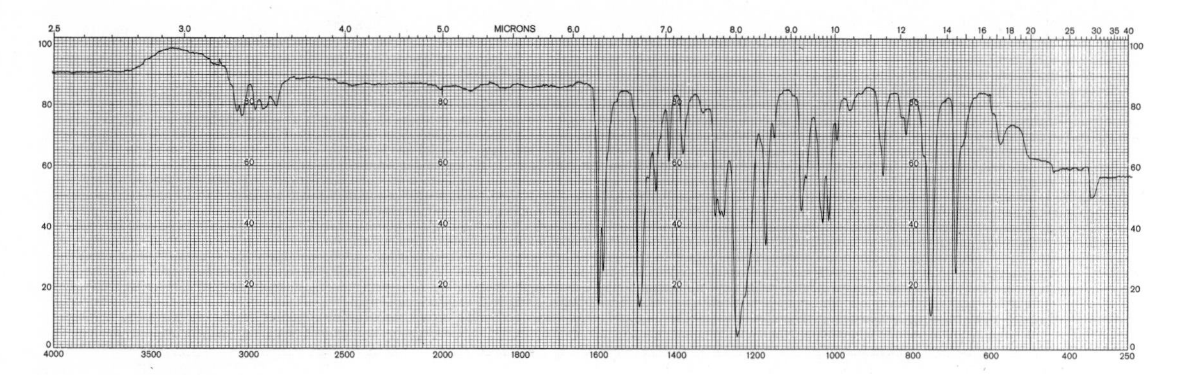

INFRARED SPECTRUM: Liquid film

ULTRAVIOLET DATA: λ_{max}^{EtOH} 219 nm (ϵ 7,700); 264(sh) nm (ϵ 1,100); 270 nm (ϵ 1,500); 277 nm (ϵ 1,230).

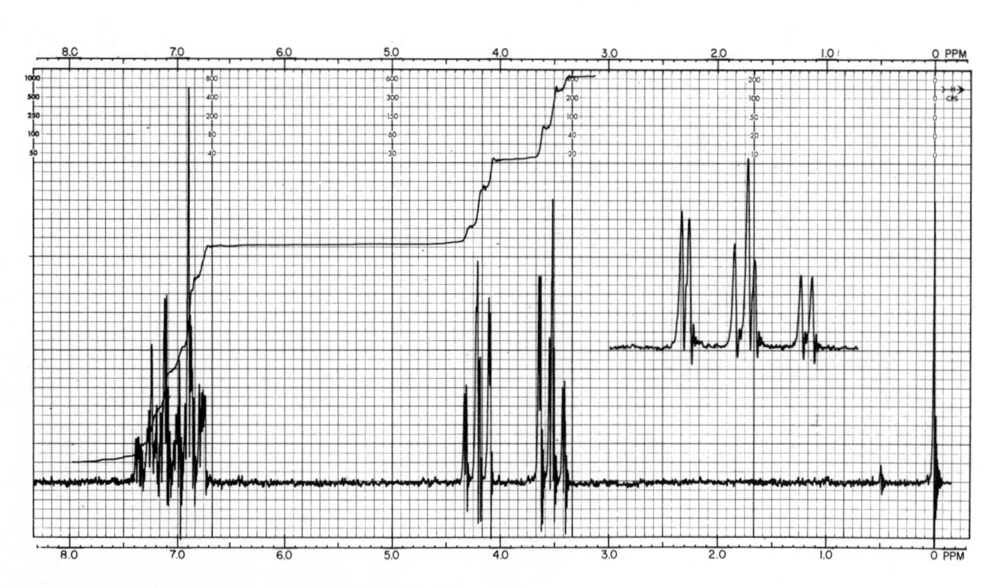

NUCLEAR MAGNETIC RESONANCE SPECTRUM:
15% in carbon tetrachloride, 500 Hz sweep width;
Expansion: 100 Hz sweep width, 190 Hz sweep offset

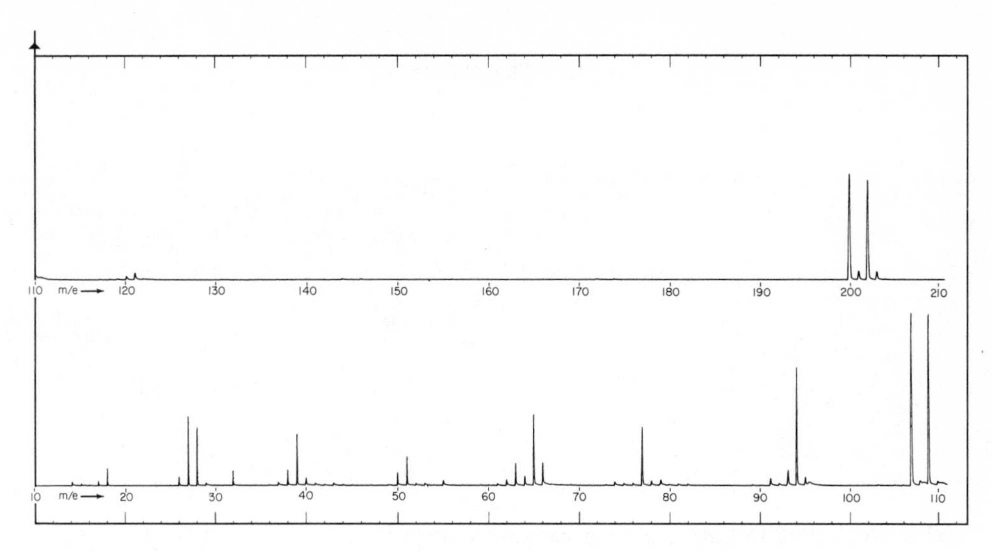

MASS SPECTRUM

MASS SPECTRAL DATA

m/e	*relative intensity*	*m/e*	*relative intensity*	*m/e*	*relative intensity*	*m/e*	*relative intensity*
14	0.7	50	4.6	78	2.7	119	1.4
15	0.4	51	10.0	79	3.3	120	3.5
16	0.5	52	0.9	80	1.1	121	5.3
17	0.7	53	1.0	81	1.1	122	1.3
18	3.3	55	2.5	82	1.0	135	0.7
26	2.7	61	1.0	89	0.6	137	0.6
27	19.7	62	2.9	91	4.0	144	1.1
28	16.5	63	8.8	92	1.5	146	1.0
29	1.1	64	3.8	93	8.0	156	0.7
32	3.3	65	29.0	94	64.0	158	0.5
37	1.2	66	10.0	95	4.7	172	1.0
38	4.6	67	0.7	107	100.0	174	1.0
39	19.4	68	0.6	108	3.4	200	82.0
40	2.9	74	1.9	109	97.0	201	7.4
41	1.0	75	1.2	110	2.8	202	80.0
42	0.5	76	1.3	117	0.7	203	7.5
43	1.0	77	26.2	118	1.1	204	0.4
44	0.4						

COMPOUND 1-8

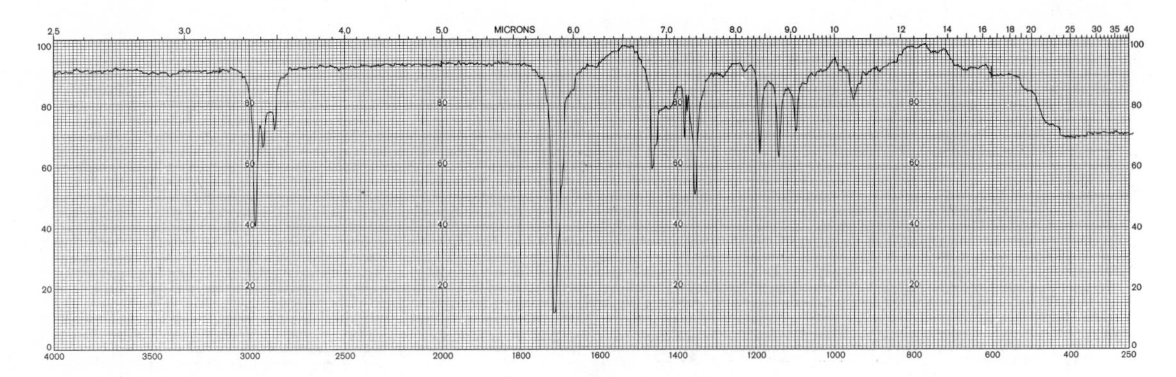

INFRARED SPECTRUM: 1.5% in carbon tetrachloride

ULTRAVIOLET DATA: λ_{max}^{EtOH} 280 nm (ϵ 21)

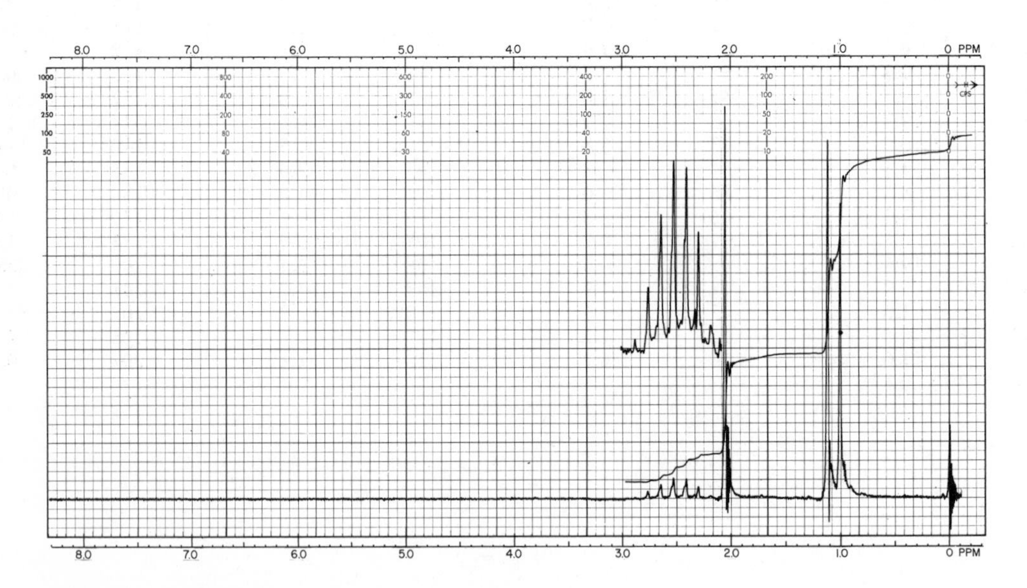

NUCLEAR MAGNETIC RESONANCE SPECTRUM:
15% in carbon tetrachloride, 500 Hz sweep width
Insert: amplitude X 10

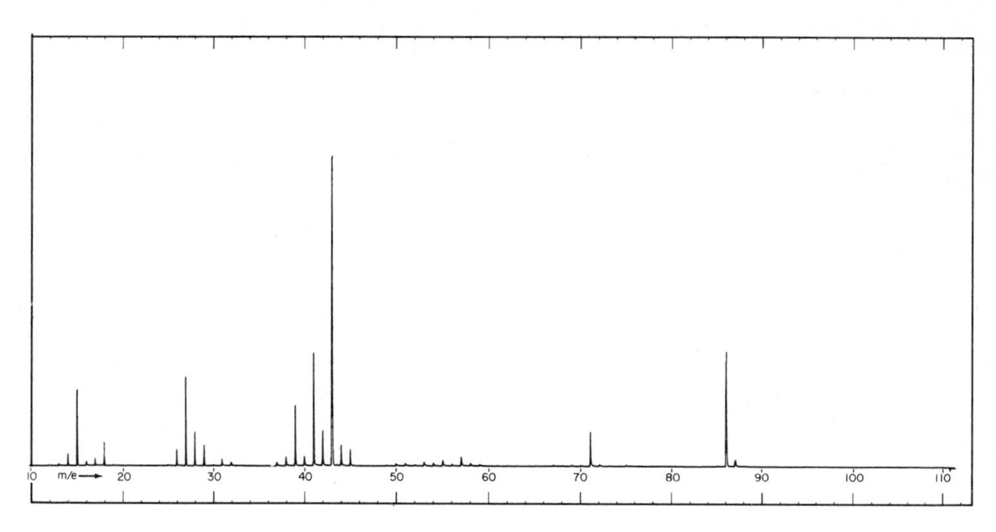

MASS SPECTRUM

MASS SPECTRAL DATA

m/e	*relative intensity*	*m/e*	*relative intensity*	*m/e*	*relative intensity*	*m/e*	*relative intensity*
13	0.2	28	4.1	42	3.7	56	0.3
14	1.5	29	2.4	43	100.0	57	1.0
15	12.8	31	0.8	44	2.5	58	0.4
16	0.5	32	0.5	45	1.8	59	0.3
17	0.8	37	0.6	50	0.2	71	3.6
18	3.8	38	1.1	51	0.3	72	0.2
25	0.2	39	7.7	53	0.5	86	13.0
26	1.9	40	1.1	54	0.4	87	1.0
27	15.8	41	15.6	55	0.7		

COMPOUND 1-9

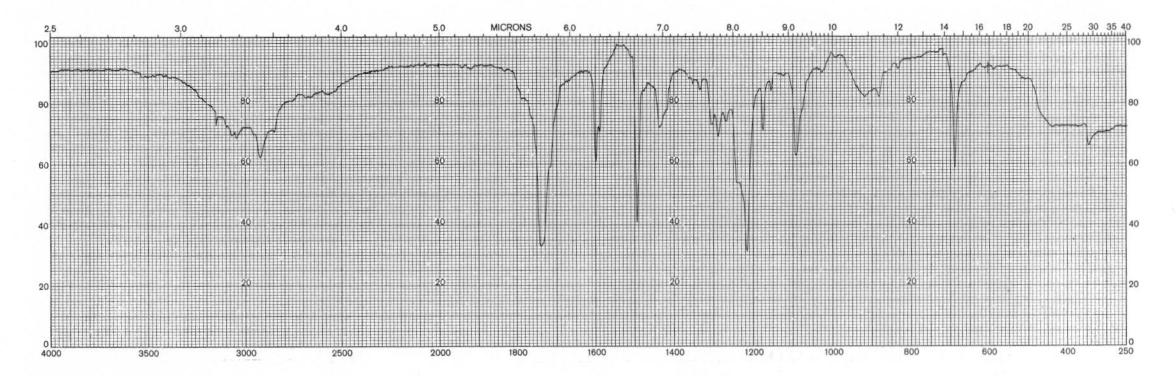

INFRARED SPECTRUM: 1.0% in carbon tetrachloride

ULTRAVIOLET DATA: λ_{max}^{EtOH} 219 nm (ϵ 6,800); 264 nm (ϵ 1,100); 270 nm (ϵ 1,500); 276 nm (ϵ 1,260).

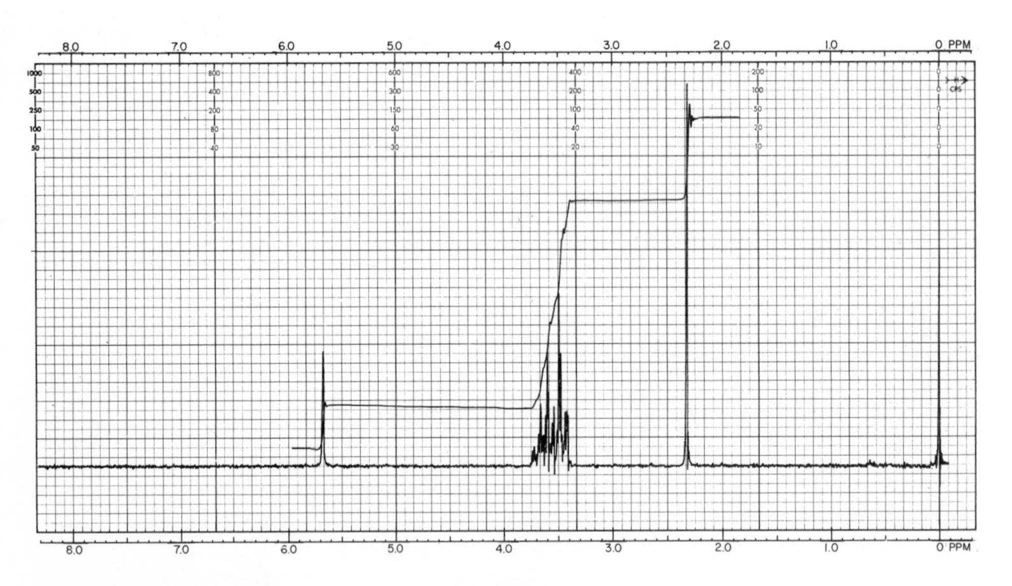

NUCLEAR MAGNETIC RESONANCE SPECTRUM:
15% in deuterochloroform, 1000 Hz sweep width

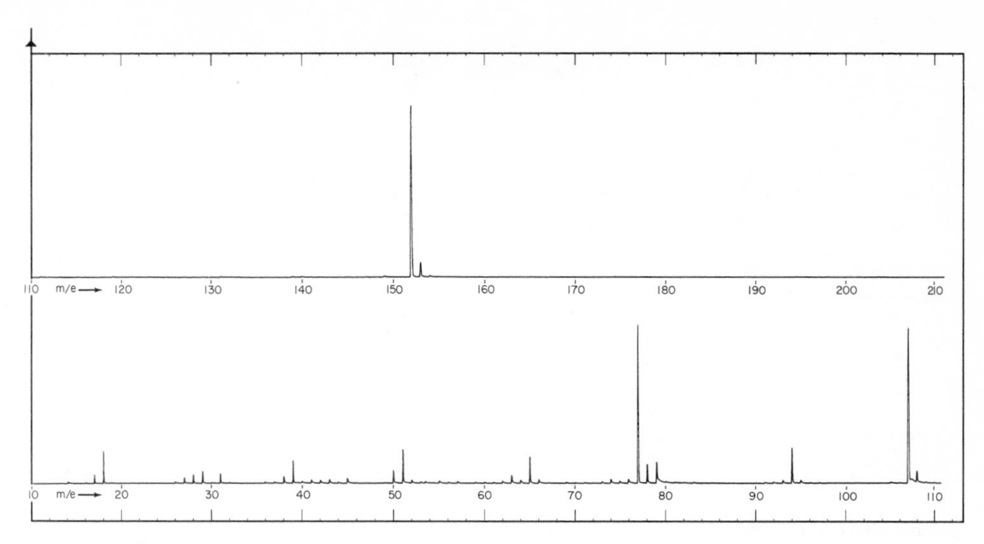

MASS SPECTRUM

MASS SPECTRAL DATA

m/e	*relative intensity*	*m/e*	*relative intensity*	*m/e*	*relative intensity*	*m/e*	*relative intensity*
17	1.3	51	15.3	74	4.1	115	2.1
18	6.6	52	2.0	75	1.5	116	1.2
27	2.1	53	1.4	76	2.2	117	0.9
28	4.4	53.5	1.1	77	81.0	119	1.2
29	2.6	54	0.7	78	9.3	128	1.3
30	0.8	55	2.3	79	10.3	129	1.0
31	2.0	57	3.5	88	1.6	131	3.8
32	0.9	60	1.5	91	2.5	139	1.7
37	1.0	61	1.2	92	1.4	140	1.6
38	2.8	62	1.3	93	1.7	141	1.1
39	10.0	63	4.1	94	18.4	142	0.8
40	1.1	64	1.9	95	2.4	143	0.7
41	3.0	65	12.4	105	1.9	147	1.2
42	1.9	66	2.8	107	82.0	149	1.1
43	1.8	67	1.0	108	6.7	152	100.0
44	1.1	69	0.9	109	1.0	153	8.9
45	2.8	71	0.6	111	1.1	154	1.2
50	5.5	73	1.8				

COMPOUND 1-10

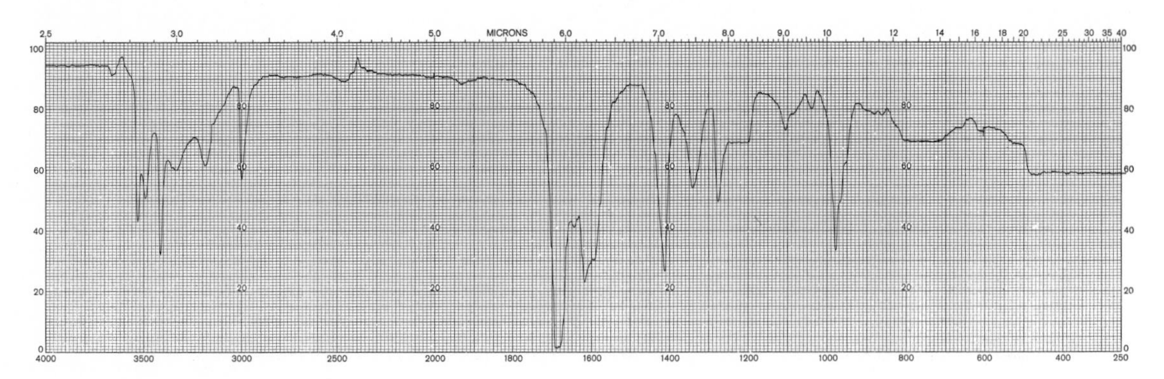

INFRARED SPECTRUM: 1.4% in chloroform

ULTRAVIOLET DATA: no λ_{max} above 205 nm

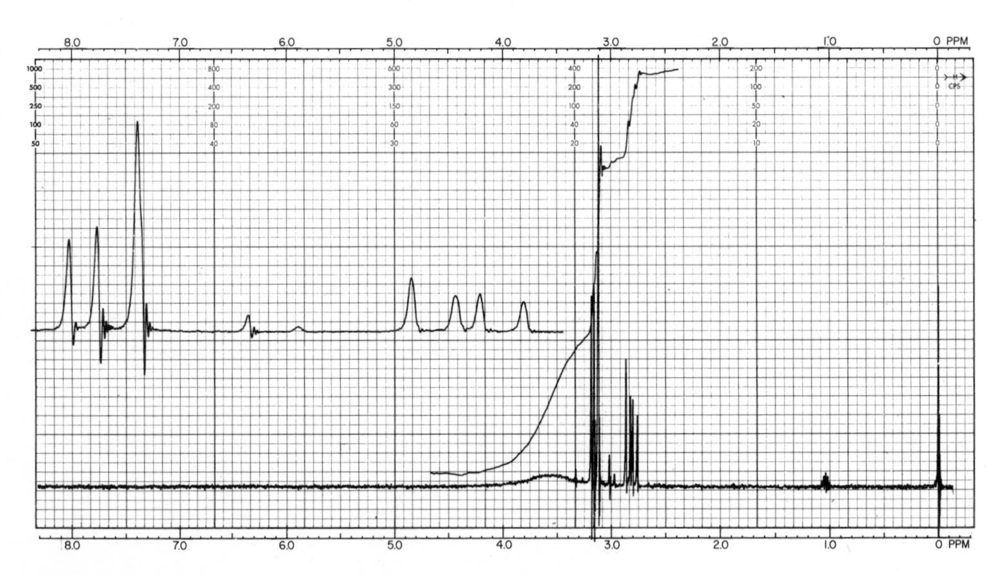

NUCLEAR MAGNETIC RESONANCE SPECTRUM:
15% in acetone-d_6, 1000 Hz sweep width
Expansion: 100 Hz sweep width, 285 Hz sweep offset

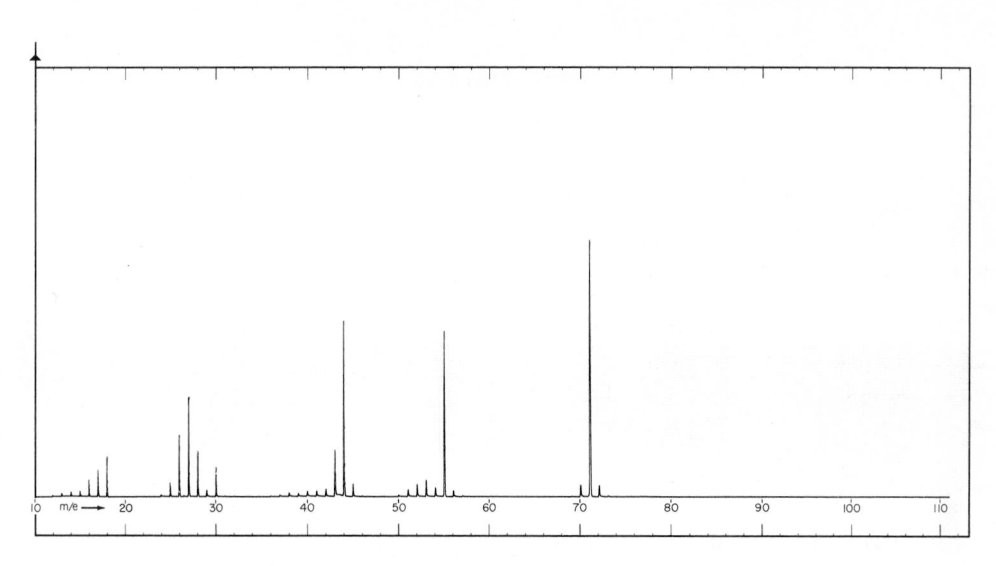

MASS SPECTRUM

MASS SPECTRAL DATA

m/e	*relative intensity*	*m/e*	*relative intensity*	*m/e*	*relative intensity*	*m/e*	*relative intensity*
12	0.3	27	47.0	42	2.2	54	3.0
13	0.5	28	11.8	43	13.0	55	58.0
14	0.8	29	1.6	44	64.0	56	1.9
15	0.9	30	6.6	45	3.4	57	0.3
16	2.7	36	0.2	46	0.2	69	0.2
17	4.8	37	0.5	50	0.5	70	4.0
18	7.8	38	1.0	51	2.0	71	100.0
24	0.4	39	0.9	52	3.7	72	3.7
25	2.6	40	1.6	53	5.4	73	0.3
26	14.9	41	1.6				

COMPOUND 1-11

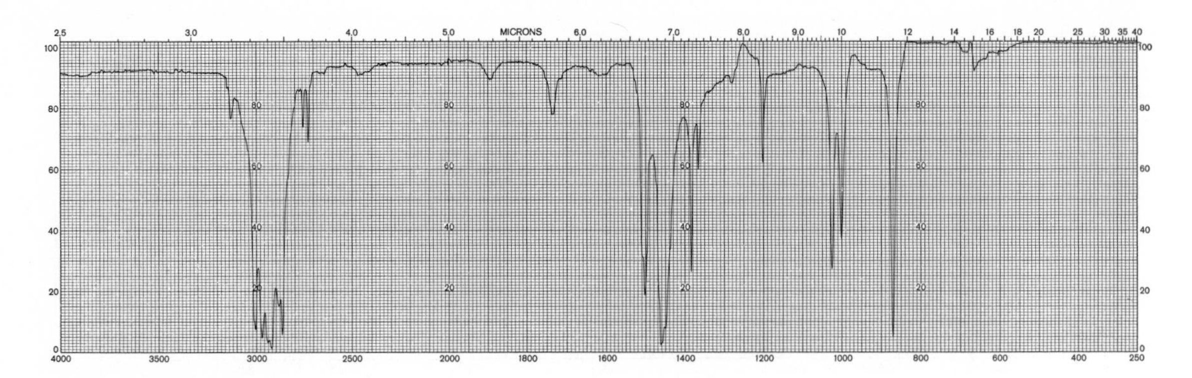

INFRARED SPECTRUM: 9.9% in carbon tetrachloride

ULTRAVIOLET DATA: λ_{max}^{EtOH} 258(sh) nm (ϵ 400); 268 nm (ϵ 630).

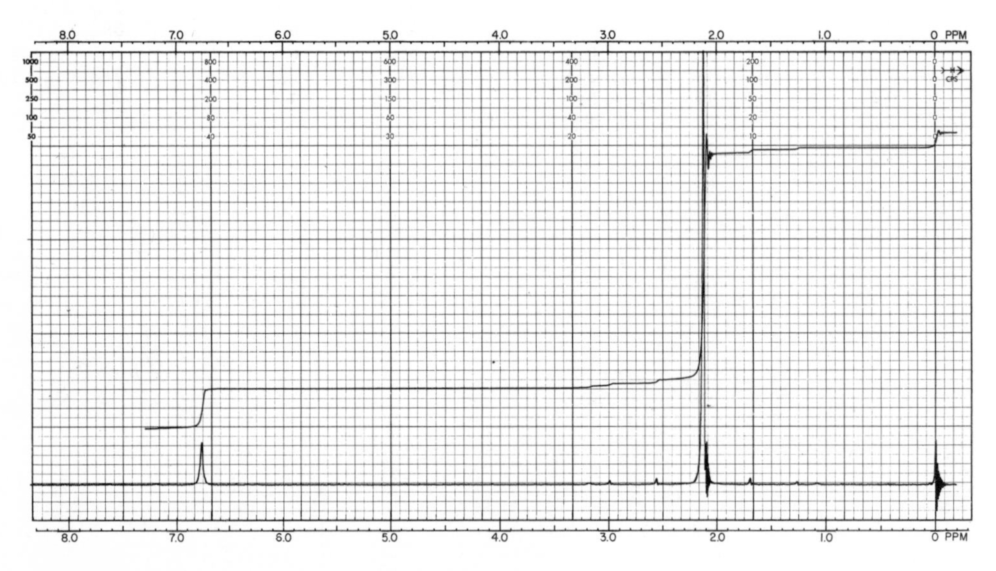

NUCLEAR MAGNETIC RESONANCE SPECTRUM:
15% in carbon tetrachloride, 500 Hz sweep width

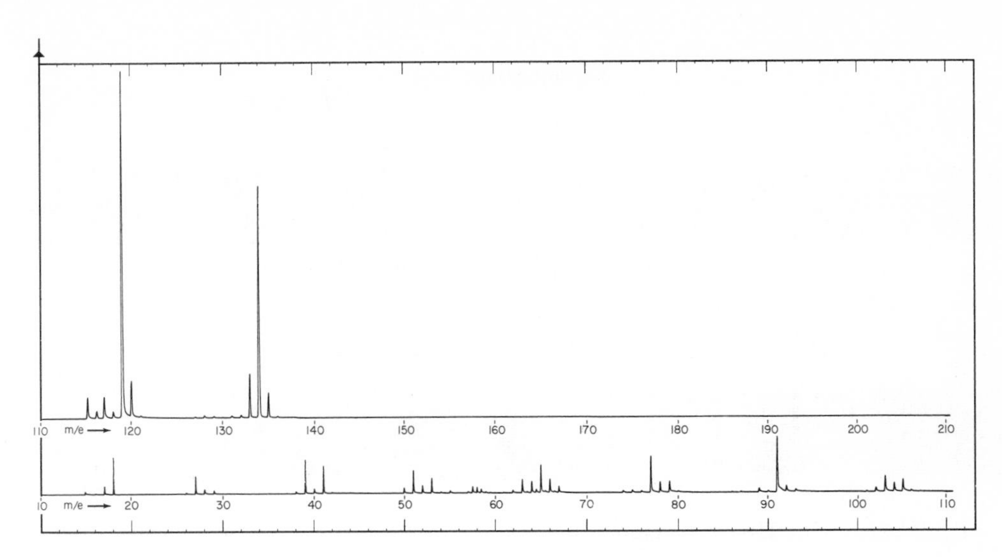

MASS SPECTRUM

MASS SPECTRAL DATA

m/e	*relative intensity*	*m/e*	*relative intensity*	*m/e*	*relative intensity*	*m/e*	*relative intensity*
15	0.5	57.5	1.0	77	5.9	115	6.0
17	0.6	58	0.9	78	2.0	116	2.3
18	2.5	58.5	0.8	79	2.1	117	6.3
27	1.9	59	0.2	80	0.4	118	2.2
28	0.7	62	0.6	87	0.2	119	100.0
29	0.6	63	2.0	89	0.8	120	11.1
38	0.3	63.5	0.3	90	0.4	121	0.8
39	4.6	64	1.8	91	10.2	127	0.3
40	0.7	64.5	0.7	92	1.3	128	0.8
41	3.2	65	4.3	93	0.7	129	0.6
50	0.9	65.5	0.6	101	0.3	131	0.8
51	3.1	66	2.0	102	1.0	132	1.0
52	1.2	66.5	0.3	103	3.0	133	13.4
53	2.3	67	1.2	104	1.7	134	69.0
54	0.2	74	0.4	105	2.4	135	8.1
55	0.4	75	0.5	106	0.5	136	0.5
57	0.3	76	0.4	107	0.2		

COMPOUND 1-12

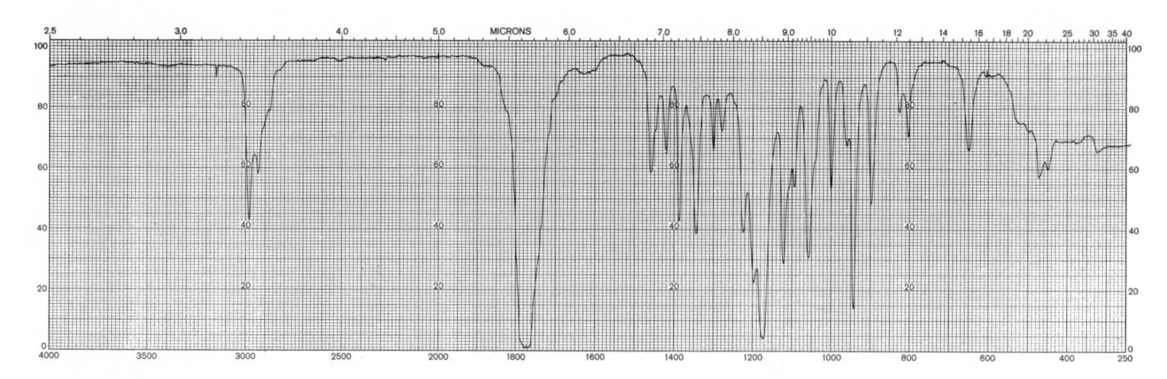

INFRARED SPECTRUM: Liquid film

ULTRAVIOLET DATA: λ_{max}^{EtOH} 209 nm (ϵ 43)

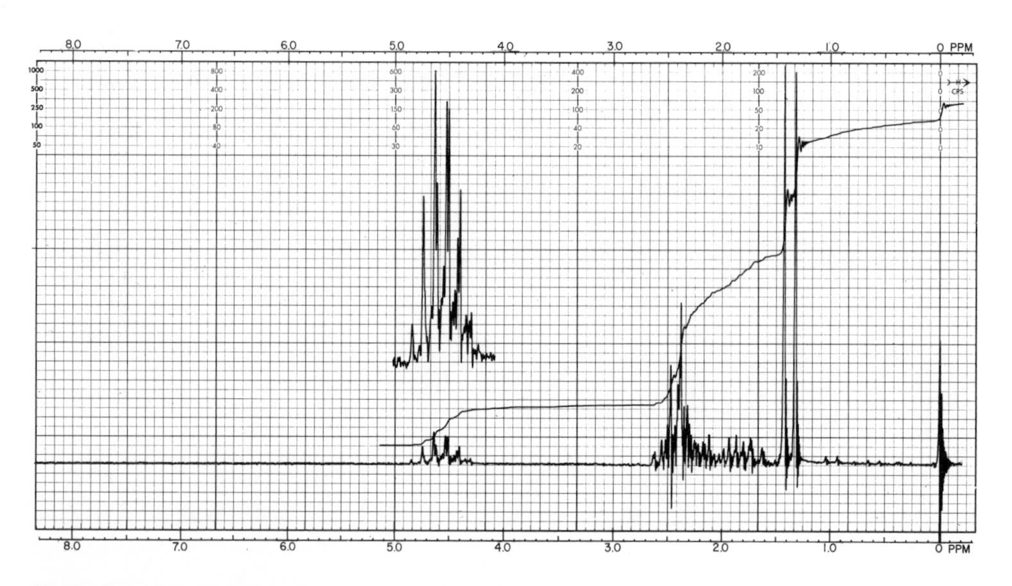

NUCLEAR MAGNETIC RESONANCE SPECTRUM:
15% in carbon tetrachloride, 500 Hz sweep width
Insert: Amplitude X 10

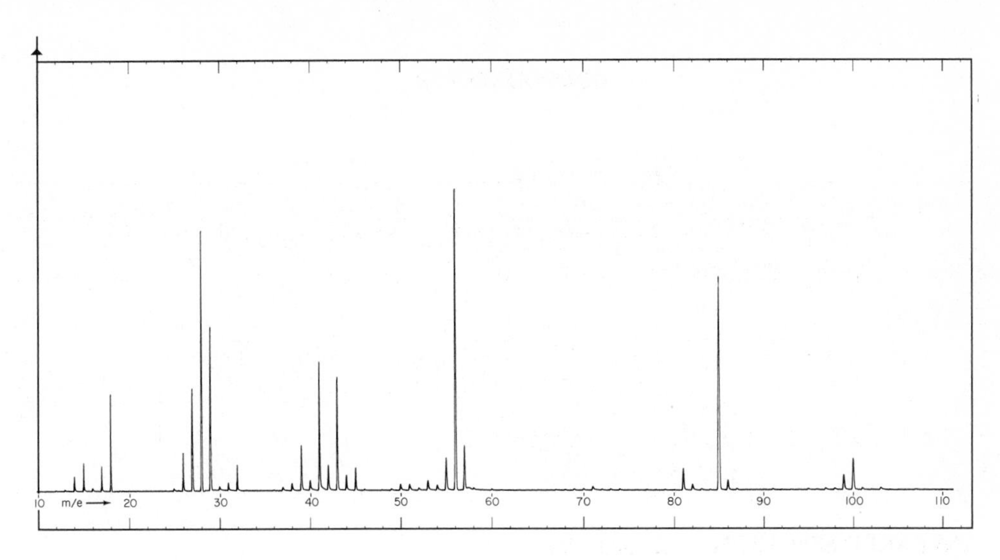

MASS SPECTRUM

MASS SPECTRAL DATA

m/e	*relative intensity*	*m/e*	*relative intensity*	*m/e*	*relative intensity*	*m/e*	*relative intensity*
14	2.7	32	7.0	50	1.7	83	0.7
15	4.8	37	1.0	51	1.7	85	74.0
16	1.0	38	2.0	52	0.7	86	3.6
17	4.4	39	11.7	53	3.1	87	0.3
18	19.6	40	3.0	54	1.6	95	0.5
25	0.5	41	33.9	55	9.3	96	0.4
26	1.8	42	6.8	56	100.0	97	0.8
27	23.3	43	29.8	57	13.8	98	0.6
28	79.0	44	4.3	71	1.5	99	5.4
29	40.8	45	6.1	74	0.6	100	10.3
30	1.2	46	0.6	81	6.8	101	0.8
31	2.0	49	0.5	82	2.2		

COMPOUND 1-13

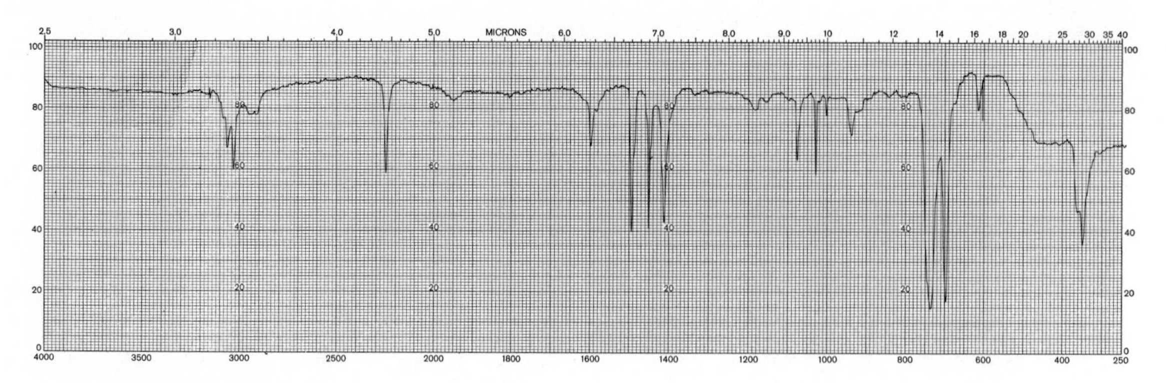

INFRARED SPECTRUM: Liquid film

ULTRAVIOLET DATA: λ_{max}^{EtOH} 206 nm (ϵ 7,300); 246 nm (ϵ 90); 251 nm (ϵ 135); 257 nm (ϵ 175); 263 nm (ϵ 135).

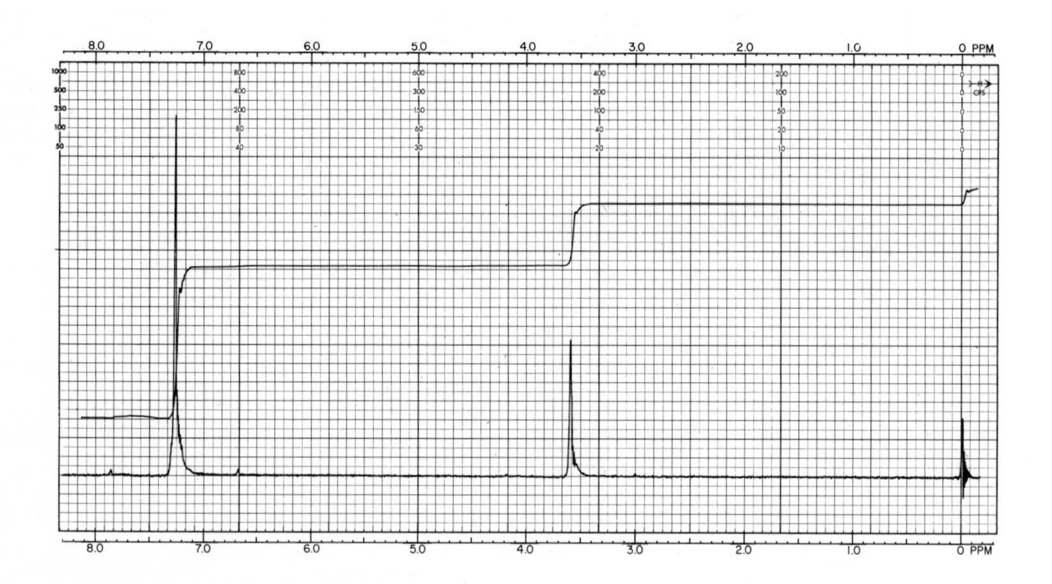

NUCLEAR MAGNETIC RESONANCE SPECTRUM:
15% in carbon tetrachloride, 500 Hz sweep width

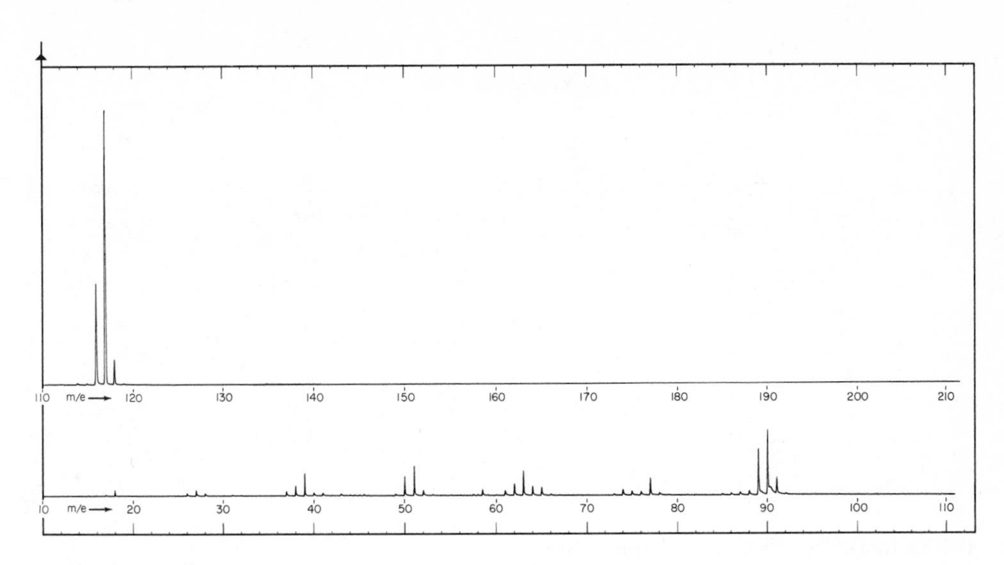

MASS SPECTRUM

MASS SPECTRAL DATA

m/e	relative intensity	m/e	relative intensity	m/e	relative intensity	m/e	relative intensity
18	0.5	50	5.1	64	3.0	88	1.7
26	0.7	51	7.4	65	2.7	89	15.5
27	1.3	52	1.5	66	0.4	90	21.6
28	0.5	53	0.3	73	0.5	91	6.2
37	1.2	57.5	0.5	74	1.9	92	0.8
38	2.2	58	0.3	75	1.6	114	0.8
39	5.6	58.5	1.7	76	1.3	115	0.7
40	0.9	59	0.4	77	5.3	116	35.0
41	0.7	60	0.2	78	0.9	117	100.0
43	0.5	61	1.6	85	0.5	118	10.1
45.5	0.4	62	3.4	86	0.8	119	0.5
49	0.5	63	7.0	87	1.3		

COMPOUND 1-14

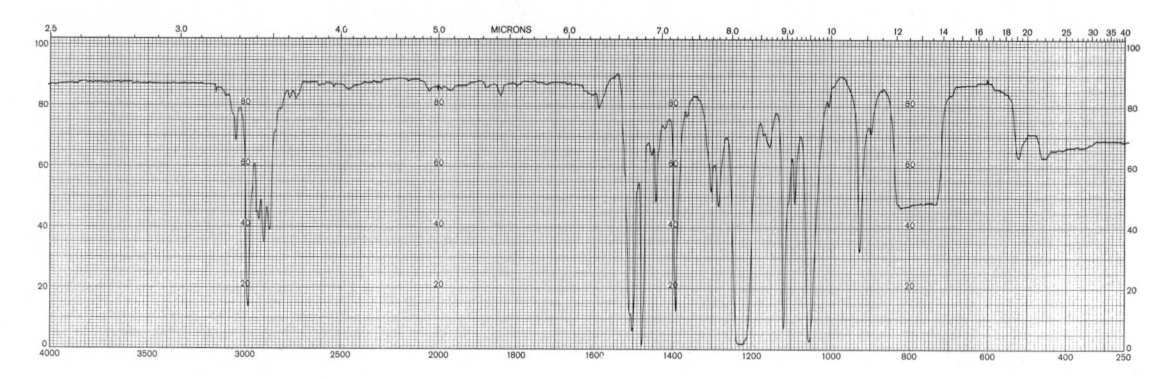

INFRARED SPECTRUM: Liquid film

ULTRAVIOLET DATA: λ_{max}^{EtOH} 291 nm (ϵ 2,400)

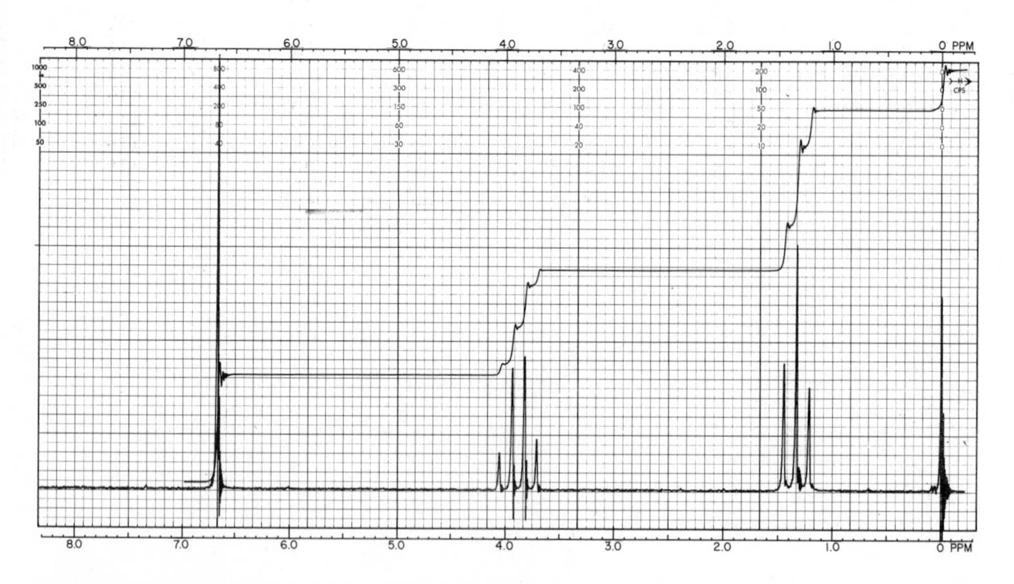

NUCLEAR MAGNETIC RESONANCE SPECTRUM:
15% in carbon tetrachloride, 500 Hz sweep width

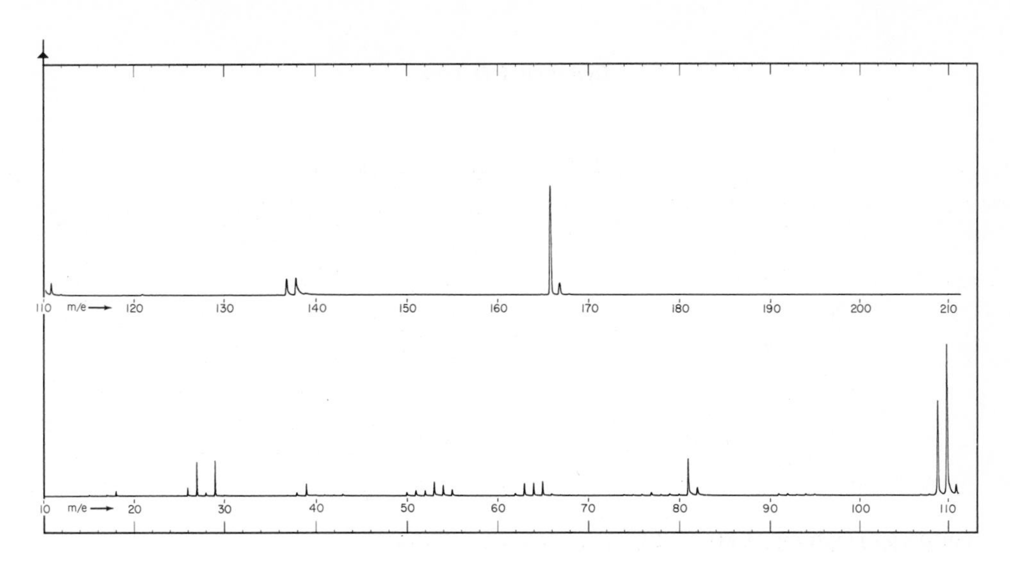

MASS SPECTRUM

MASS SPECTRAL DATA

m/e	*relative intensity*	*m/e*	*relative intensity*	*m/e*	*relative intensity*	*m/e*	*relative intensity*
18	1.2	53	6.4	77	1.8	109	62.0
26	4.1	54	5.1	78	0.3	110	100.0
27	15.6	55	3.1	79	1.1	111	7.2
28	1.7	56	0.3	80	0.8	112	0.8
29	16.3	61	0.3	81	19.0	121	1.0
38	1.8	62	1.3	82	4.2	137	6.7
39	6.1	63	6.2	91	1.0	138	7.0
40	0.4	64	6.4	92	1.1	139	1.0
41	0.5	65	7.0	93	0.6	151	0.6
43	0.8	66	1.3	94	0.9	165	0.5
50	1.9	74	0.5	95	0.8	166	49.0
51	2.7	75	0.5	107	0.4	167	5.5
52	2.8	76	0.7	108	0.7	168	0.7

COMPOUND 1-15

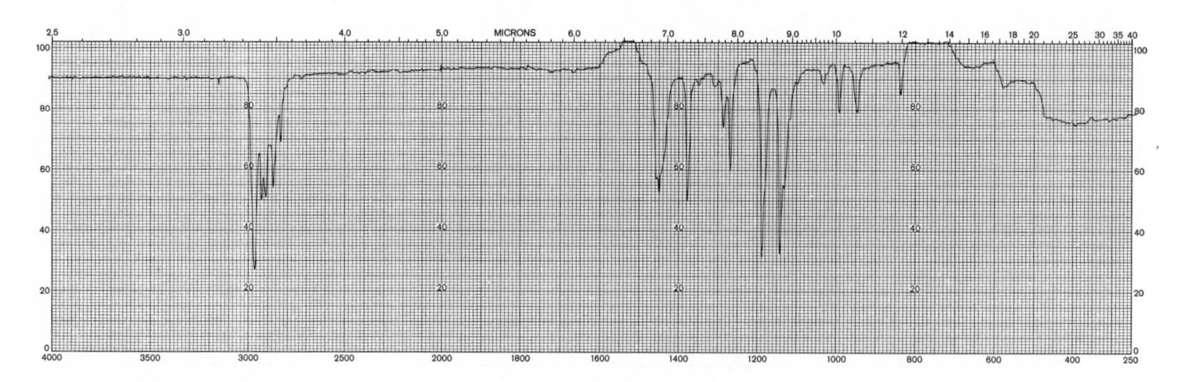

INFRARED SPECTRUM: 4.8% in carbon tetrachloride

ULTRAVIOLET DATA: λ_{max}^{Hexane} 261 nm (ϵ 550)

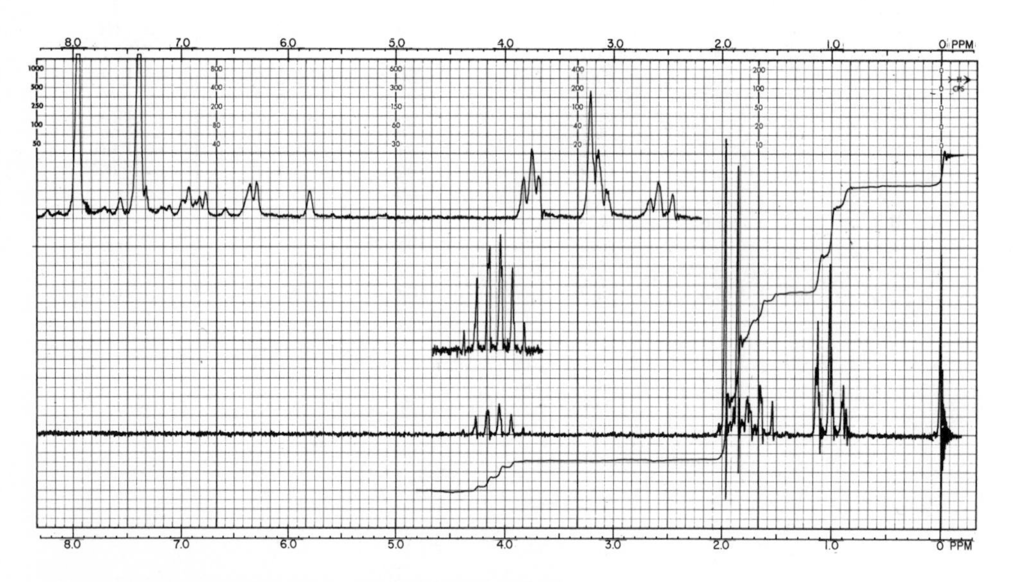

NUCLEAR MAGNETIC RESONANCE SPECTRUM:
15% in carbon tetrachloride, 500 Hz sweep width
Expansion: 100 Hz sweep width, 23 Hz sweep offset
Insert: Amplitude X 4

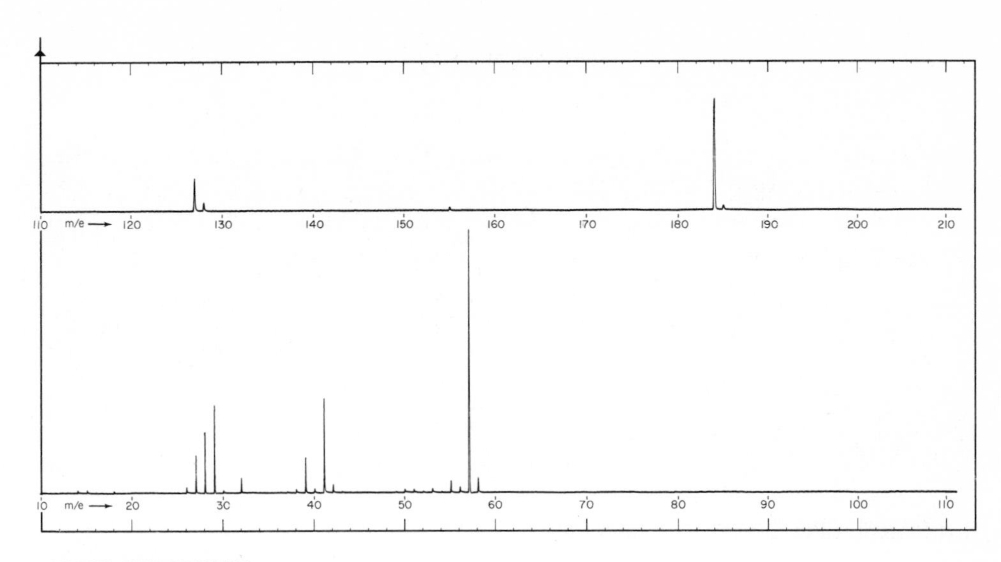

MASS SPECTRUM

MASS SPECTRAL DATA

m/e	*relative intensity*	*m/e*	*relative intensity*	*m/e*	*relative intensity*	*m/e*	*relative intensity*
14	1.2	30	1.0	49	0.5	57	100.0
15	0.6	32	8.2	50	1.5	58	6.6
16	0.6	37	0.6	51	1.5	127	21.0
18	0.9	38	1.4	52	0.7	128	6.2
26	2.4	39	13.5	53	2.1	141	1.1
27	12.3	40	1.9	54	0.6	155	2.9
28	34.0	41	40.0	55	5.7	184	82.0
29	43.0	42	3.5	56	2.9	185	3.7

COMPOUND 1-16

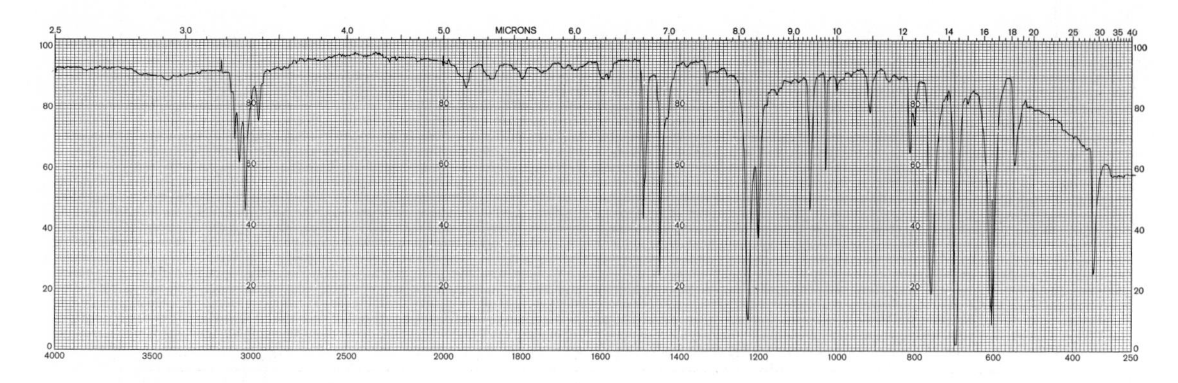

INFRARED SPECTRUM: Liquid film

ULTRAVIOLET DATA: λ_{max}^{Hexane} 225 nm (ϵ 7,400).

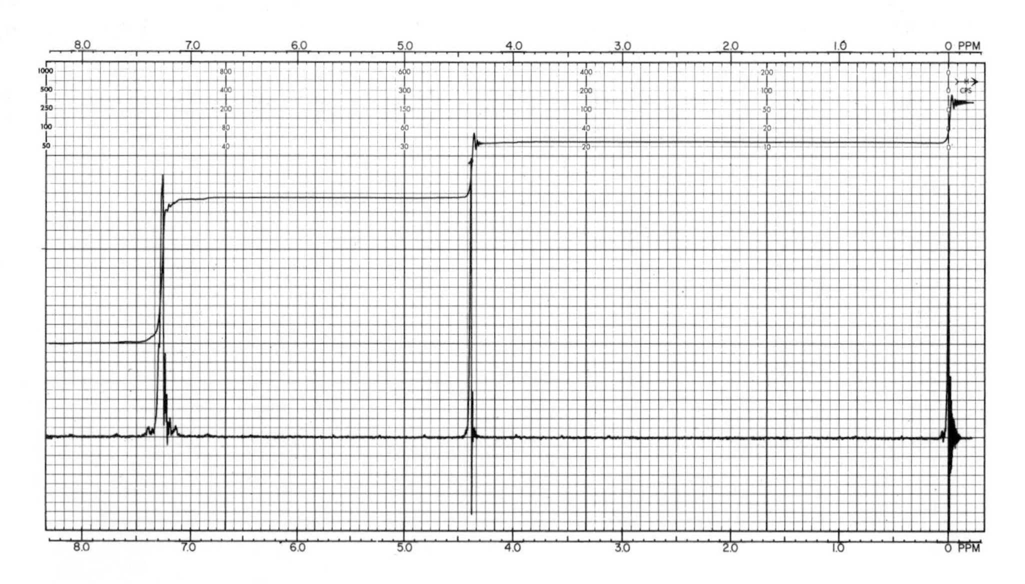

NUCLEAR MAGNETIC RESONANCE SPECTRUM
15% in carbon tetrachloride, 500 Hz sweep width

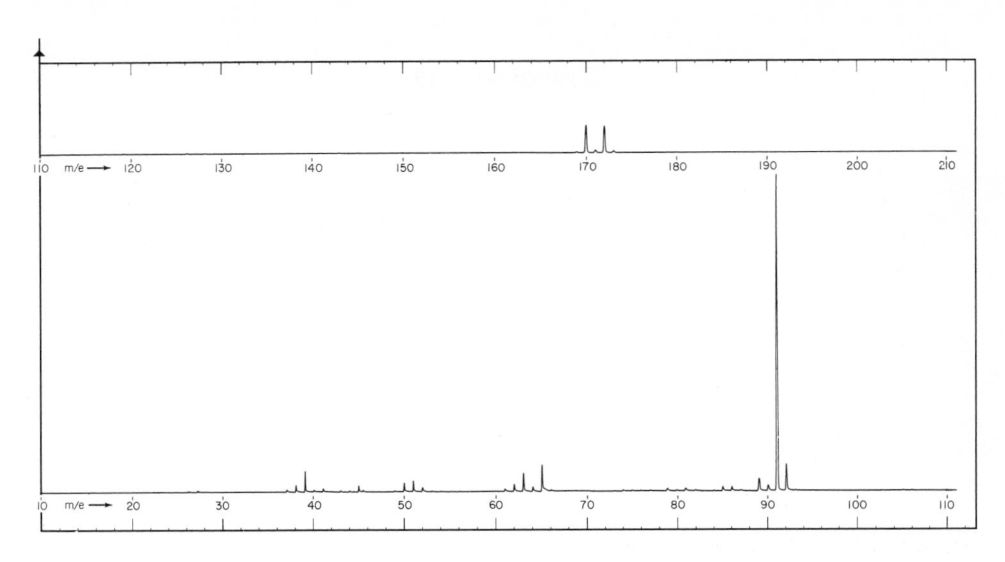

MASS SPECTRUM

MASS SPECTRAL DATA

m/e	*relative intensity*	*m/e*	*relative intensity*	*m/e*	*relative intensity*	*m/e*	*relative intensity*
26	1.1	49	0.7	75	0.6	89	4.8
27	1.3	50	3.6	76	0.6	90	2.5
37	1.1	51	4.8	77	0.4	91	100.0
38	2.7	52	2.1	79	1.4	92	10.1
39	7.9	61	1.4	80	0.6	93	0.6
40	0.9	62	2.7	81	1.5	169	0.6
41	1.8	63	6.8	82	0.7	170	10.3
43	0.8	64	2.3	85	2.1	171	1.5
44	0.8	65	9.8	86	1.7	172	9.9
45	2.5	66	0.9	87	0.6	173	1.1
45.5	0.9	74	0.7				

COMPOUND 1-17

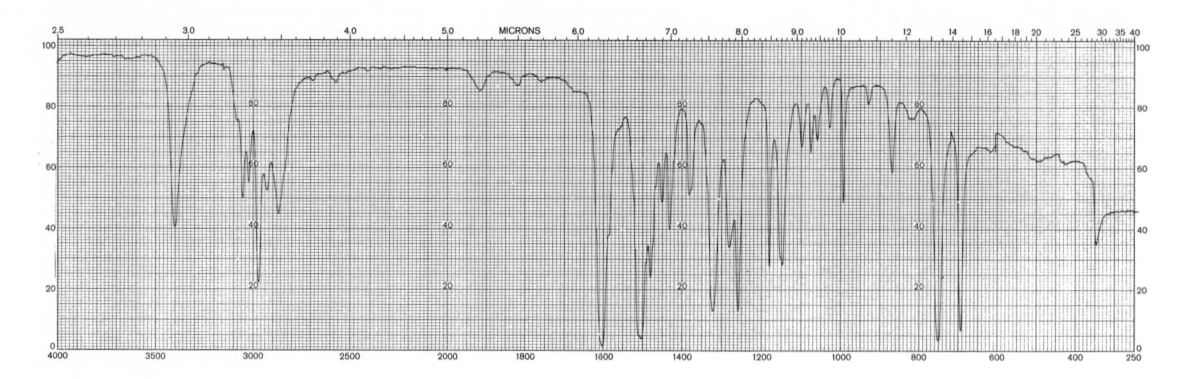

INFRARED SPECTRUM: Liquid film

ULTRAVIOLET DATA: λ_{max}^{Hexane} 245 nm (ϵ 13,000); 294 nm (ϵ 2,200).

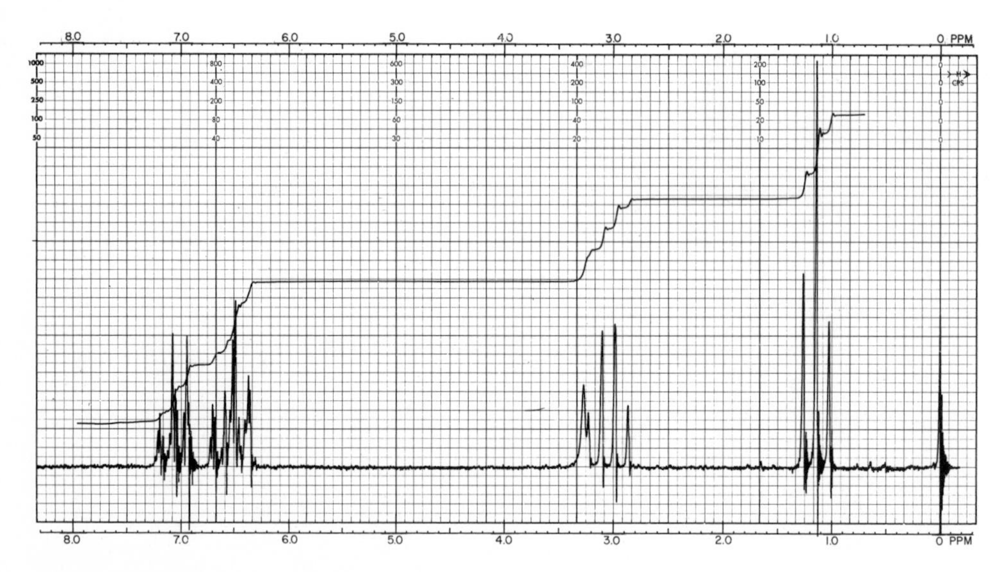

NUCLEAR MAGNETIC RESONANCE SPECTRUM:
15% in carbon tetrachloride, 500 Hz sweep width

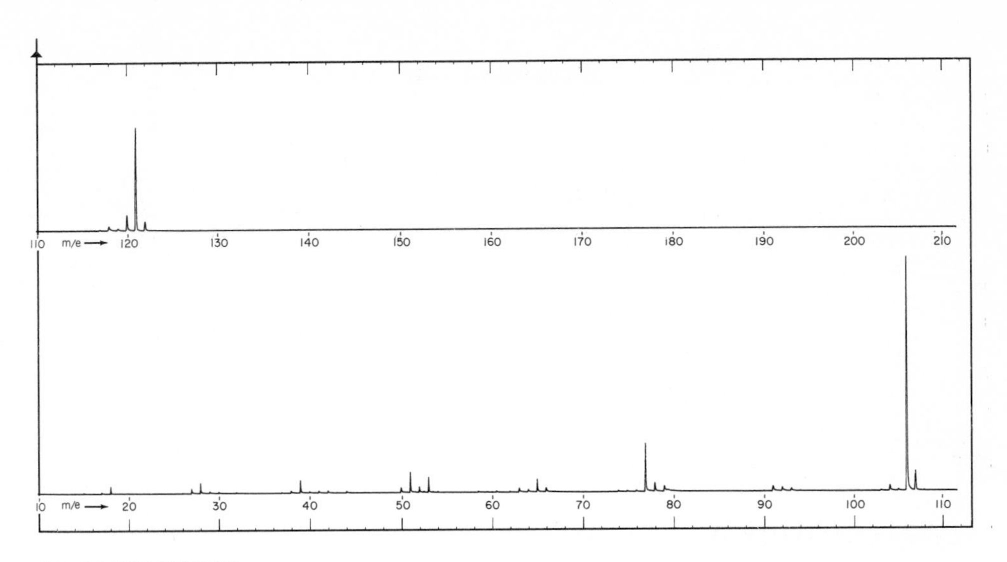

MASS SPECTRUM

MASS SPECTRAL DATA

m/e	*relative intensity*	*m/e*	*relative intensity*	*m/e*	*relative intensity*	*m/e*	*relative intensity*
15	0.5	44	1.1	64	1.5	93	1.6
17	0.7	50	3.2	65	6.1	94	0.3
18	3.4	51	8.6	66	2.1	103	0.6
26	0.5	52	3.2	67	0.5	104	2.8
27	2.8	52.5	0.8	74	0.7	105	1.1
28	4.8	53	7.4	75	0.7	106	100.0
29	1.1	53.5	0.5	76	1.0	107	8.4
30	0.9	54	0.5	77	20.0	108	0.4
32	0.7	58.5	0.8	78	4.1	117	0.6
37	0.3	59	0.2	79	2.9	118	2.2
38	1.3	59.5	0.4	80	0.8	119	1.1
39	5.4	60	0.1	89	0.3	120	7.2
40	1.0	60.5	0.8	90	0.3	121	45.0
41	1.2	61	0.2	91	2.4	122	4.2
42	1.3	62	0.5	92	2.1	123	0.1
43	0.4	63	2.3				

COMPOUND 1-18

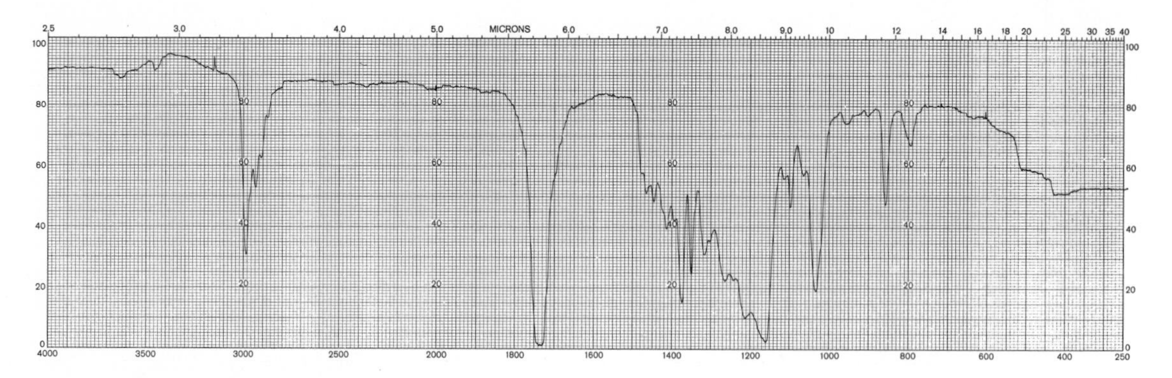

INFRARED SPECTRUM: Liquid film

ULTRAVIOLET DATA: λ_{max}^{EtOH} 209 nm (ϵ 105).

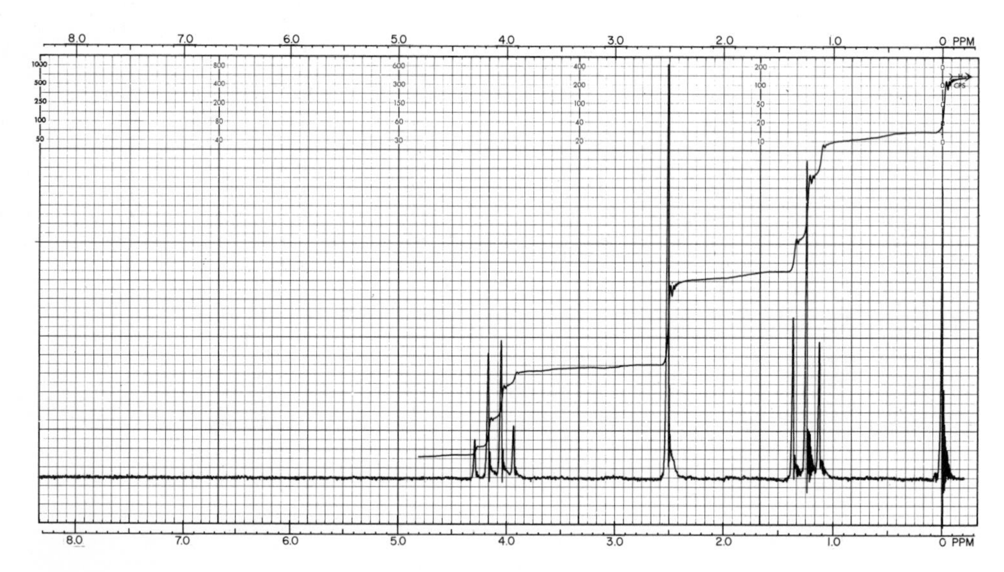

NUCLEAR MAGNETIC RESONANCE SPECTRUM:
15% in carbon tetrachloride, 500 Hz sweep width

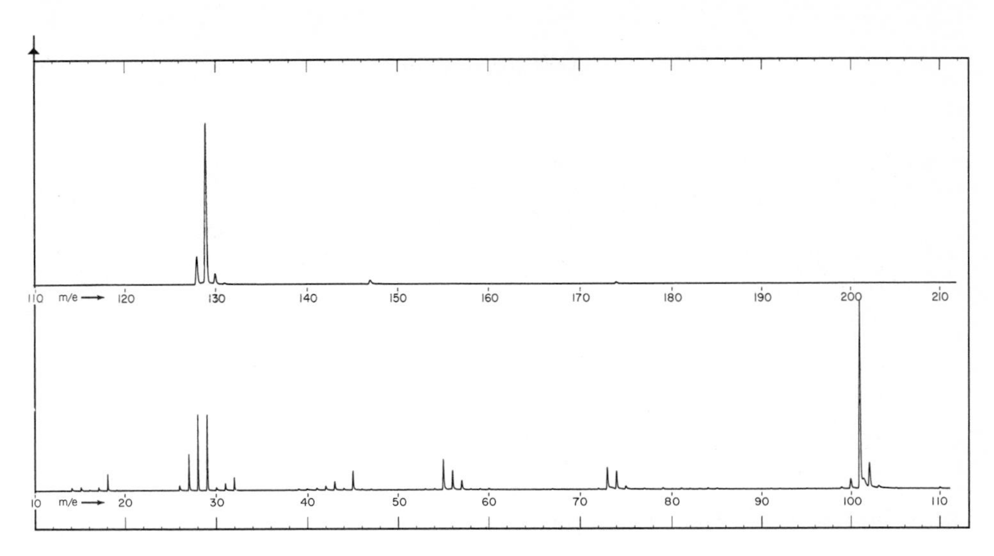

MASS SPECTRUM

MASS SPECTRAL DATA

m/e	*relative intensity*	*m/e*	*relative intensity*	*m/e*	*relative intensity*	*m/e*	*relative intensity*
14	0.5	41	0.8	59	0.4	100	4.7
15	1.2	42	1.9	60	0.7	101	100.0
16	0.4	43	3.6	71	0.2	102	13.2
17	0.7	44	0.7	72	0.5	103	1.7
18	2.8	45	8.2	73	10.6	110	0.5
26	2.1	46	0.4	74	8.8	128	13.9
27	13.5	47	0.4	75	1.8	129	83.0
28	19.6	52	0.3	79	0.5	130	6.0
29	31.0	53	0.4	81	0.4	131	0.9
30	1.4	54	0.5	82	0.4	147	2.3
31	2.4	55	13.1	83	0.2	148	0.2
32	2.1	56	8.5	84	0.6	174	1.0
39	0.4	57	4.2	85	0.4	175	0.1
40	0.3	58	0.3	99	0.9		

COMPOUND 1-19

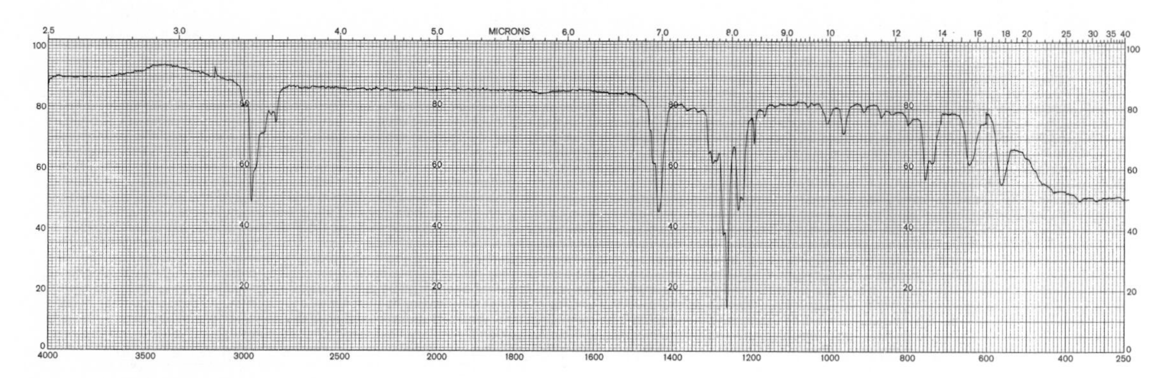

INFRARED SPECTRUM: Liquid film

ULTRAVIOLET DATA: no λ_{max} above 205 nm

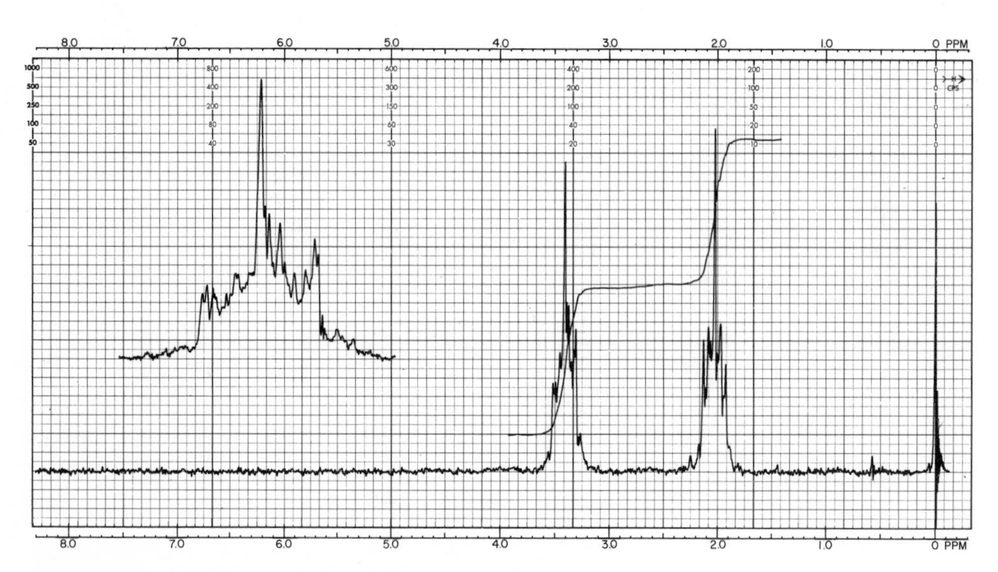

NUCLEAR MAGNETIC RESONANCE SPECTRUM:
15% in carbon tetrachloride, 500 Hz sweep width
Expansion: 100 Hz sweep width, 130 Hz sweep offset

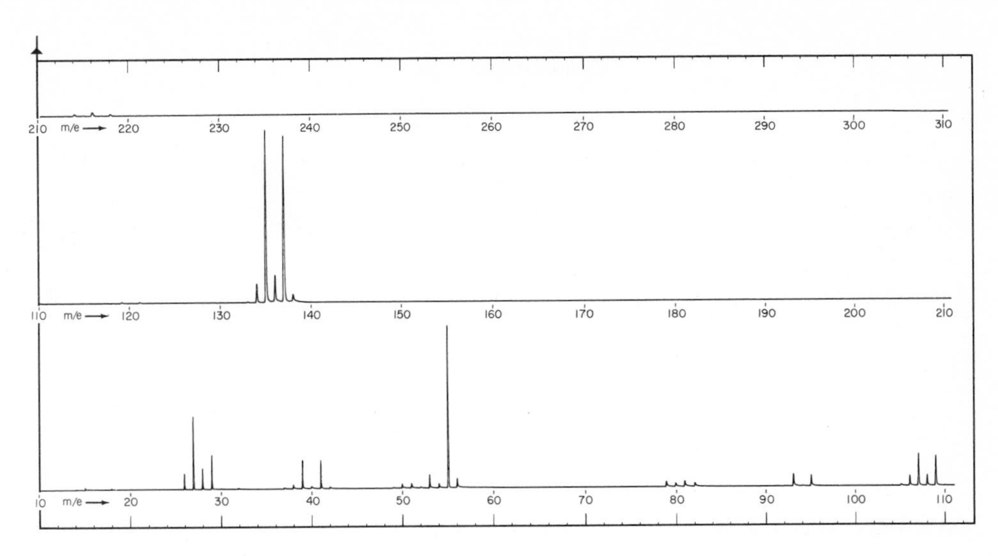

MASS SPECTRUM

MASS SPECTRAL DATA

m/e	*relative intensity*
14	0.3
15	0.6
18	0.4
25	0.3
26	4.9
27	25.9
28	7.3
29	12.3
30	0.4
32	0.4
37	0.7
38	1.6
39	11.6
40	1.1
41	10.4

m/e	*relative intensity*
42	0.9
43	0.2
49	0.5
50	1.9
51	2.0
52	0.7
53	5.6
54	2.1
55	79.0
56	4.6
57	0.2
67	0.5
68	0.4
79	3.0
80	2.5

m/e	*relative intensity*
81	3.5
82	2.5
83	0.4
91	0.4
92	0.5
93	6.4
94	0.6
95	6.0
96	0.2
104	0.3
105	1.2
106	5.9
107	17.6
108	6.3
109	16.8

m/e	*relative intensity*
110	0.4
117	0.2
119	1.1
121	0.9
123	0.8
133	0.7
134	11.3
135	100.0
136	16.8
137	97.0
138	4.9
214	1.3
216	2.6
218	1.2

COMPOUND 1-20

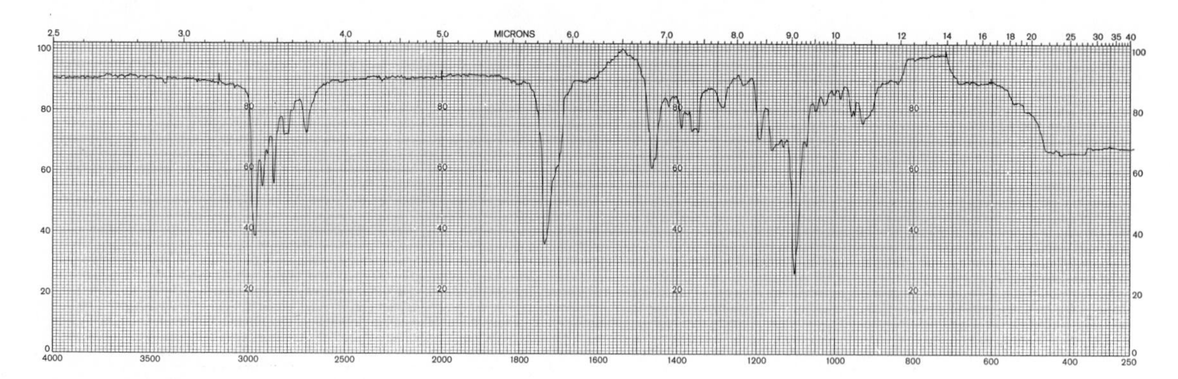

INFRARED SPECTRUM: 2.2% in carbon tetrachloride

ULTRAVIOLET DATA: $\lambda_{max}^{H_2O}$ 286 nm (ϵ 25).

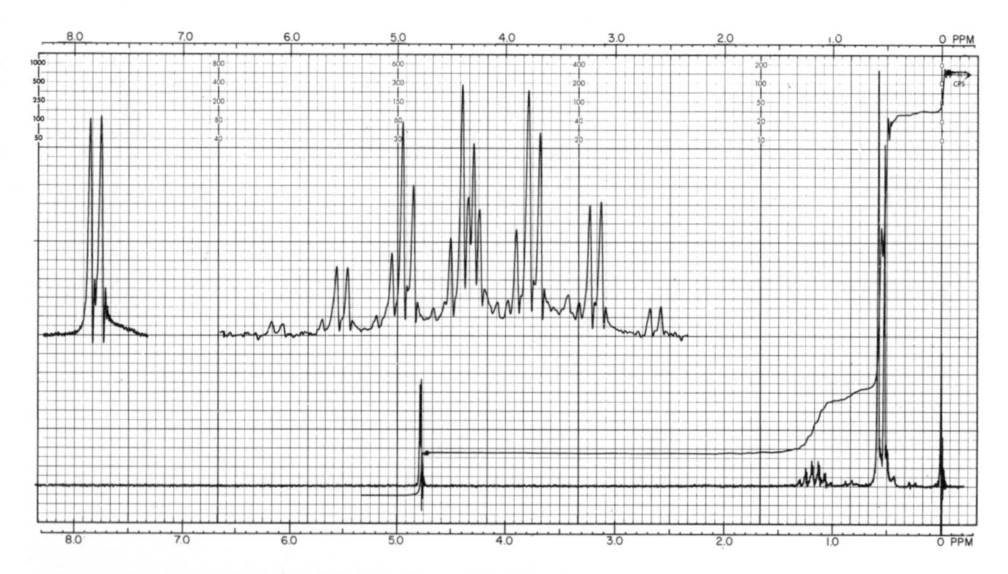

NUCLEAR MAGNETIC RESONANCE SPECTRUM:
15% in carbon tetrachloride, 1000 Hz sweep width
Expansions: (left) 100 Hz sweep width, 480 Hz sweep offset
(right) 100 Hz sweep width, 90 Hz sweep offset

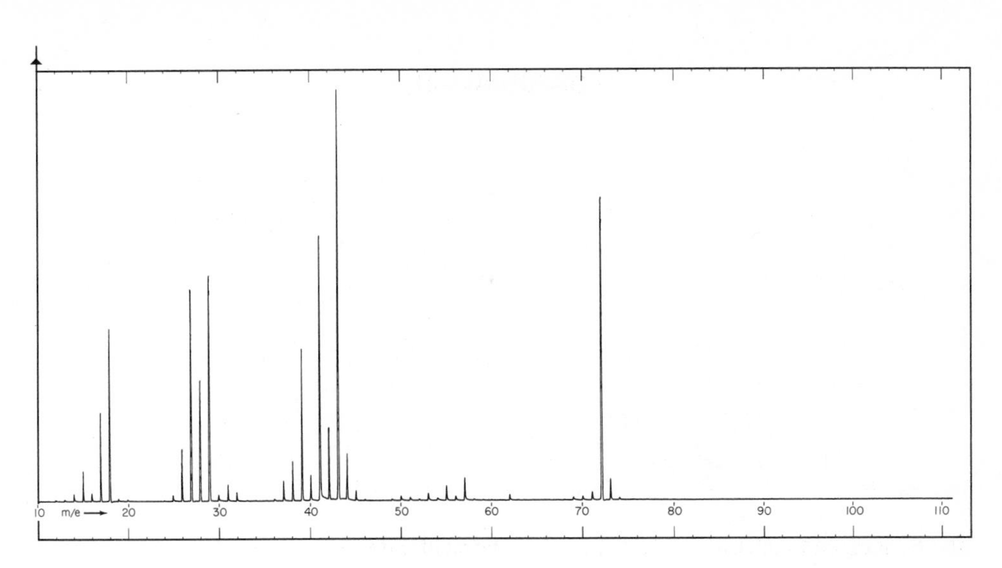

MASS SPECTRUM

MASS SPECTRAL DATA

m/e	*relative intensity*	*m/e*	*relative intensity*	*m/e*	*relative intensity*	*m/e*	*relative intensity*
12	0.2	28	15.9	41	69.0	54	0.4
13	0.3	29	36.0	42	9.6	55	1.9
14	0.7	30	1.1	43	100.0	56	0.7
15	2.9	31	1.5	44	6.3	57	3.1
16	0.6	32	0.4	45	1.0	69	0.4
17	3.9	36	0.3	49	0.2	70	0.4
18	20.2	37	2.5	50	0.6	71	1.1
25	0.7	38	4.5	51	0.4	72	63.0
26	5.9	39	23.0	52	0.1	73	3.3
27	52.0	40	3.4	53	1.1	74	0.3

COMPOUND 1-21

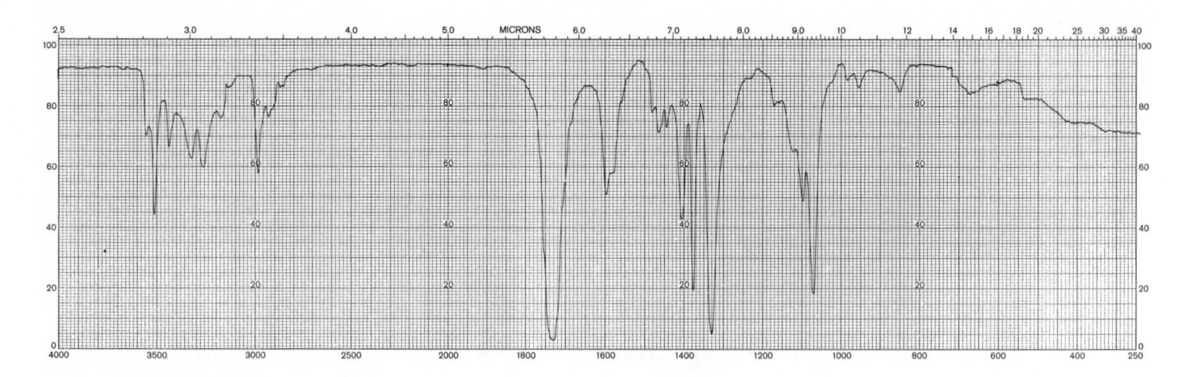

INFRARED SPECTRUM: 1.2% in carbon tetrachloride

ULTRAVIOLET DATA: no λ_{max} above 205 nm

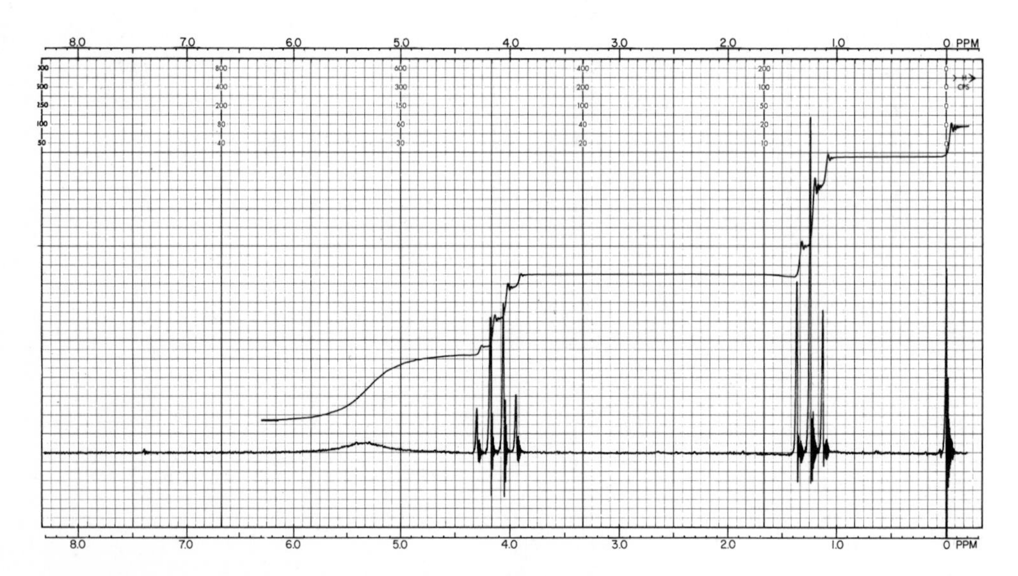

NUCLEAR MAGNETIC RESONANCE SPECTRUM:
15% in carbon tetrachloride

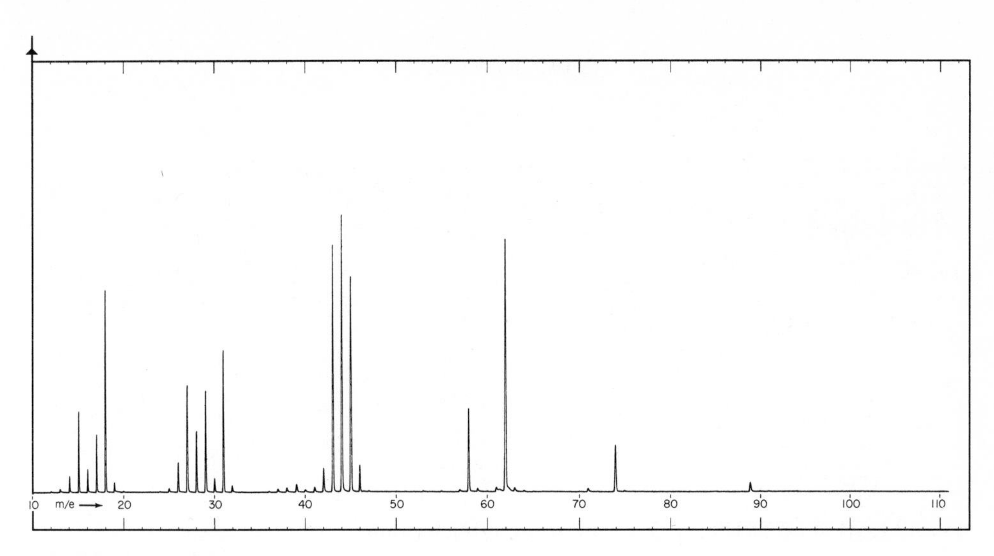

MASS SPECTRUM

MASS SPECTRAL DATA

m/e	*relative intensity*	*m/e*	*relative intensity*	*m/e*	*relative intensity*	*m/e*	*relative intensity*
12	0.8	27	38.0	40	1.0	59	1.0
13	0.9	28	21.0	41	1.6	61	2.2
14	3.1	29	38.0	42	5.4	62	95.0
15	12.1	30	5.8	43	48.0	63	2.5
16	7.6	31	50.0	44	100.0	64	0.4
17	12.9	32	2.0	45	78.0	71	1.7
18	46.0	33	0.1	46	10.8	74	18.0
19	3.6	37	0.8	47	0.7	75	0.9
25	1.0	38	1.0	57	0.8	89	4.0
26	9.1	39	1.5	58	11.8	90	0.2

COMPOUND 1-22

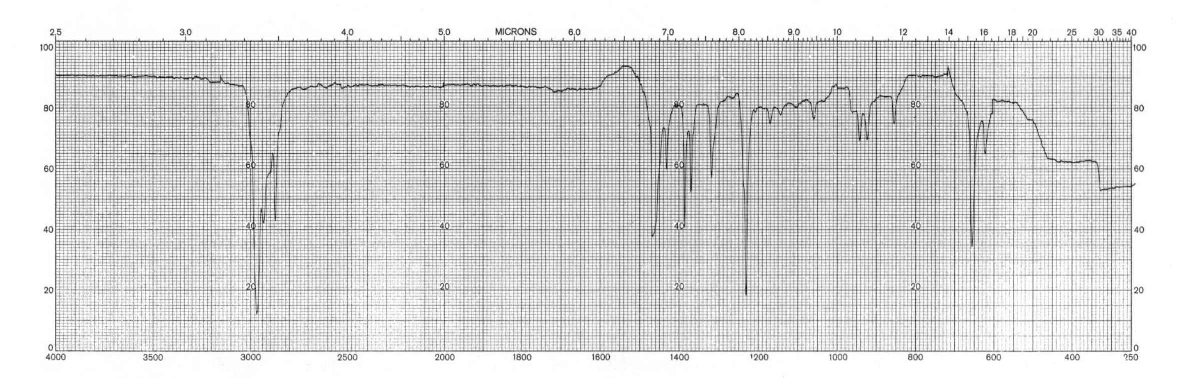

INFRARED SPECTRUM: 5.1% in carbon tetrachloride

ULTRAVIOLET DATA: no λ_{max}^{EtOH} above 205 nm

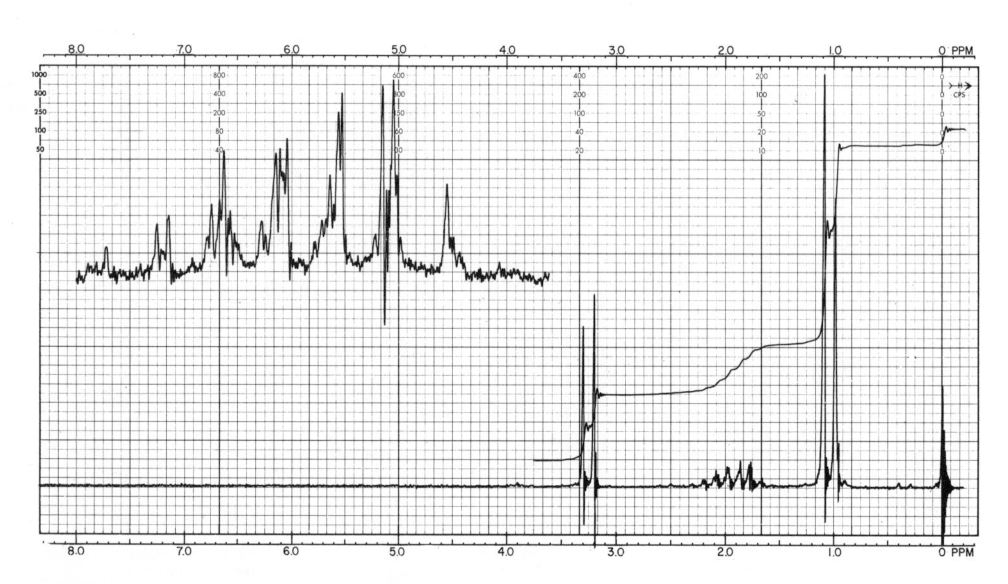

NUCLEAR MAGNETIC RESONANCE SPECTRUM:
15% in carbon tetrachloride, 500 Hz sweep width
Expansion: 100 Hz sweep width, 45 Hz sweep offset

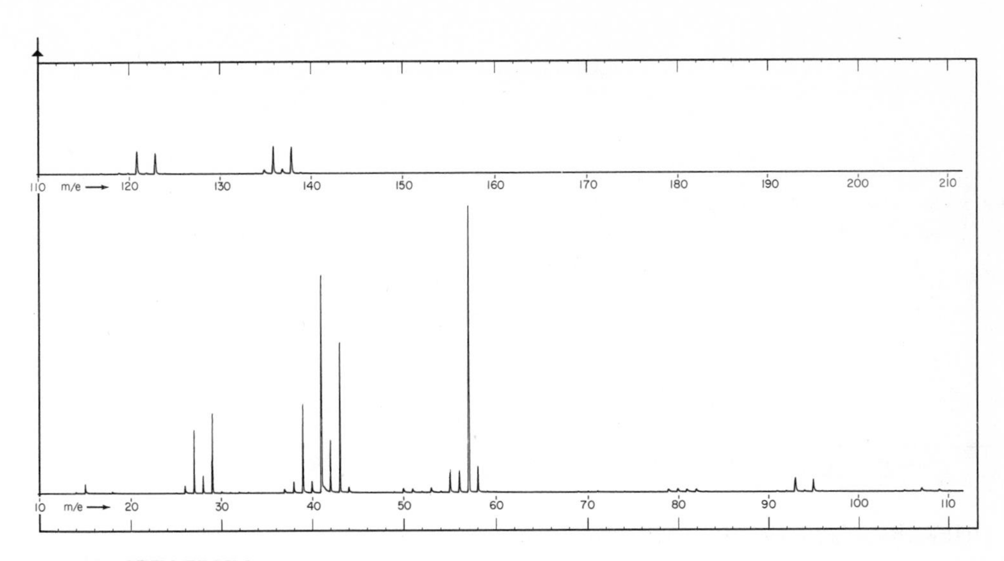

MASS SPECTRUM

MASS SPECTRAL DATA

m/e	relative intensity	m/e	relative intensity	m/e	relative intensity	m/e	relative intensity
14	0.5	42	9.5	79	0.8	118	0.2
15	2.1	43	33.0	80	1.0	119	0.2
17	0.6	44	1.5	81	0.8	120	0.3
26	1.9	49	0.5	82	0.9	121	4.1
27	16.0	50	0.9	93	2.5	122	0.6
28	3.7	51	1.1	94	0.5	123	4.0
29	19.5	52	0.2	95	2.4	124	0.2
30	0.5	53	1.1	96	0.1	135	1.0
37	1.0	54	0.2	105	0.5	136	5.2
38	2.5	55	4.4	107	0.7	137	1.0
39	16.6	56	3.9	109	0.6	138	5.1
40	2.3	57	100.0	117	0.3	139	0.3
41	50.0	58	4.7				

COMPOUND 1-23

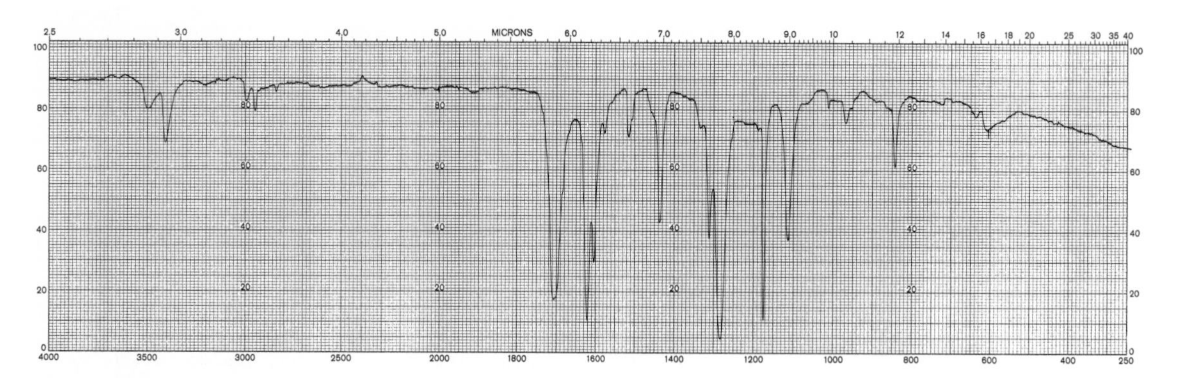

INFRARED SPECTRUM: 1.1% in chloroform

ULTRAVIOLET DATA: λ_{max}^{EtOH} 294 nm (ϵ 20,000).

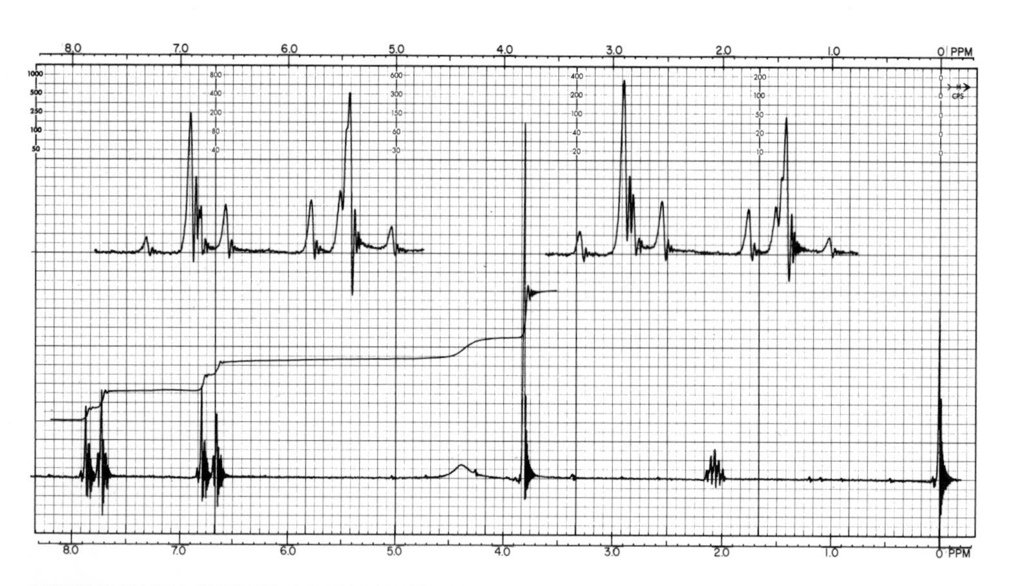

NUCLEAR MAGNETIC RESONANCE SPECTRUM:
15% in acetone-d_6, 500 Hz sweep width
Expansions: (left) 50 Hz sweep width, 430 Hz sweep offset
(right) 50 Hz sweep width, 390 Hz sweep offset

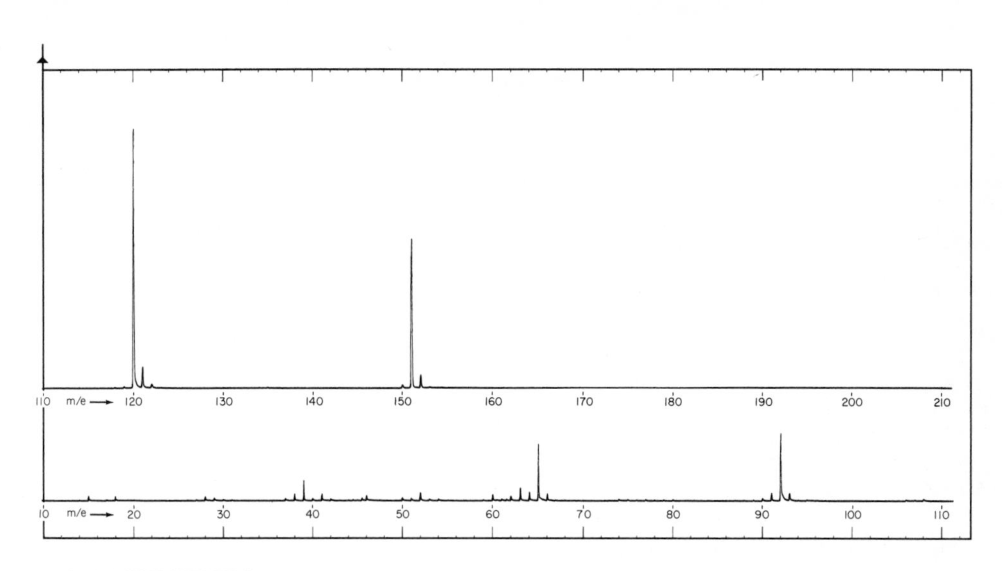

MASS SPECTRUM

MASS SPECTRAL DATA

m/e	*relative intensity*	*m/e*	*relative intensity*	*m/e*	*relative intensity*	*m/e*	*relative intensity*
15	1.1	50	1.0	67	0.2	95	0.3
18	0.7	51	0.8	74	0.5	106	0.6
27	0.4	52	1.9	75	0.4	107	0.3
28	1.2	53	0.5	76	0.3	108	0.9
29	0.8	54	0.5	77	0.3	118	0.3
30	0.3	59	0.4	78	0.2	119	0.9
31	0.3	60	1.9	79	0.4	120	100.0
37	0.6	60.5	0.2	80	0.3	121	9.6
38	1.6	61	0.6	89	0.3	122	2.1
39	4.8	61.5	0.4	90	1.0	135	0.3
40	0.7	62	1.5	91	3.1	150	1.7
41	1.8	63	3.6	92	22.0	151	68.0
42	0.6	64	2.8	93	2.9	152	6.5
45.5	0.7	65	14.8	94	0.3	153	0.6
46	1.4	66	2.4				

COMPOUND 1-24

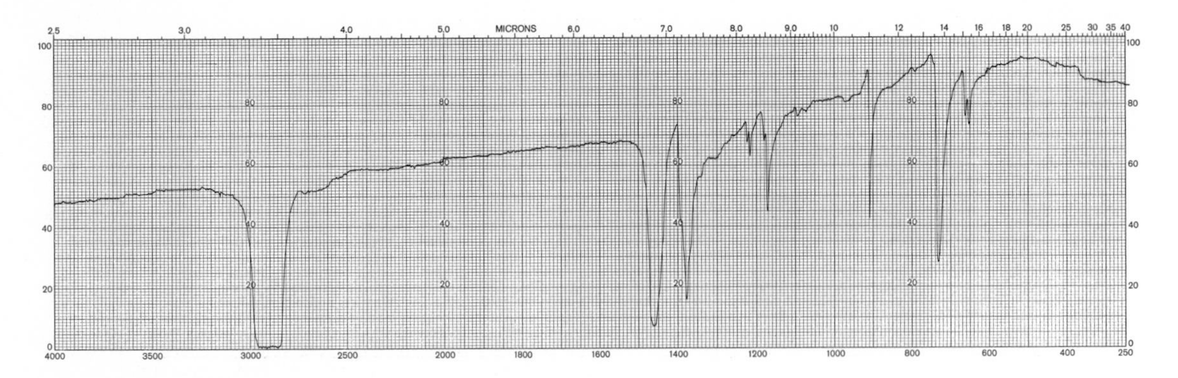

INFRARED SPECTRUM: Nujol mull

ULTRAVIOLET DATA: λ_{max}^{EtOH} 240 nm (ϵ 1,450)

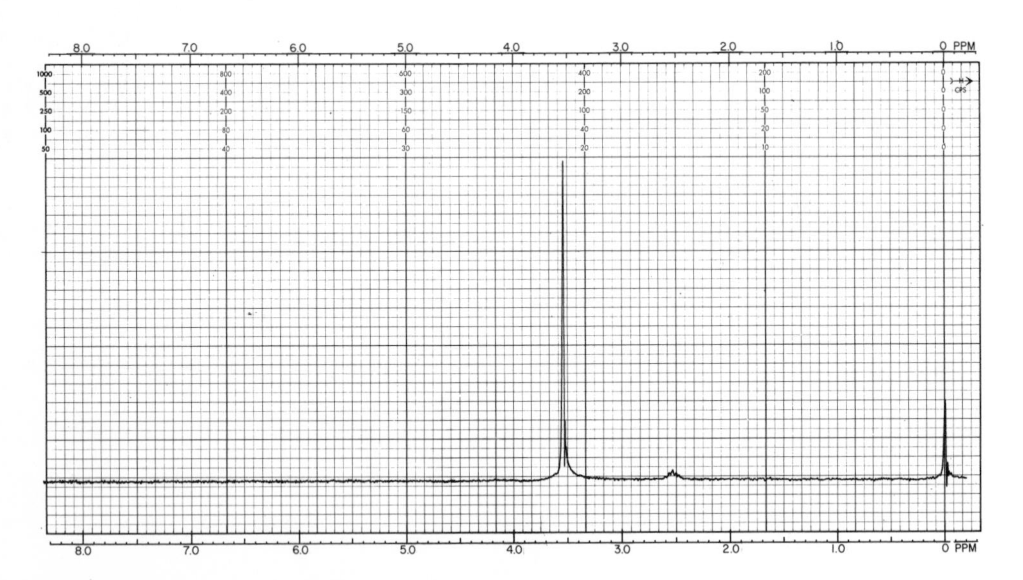

NUCLEAR MAGNETIC RESONANCE SPECTRUM:
15% in DMSO-d_6, 500 Hz sweep width, 80°

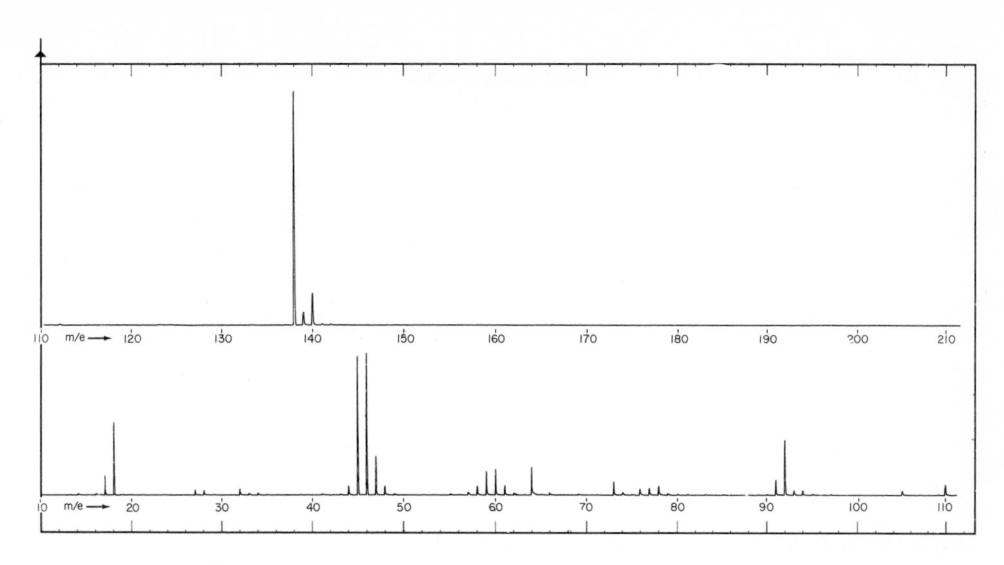

MASS SPECTRUM

MASS SPECTRAL DATA

m/e	*relative intensity*	*m/e*	*realtive intensity*	*m/e*	*relative intensity*	*m/e*	*relative intensity*
16	0.5	46	57.0	62	0.8	92	22.6
17	3.6	47	12.4	64	9.3	93	1.9
18	13.5	48	3.0	66	1.0	94	2.0
27	1.5	49	0.7	73	5.0	105	2.1
28	1.5	55	0.7	74	1.3	110	5.0
32	1.7	57	1.2	76	2.7	138	100.0
33	0.6	58	3.2	77	2.6	139	5.6
34	0.9	59	7.7	78	3.6	140	13.6
44	3.2	60	8.4	79	0.7	141	0.7
45	47.4	61	3.4	91	6.5	142	0.7

COMPOUND 1-25

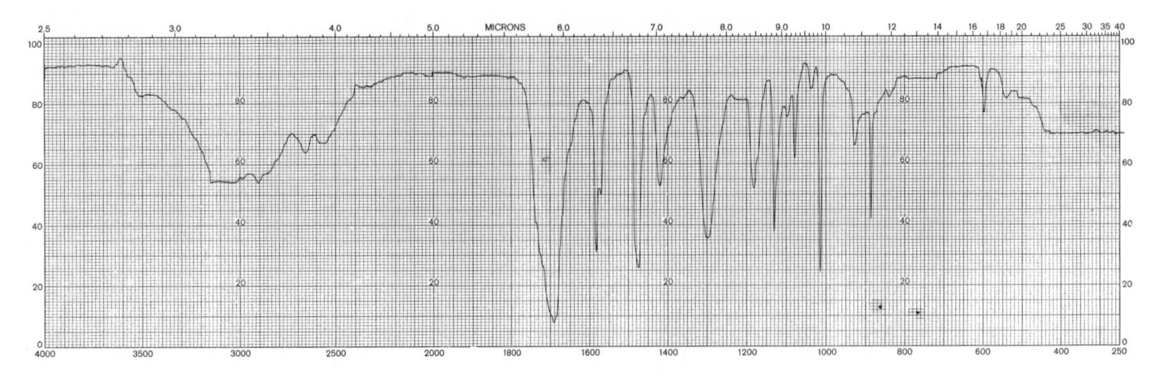

INFRARED SPECTRUM: 1.2% in carbon tetrachloride

ULTRAVIOLET DATA: λ_{max}^{EtOH} 248 nm (ϵ 25,800)

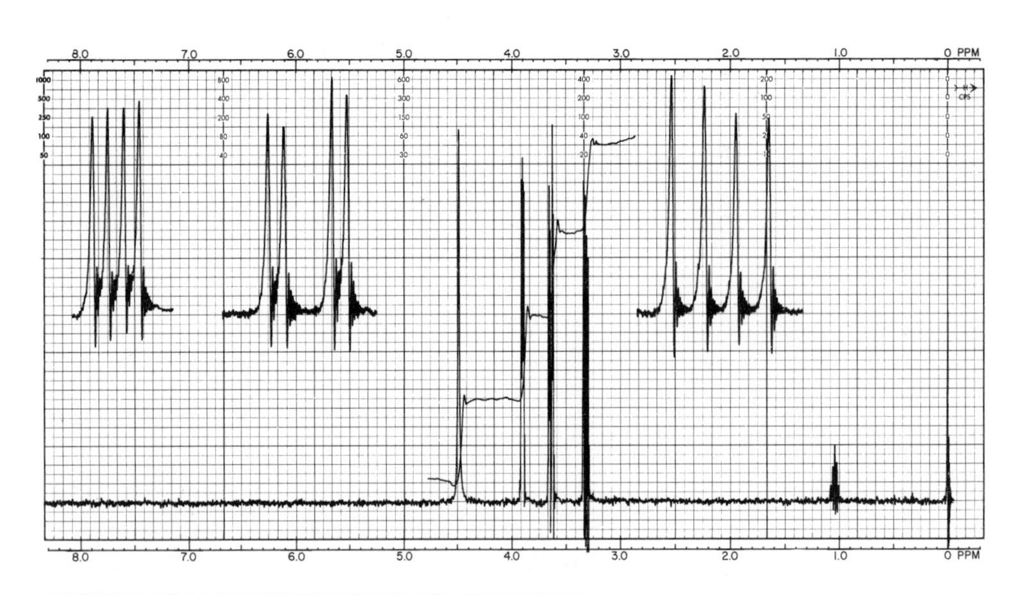

NUCLEAR MAGNETIC RESONANCE SPECTRUM:
15% in acetone-d_6, 1000 Hz sweep width
Expansions: (left) 50 Hz sweep width, 421 Hz sweep offset
(center) 50 Hz sweep width, 400 Hz sweep offset
(right) 50 Hz sweep width, 385 Hz sweep offset

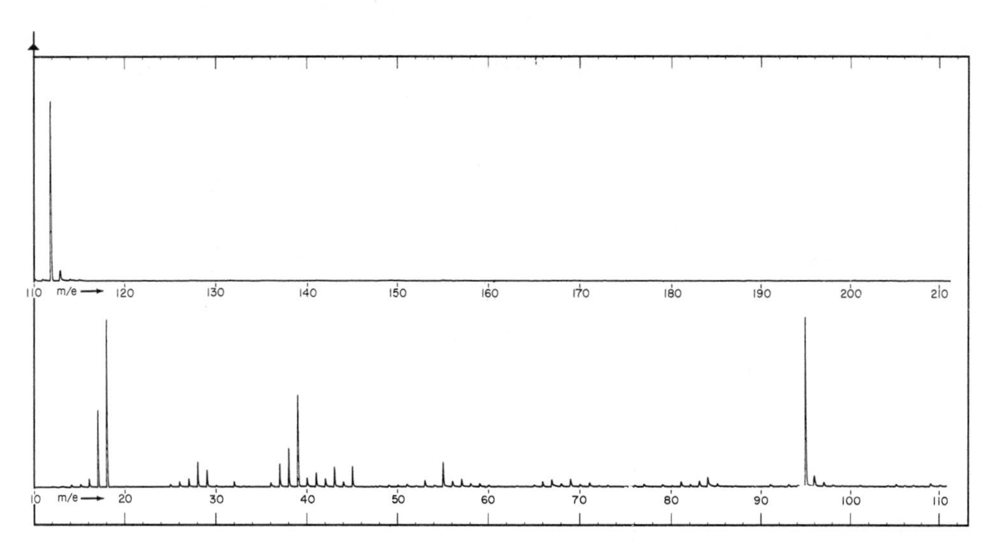

MASS SPECTRUM

MASS SPECTRAL DATA

m/e	*relative intensity*	*m/e*	*relative intensity*	*m/e*	*relative intensity*	*m/e*	*relative intensity*
17	12.4	39	43.5	56	2.9	83	2.2
18	48.3	40	4.2	57	2.2	84	5.2
26	2.6	41	4.6	66	2.6	85	1.3
27	2.8	42	4.1	67	3.5	95	87.0
28	9.2	43	6.0	68	2.0	96	5.8
29	8.0	44	2.4	69	2.6	97	1.8
32	1.3	45	9.6	71	1.4	112	100.0
36	2.0	53	3.3	81	1.6	113	6.1
37	11.0	55	10.4	82	1.1	114	1.1
38	16.5						

COMPOUND 1-26

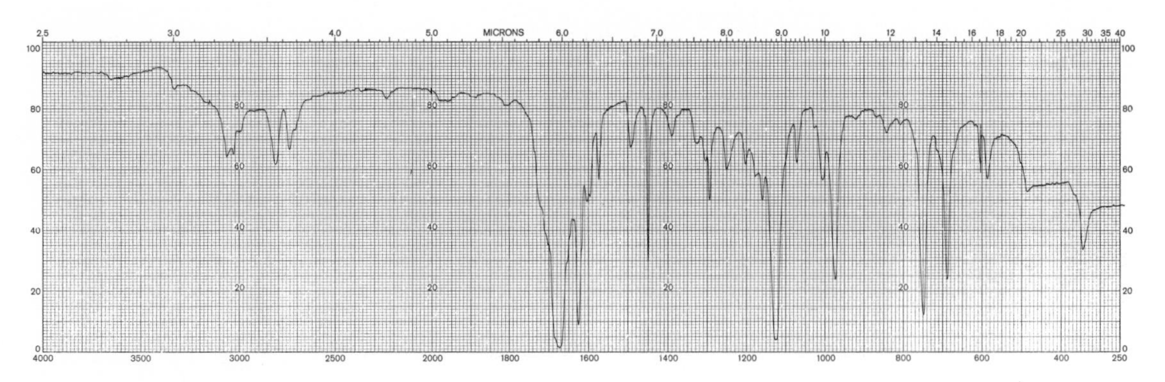

INFRARED SPECTRUM: Liquid film

ULTRAVIOLET DATA: λ_{max}^{EtOH} 287 nm (ϵ 22,000)

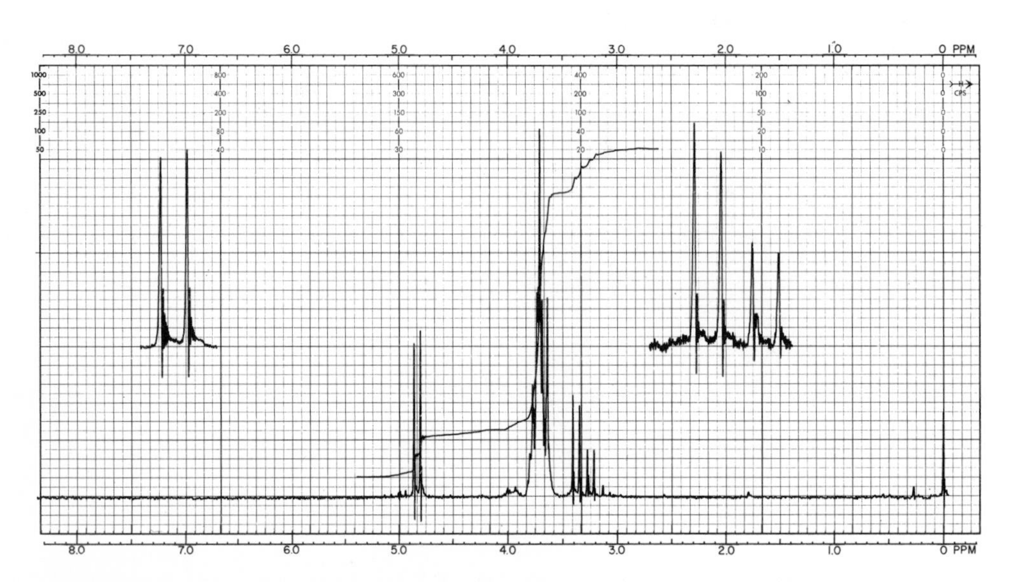

NUCLEAR MAGNETIC RESONANCE SPECTRUM:
15% in carbon tetrachloride, 1000 Hz sweep width
Expansions: (left) 250 Hz sweep width, 365 Hz sweep offset
(right) 250 Hz sweep width, 340 Hz sweep offset

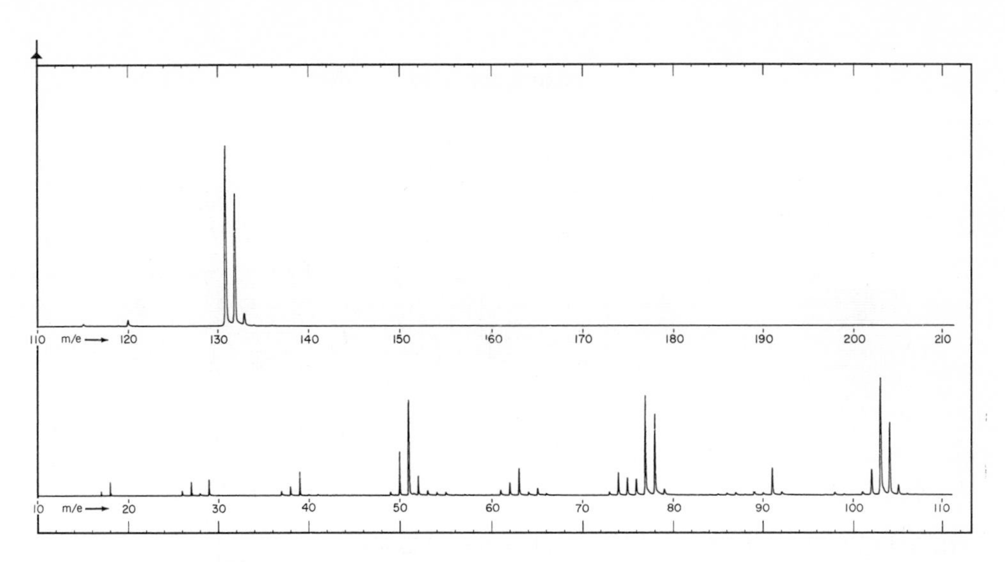

MASS SPECTRUM

MASS SPECTRAL DATA

m/e	relative intensity	m/e	relative intensity	m/e	relative intensity	m/e	relative intensity
17	0.9	52	9.9	76	8.5	99	0.5
18	3.1	53	2.3	77	52.0	101	2.1
26	2.1	54	1.4	78	41.0	102	14.2
27	5.7	55	1.4	79	3.0	103	62.0
28	1.3	61	2.4	85	0.9	104	39.0
29	6.5	62	5.5	86	1.1	105	5.5
37	2.2	63	12.5	87	1.3	106	0.8
38	4.0	64	1.5	88	0.5	115	1.2
39	10.2	65	3.1	89	2.5	116	0.5
40	0.6	65.5	0.6	90	0.9	120	2.7
41	0.7	66	1.0	91	11.0	131	100.0
49	1.9	73	1.7	92	1.6	132	72.0
50	21.1	74	11.3	97	0.4	133	7.4
51	45.0	75	9.1	98	1.5	134	0.8
51.5	1.5						

COMPOUND 1-27

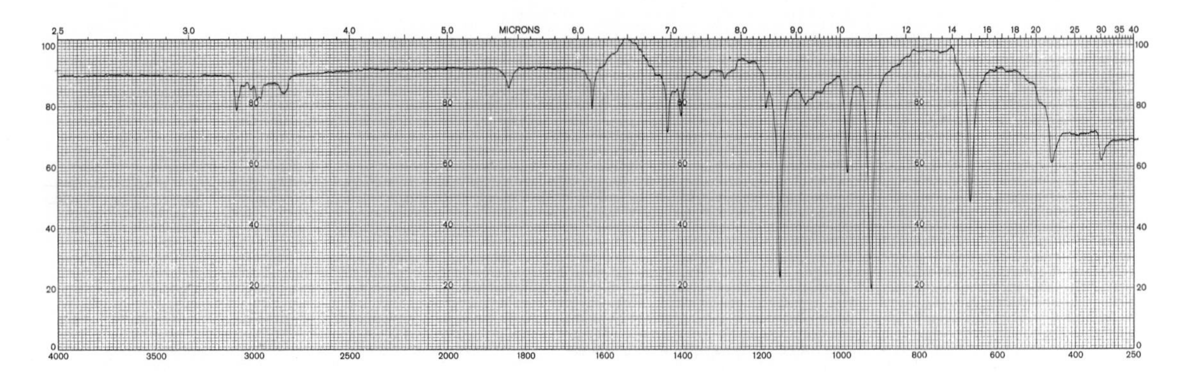

INFRARED SPECTRUM: 5.3% in carbon tetrachloride

ULTRAVIOLET DATA: λ_{max}^{EtOH} 219 nm (ϵ 2,050); 268 nm (ϵ 770).

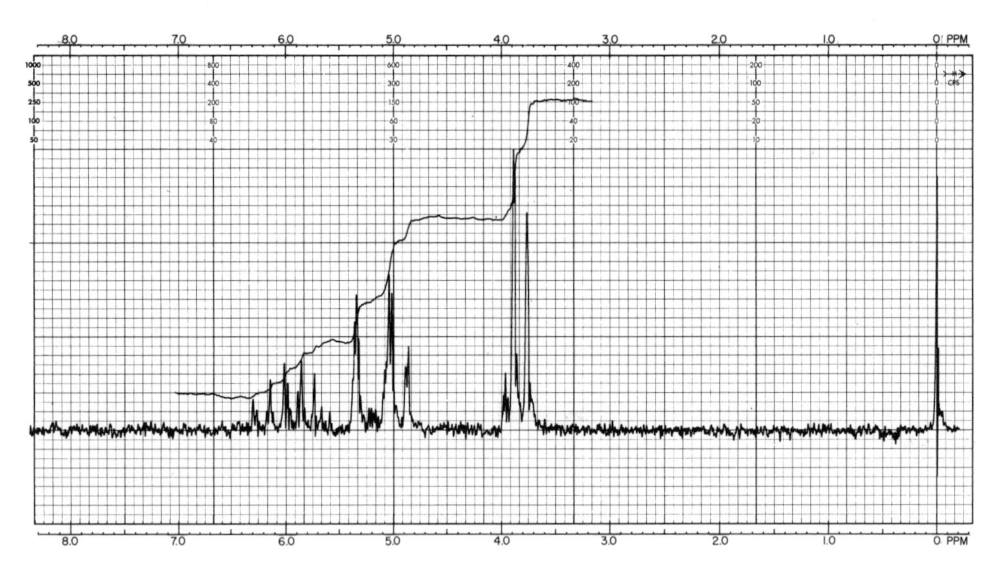

NUCLEAR MAGNETIC RESONANCE SPECTRUM:
15% in carbon tetrachloride, 500 Hz sweep width

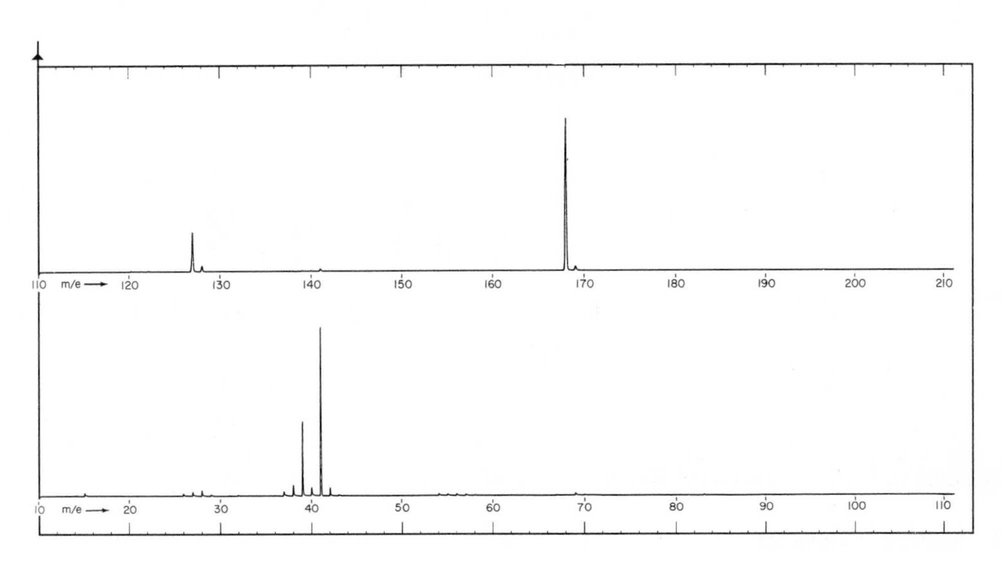

MASS SPECTRUM

MASS SPECTRAL DATA

m/e	*relative intensity*	*m/e*	*relative intensity*	*m/e*	*relative intensity*	*m/e*	*relative intensity*
14	1.8	29	1.6	40	6.3	139	0.8
15	3.0	32	5.9	41	92.0	140	0.8
25	0.7	36	0.6	42	6.8	141	2.3
26	2.6	37	4.0	43	1.0	168	100.0
27	4.0	38	8.1	127	27.0	169	3.5
28	24.0	39	56.0	128	4.8		

COMPOUND 1-28

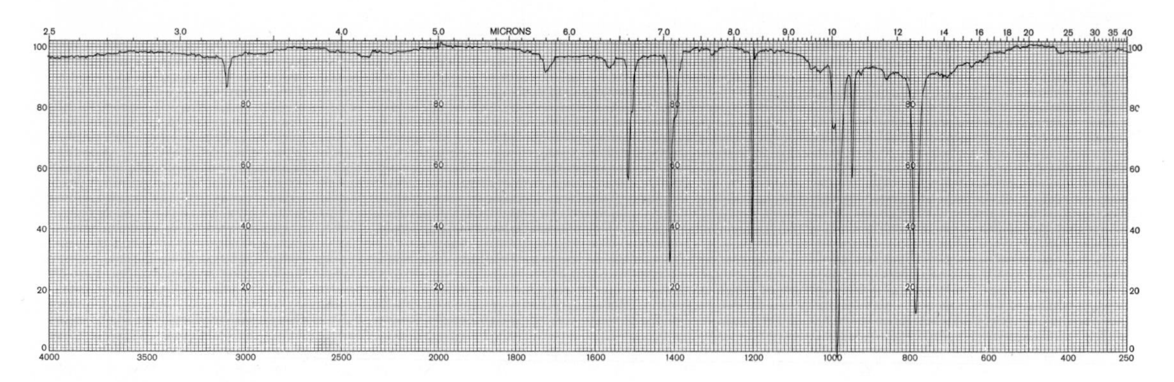

INFRARED SPECTRUM: Liquid film

ULTRAVIOLET DATA: λ_{max}^{Hexane} 252 nm (ϵ 8,900).

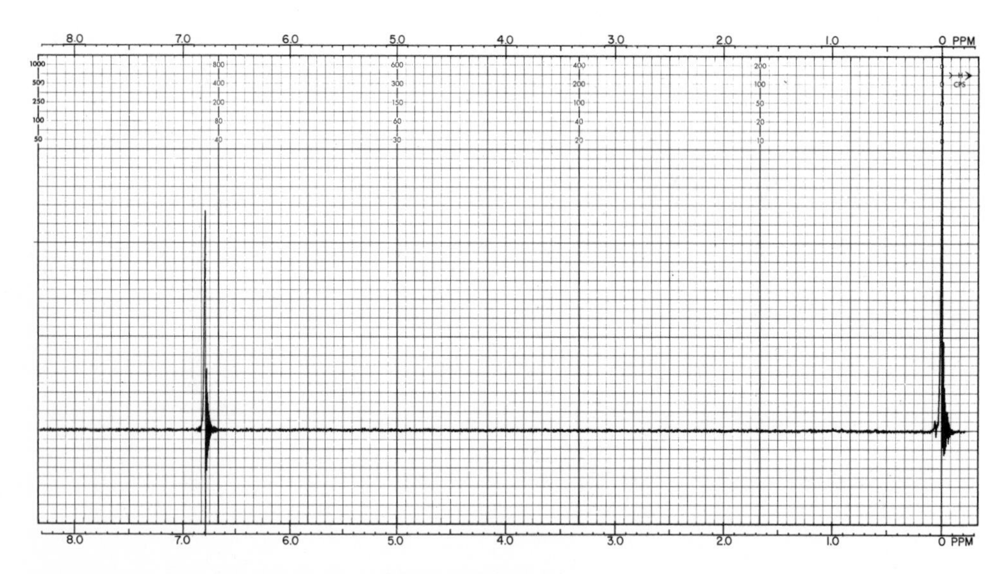

NUCLEAR MAGNETIC RESONANCE SPECTRUM:
15% in carbon tetrachloride, 500 Hz sweep width

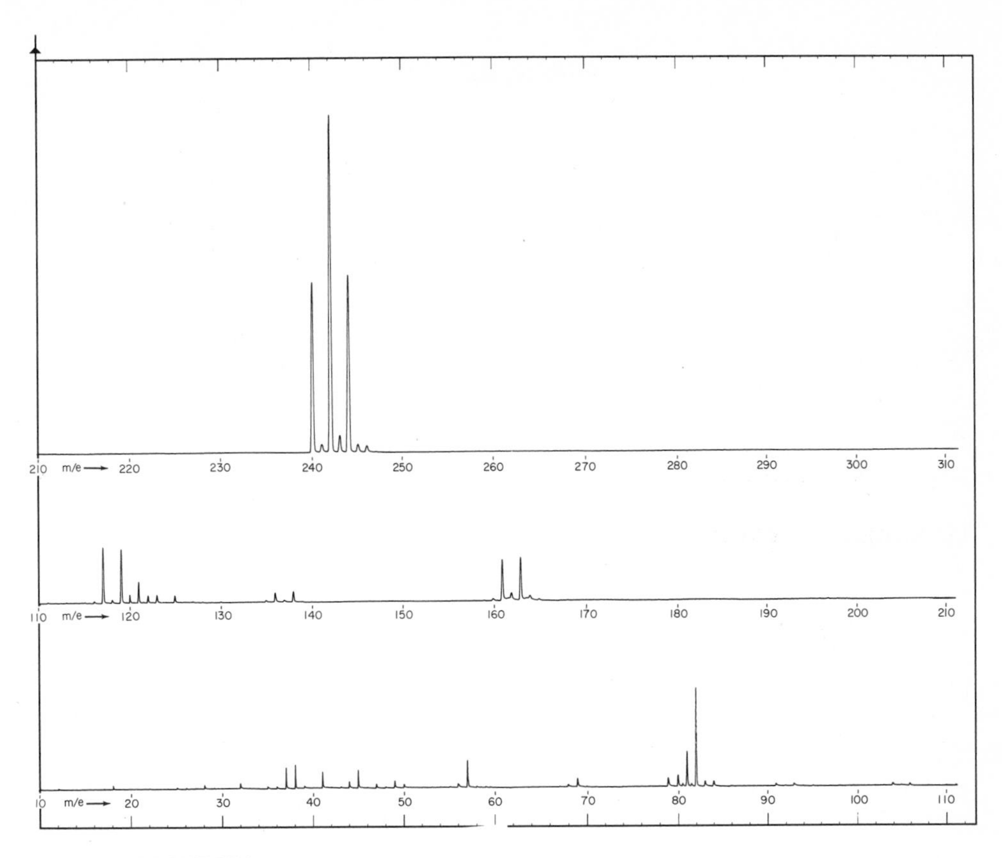

MASS SPECTRUM

MASS SPECTRAL DATA

m/e	*relative intensity*	*m/e*	*relative intensity*	*m/e*	*relative intensity*	*m/e*	*relative intensity*
26	0.5	56	0.9	91	0.8	137	0.8
28	0.7	57	5.4	93	0.8	138	2.9
32	0.9	58	0.4	104	1.0	160	0.9
35	0.6	59	0.5	106	1.0	161	11.8
36	0.7	59.5	0.4	116	0.9	162	2.2
37	3.4	68	0.7	117	15.2	163	12.8
38	3.7	69	2.1	118	1.2	164	2.0
39	0.6	70	0.4	119	14.2	165	0.8
41	3.8	79	2.4	120	2.3	240	52.0
43	0.7	80	2.7	121	5.4	241	2.6
44	1.3	80.5	0.7	122	2.2	242	100.0
45	3.4	81	8.7	123	2.2	243	5.4
47	1.0	81.5	1.0	125	2.0	244	55.0
48	0.5	82	24.0	135	0.8	245	2.8
49	1.8	83	1.8	136	2.6	246	2.1
50	0.9	84	1.5				

COMPOUND 1-29

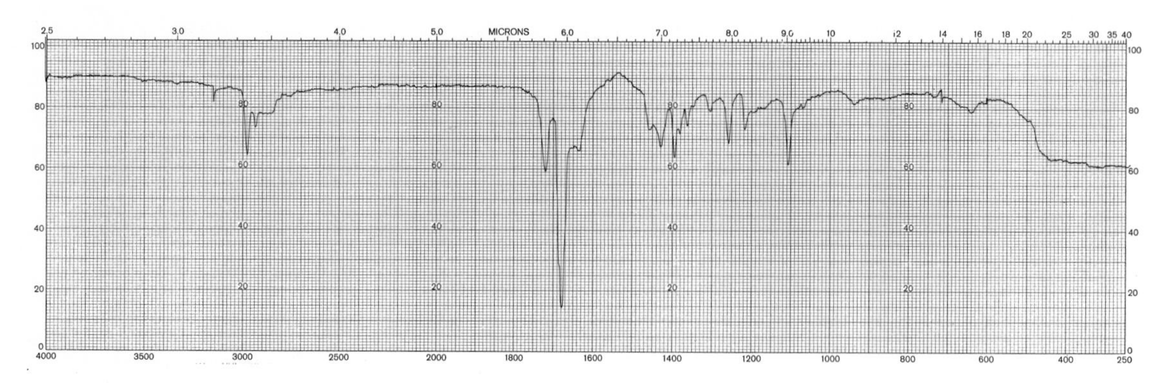

INFRARED SPECTRUM: 1.4% in carbon tetrachloride

ULTRAVIOLET DATA: λ_{max}^{EtOH} 202 nm (ϵ 7,050).

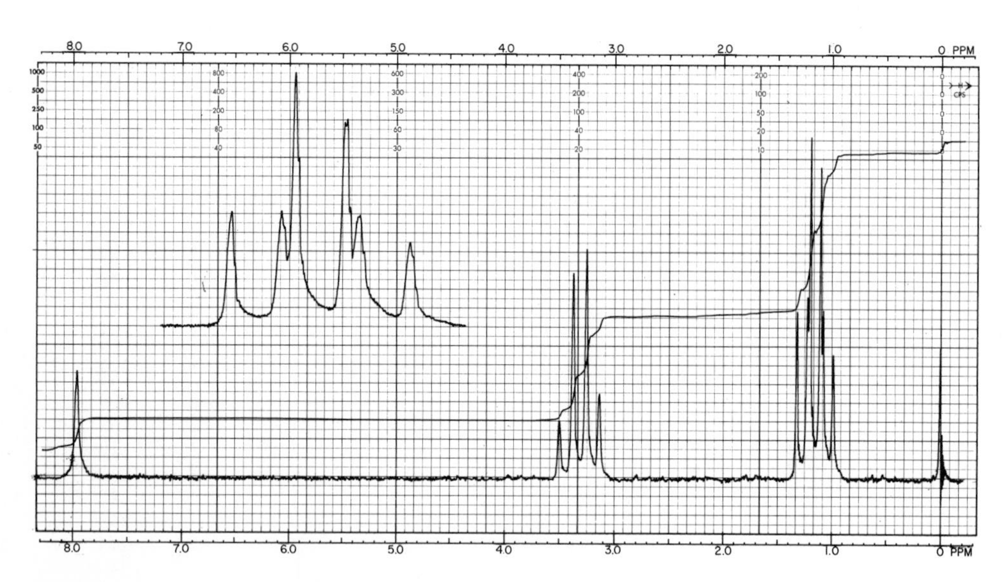

NUCLEAR MAGNETIC RESONANCE SPECTRUM:
15% in carbon tetrachloride, 500 Hz sweep width
Expansion: 100 Hz sweep width, 0 Hz sweep offset

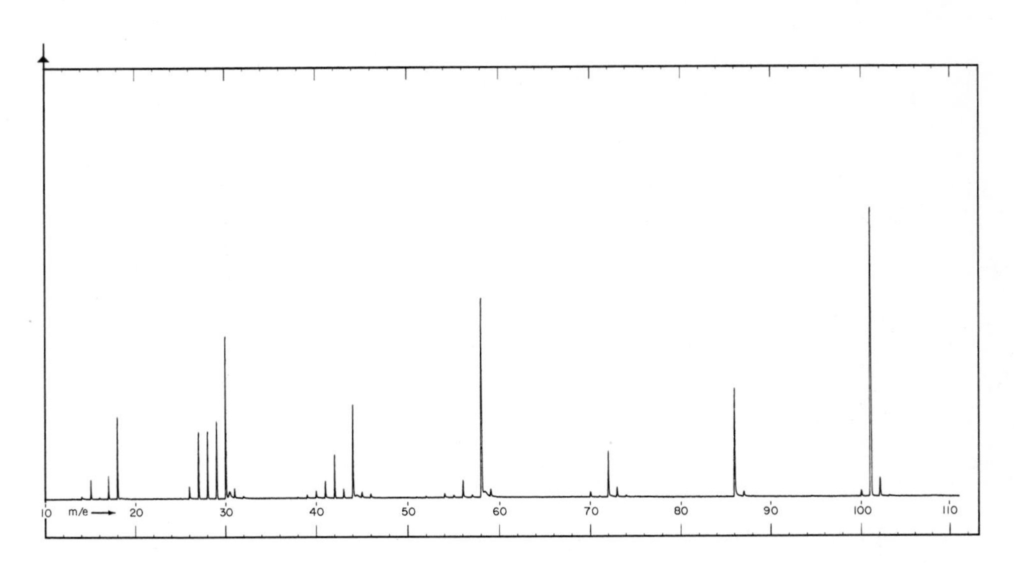

MASS SPECTRUM

MASS SPECTRAL DATA

m/e	*relative intensity*	*m/e*	*relative intensity*	*m/e*	*relative intensity*	*m/e*	*relative intensity*
14	0.7	32	0.6	52	0.4	72	13.9
15	3.5	38	0.4	54	1.1	73	2.8
16	0.4	39	0.7	55	0.6	74	0.8
17	3.7	40	1.7	56	4.2	86	33.0
18	19.2	41	3.7	57	0.8	87	1.9
26	2.8	42	10.0	58	61.0	88	0.1
27	16.6	43	2.2	59	2.6	100	2.0
28	17.3	44	23.7	60	0.1	101	100.0
29	19.2	45	1.9	70	1.6	102	6.1
30	45.0	46	1.0	71	0.3	103	0.3
31	2.4						

COMPOUND 1-30

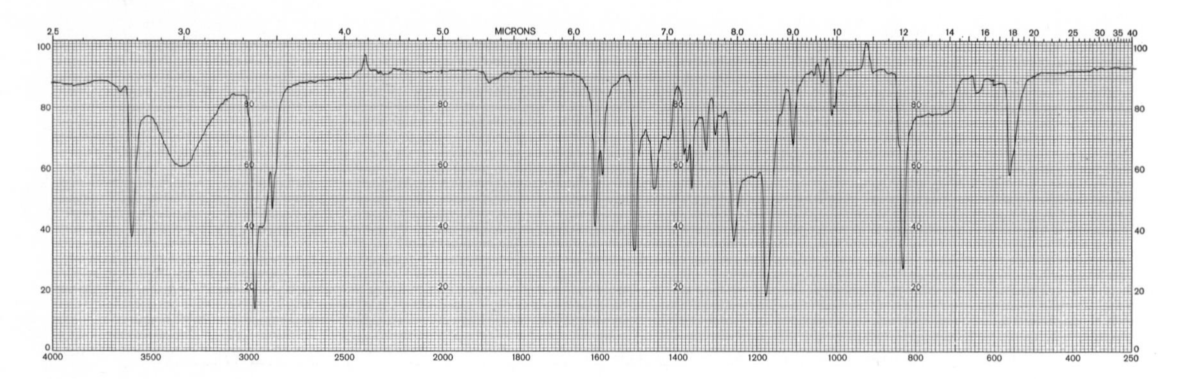

INFRARED SPECTRUM: 3.1% in chloroform

ULTRAVIOLET DATA: λ_{max}^{EtOH} 278 nm (ϵ 3,200); 284(sh) nm (ϵ 2,500).

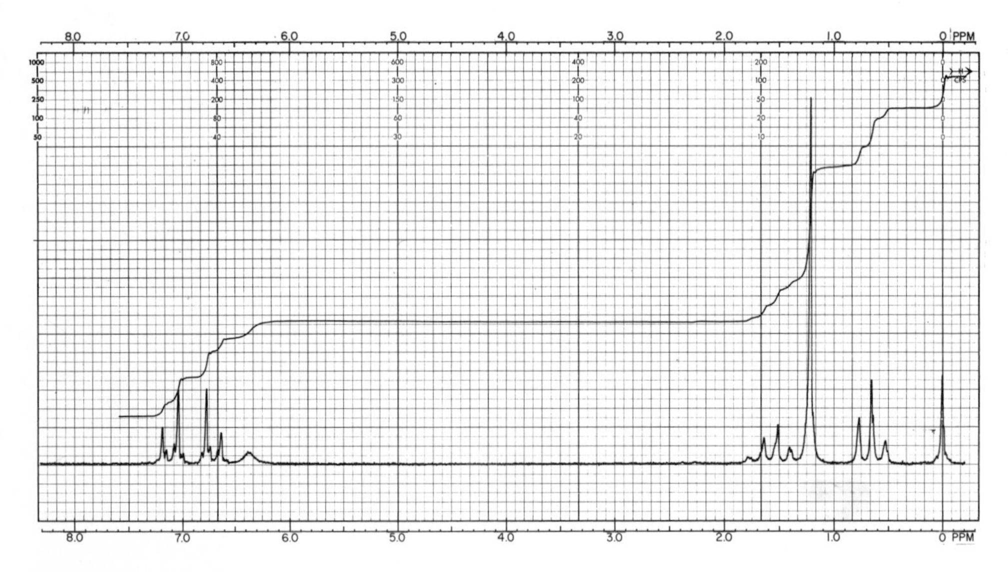

NUCLEAR MAGNETIC RESONANCE SPECTRUM:
15% in carbon tetrachloride, 500 Hz sweep width

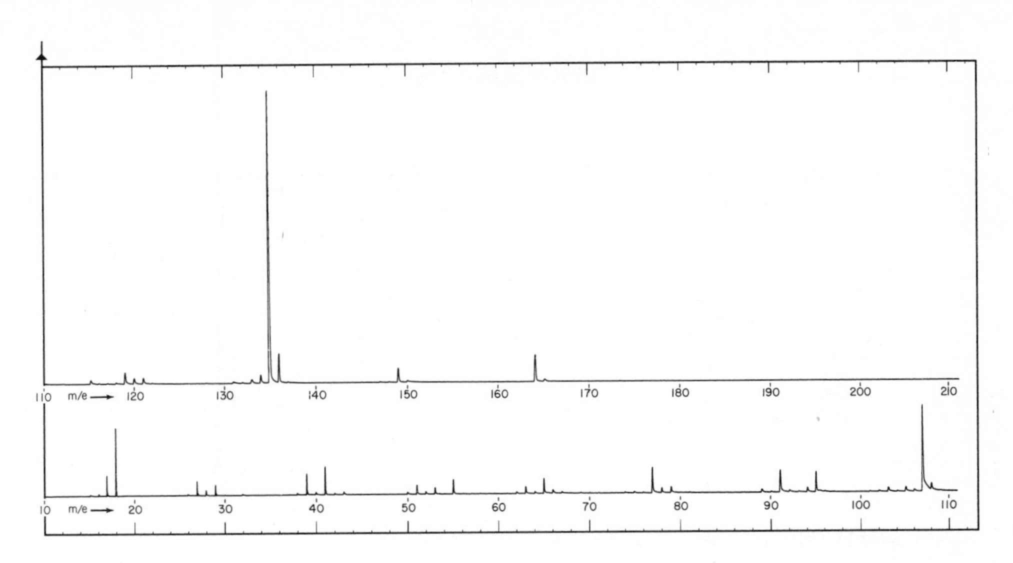

MASS SPECTRUM

MASS SPECTRAL DATA

m/e	*relative intensity*	*m/e*	*relative intensity*	*m/e*	*relative intensity*	*m/e*	*relative intensity*
15	0.2	56	0.3	92	0.8	120	2.4
17	0.6	57	0.1	93	0.3	121	2.5
18	2.3	62	0.5	94	1.5	122	0.3
26	0.2	63	1.7	95	5.6	127	0.1
27	2.4	64	0.6	96	0.5	128	0.2
28	0.8	65	3.6	101	0.3	131	0.8
29	1.9	66	1.0	102	0.3	132	0.4
38	0.4	67	0.3	103	1.3	133	1.7
39	3.9	74	0.2	104	0.3	134	3.4
40	0.6	75	0.5	105	1.5	135	100.0
41	5.2	76	0.2	106	0.7	136	10.5
42	0.4	77	6.1	107	23.4	137	0.8
43	0.7	78	1.5	108	2.8	147	0.5
50	0.7	79	1.6	109	0.4	148	0.3
51	2.0	80	0.2	115	1.5	149	5.5
52	0.7	81	0.3	116	0.5	150	0.9
53	1.8	89	1.0	117	0.5	164	9.4
54	0.2	90	0.3	118	0.6	165	1.3
55	3.5	91	5.8	119	4.2	166	0.1

COMPOUND 1-31

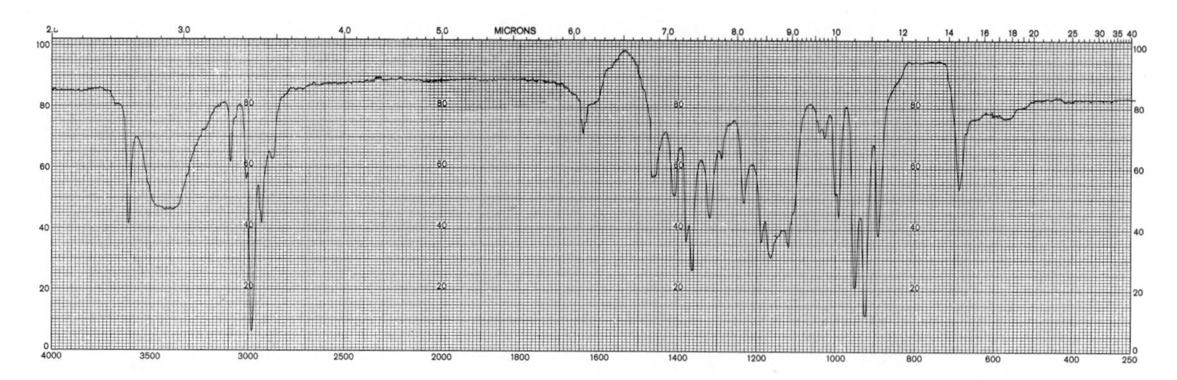

INFRARED SPECTRUM: 3.8% in carbon tetrachloride

ULTRAVIOLET DATA: no λ_{max} above 205 nm

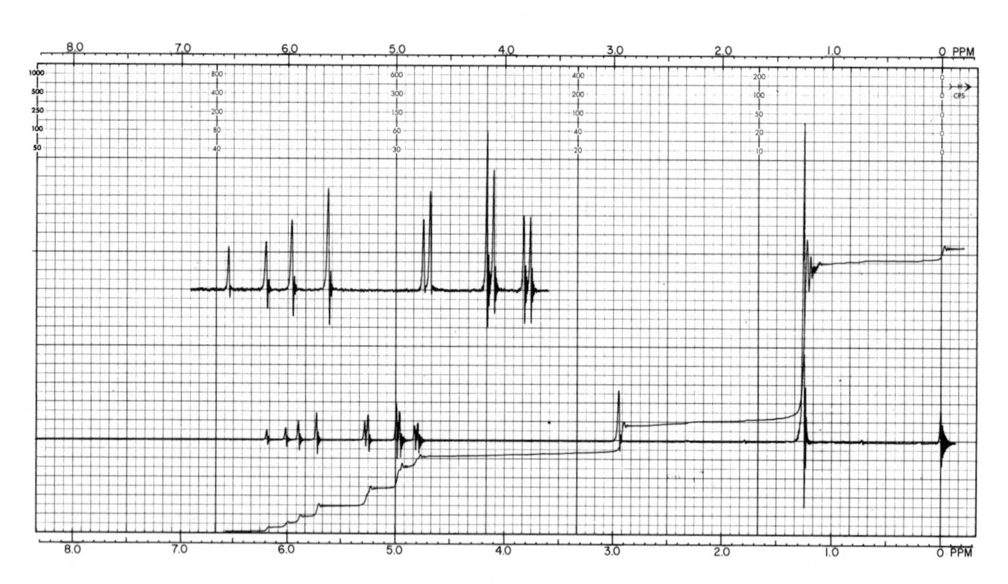

NUCLEAR MAGNETIC RESONANCE SPECTRUM:
15% in carbon tetrachloride, 500 Hz sweep width
Expansion: 250 Hz sweep width, 175 Hz sweep offset

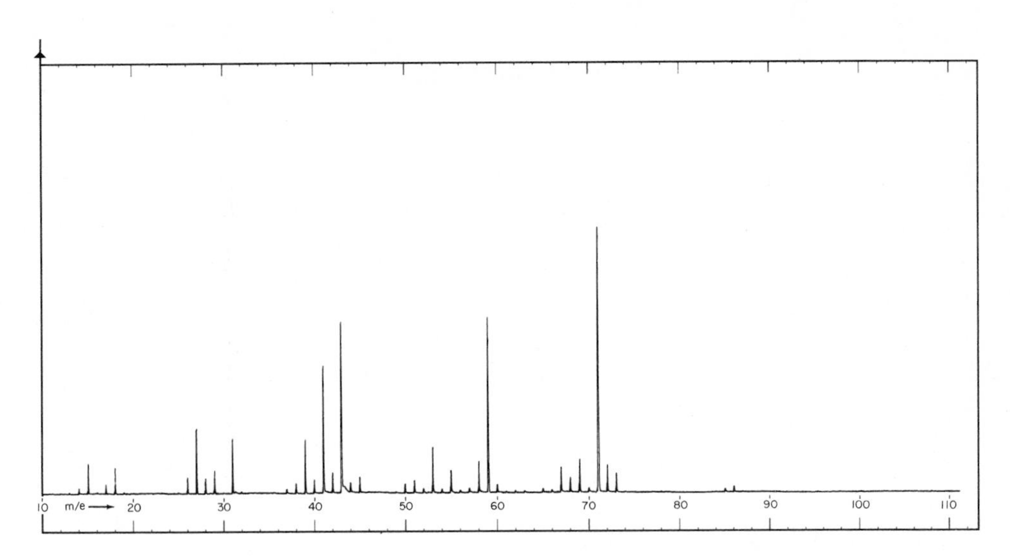

MASS SPECTRUM

MASS SPECTRAL DATA

m/e	*relative intensity*	*m/e*	*relative intensity*	*m/e*	*relative intensity*	*m/e*	*relative intensity*
14	0.9	39	10.9	53	7.6	67	4.6
15	3.5	40	2.2	54	0.9	68	2.6
17	0.8	41	21.2	55	4.9	69	5.8
18	3.8	42	4.1	56	0.8	70	0.9
26	2.6	43	46.0	57	1.1	71	100.0
27	12.7	44	1.7	58	5.2	72	5.3
28	3.1	45	2.8	59	44.0	73	3.8
29	3.9	49	0.6	60	1.5	74	0.5
31	8.0	50	2.0	61	0.1	85	1.1
37	0.7	51	2.2	65	0.8	86	1.1
38	2.0	52	0.8				

COMPOUND 1-32

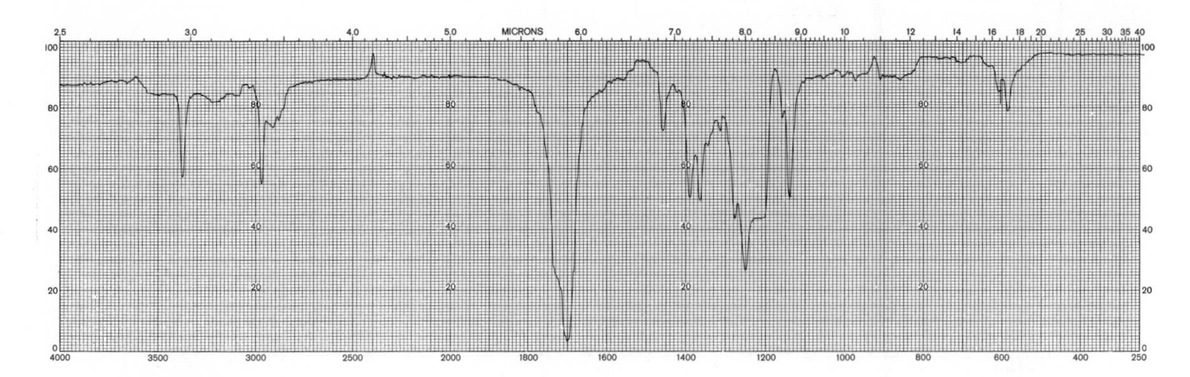

INFRARED SPECTRUM: 1.1% in chloroform

ULTRAVIOLET DATA: λ_{max}^{EtOH} 204 nm (ϵ 15,700); 230(sh) nm (ϵ 220); 255(sh) nm (ϵ 65).

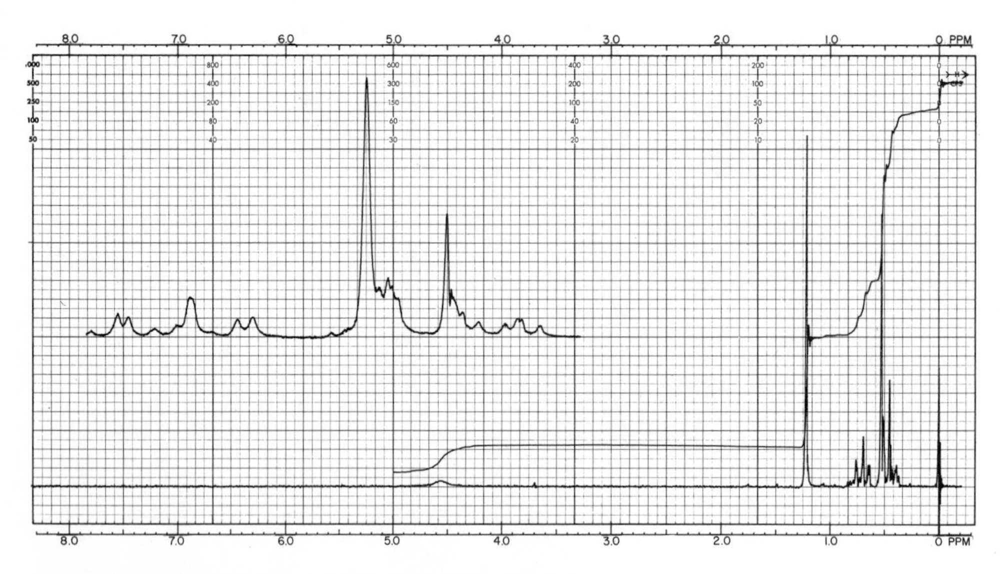

NUCLEAR MAGNETIC RESONANCE SPECTRUM:
15% in deuterochloroform, 1000 Hz sweep width
Expansion: 100 Hz sweep width, 0 Hz sweep offset

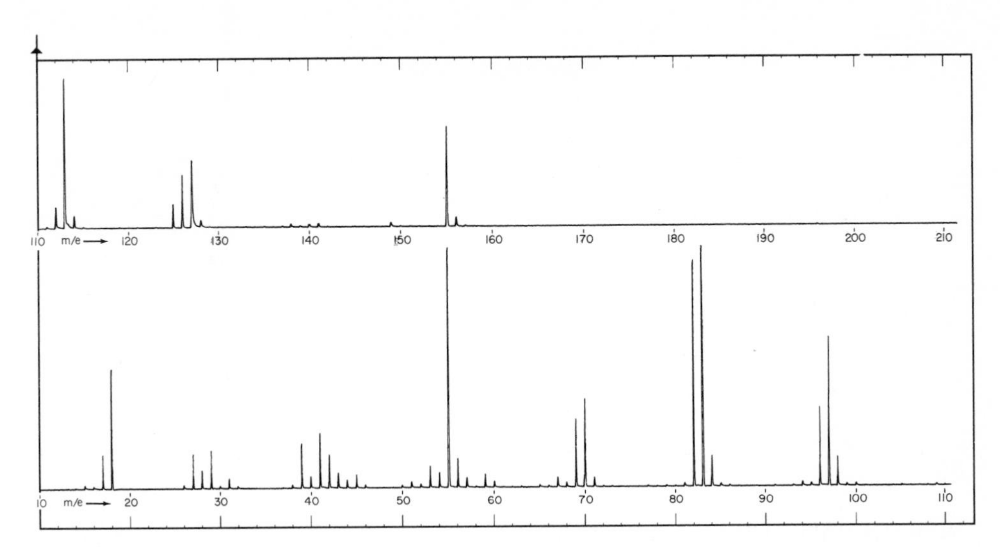

MASS SPECTRUM

MASS SPECTRAL DATA

m/e	*relative intensity*	*m/e*	*relative intensity*	*m/e*	*relative intensity*	*m/e*	*relative intensity*
14	0.9	50	2.0	72	0.8	110	1.1
15	2.3	51	3.9	73	0.5	111	1.0
16	0.4	52	2.6	77	0.5	112	11.5
17	2.7	53	14.8	79	0.6	113	89.0
18	9.8	54	10.5	80	0.5	114	7.0
26	1.9	55	99.0	81	1.7	115	0.9
27	22.7	56	19.7	82	94.0	125	13.6
28	8.8	57	5.8	83	100.0	126	29.0
29	27.0	58	0.7	84	20.9	127	41.0
30	2.2	59	8.6	85	2.0	128	4.3
31	2.4	60	3.6	86	1.0	129	0.6
37	0.5	63	0.6	94	3.1	137	0.6
38	2.3	64	0.3	95	1.8	138	1.9
39	31.0	65	1.6	96	32.0	139	1.0
40	7.4	66	1.3	97	61.0	140	3.5
41	42.0	67	6.3	98	12.3	141	3.6
42	26.0	68	3.5	99	1.8	142	0.5
43	9.1	69	42.0	100	2.3	155	50.0
44	5.2	70	55.0	108	0.4	156	4.7
45	1.7	71	4.9	109	1.0	157	0.6
46	0.7						

COMPOUND 1-33

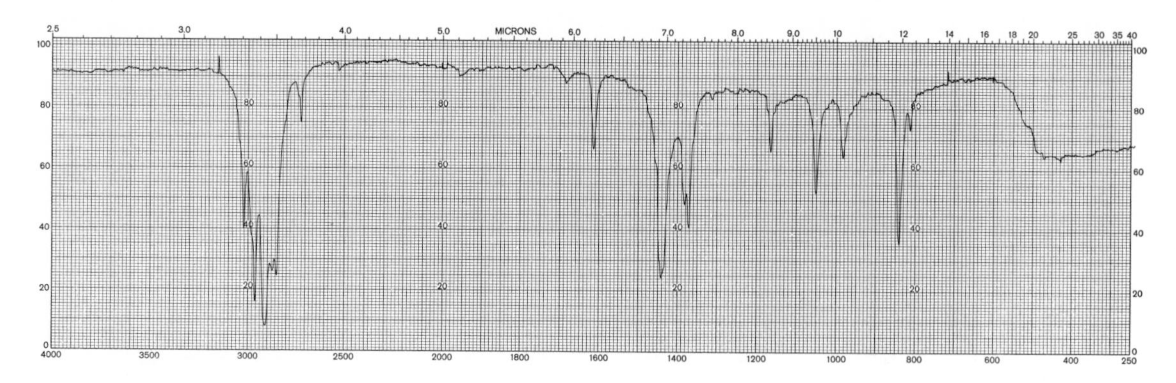

INFRARED SPECTRUM: Liquid film

ULTRAVIOLET DATA: λ_{max}^{EtOH} 241 nm (ϵ 22,400); 245(sh) nm (ϵ 21,400); 250 (sh) nm (ϵ 16,600).

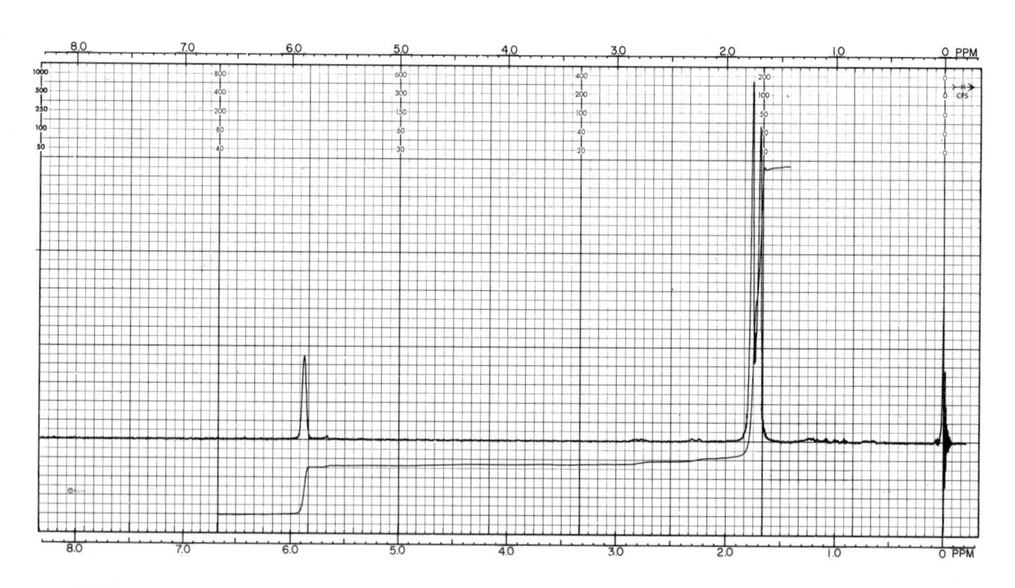

NUCLEAR MAGNETIC RESONANCE SPECTRUM:
15% in carbon tetrachloride, 500 Hz sweep width

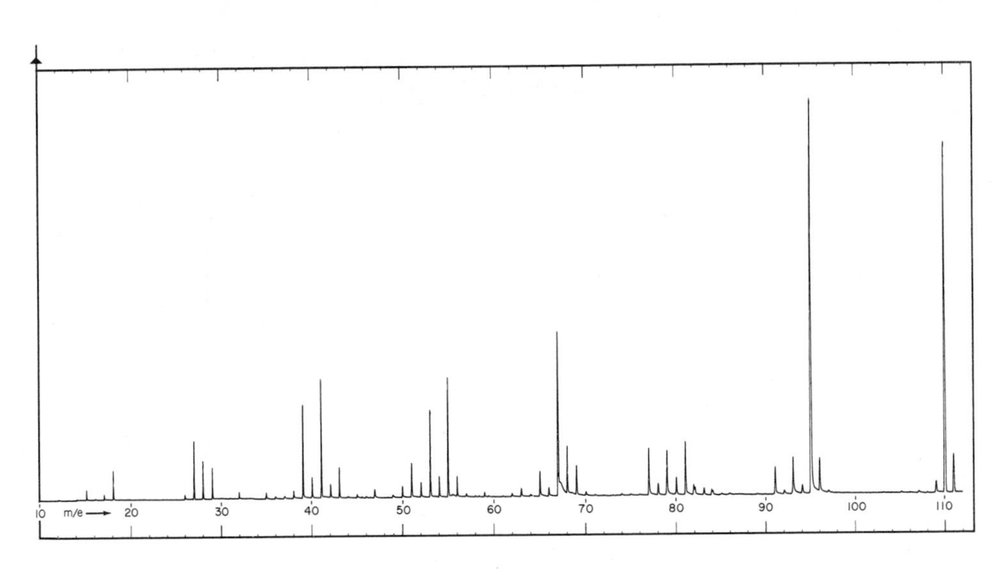

MASS SPECTRUM

MASS SPECTRAL DATA

m/e	*relative intensity*	*m/e*	*relative intensity*	*m/e*	*relative intensity*	*m/e*	*relative intensity*
15	1.2	44	0.7	62	0.8	83	2.0
17	0.6	45	0.8	63	2.5	84	1.8
18	2.7	45.5	0.6	64	0.5	85	0.6
26	0.9	46	0.6	65	6.2	91	7.3
27	9.8	47	1.7	66	2.1	92	1.1
28	6.8	49	0.8	67	40.0	93	10.3
29	5.0	50	2.5	68	12.1	94	2.5
32	0.9	51	6.9	69	7.4	95	100.0
35	1.4	52	3.0	70	1.2	96	10.2
36	0.6	53	17.8	74	0.7	97	1.1
37	0.8	54	4.7	77	12.6	107	0.7
38	1.5	55	29.0	78	3.0	108	0.5
39	19.6	56	4.7	79	11.2	109	3.2
40	4.3	57	0.9	80	4.7	110	94.0
41	26.0	58	0.5	81	14.5	111	11.1
42	2.5	58.5	0.6	82	3.2	112	0.6
43	7.6	59	1.2				

COMPOUND 1-34

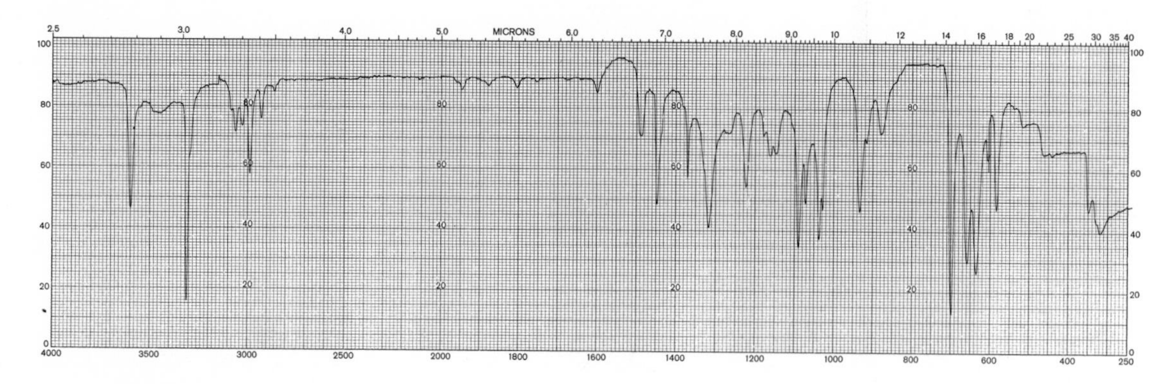

INFRARED SPECTRUM: 2.7% in carbon tetrachloride

ULTRAVIOLET DATA: λ_{max}^{EtOH} 206 nm (ϵ 8,300); 245 nm (ϵ 90); 251 nm (ϵ 130); 257 nm (ϵ 180); 263 nm (ϵ 140).

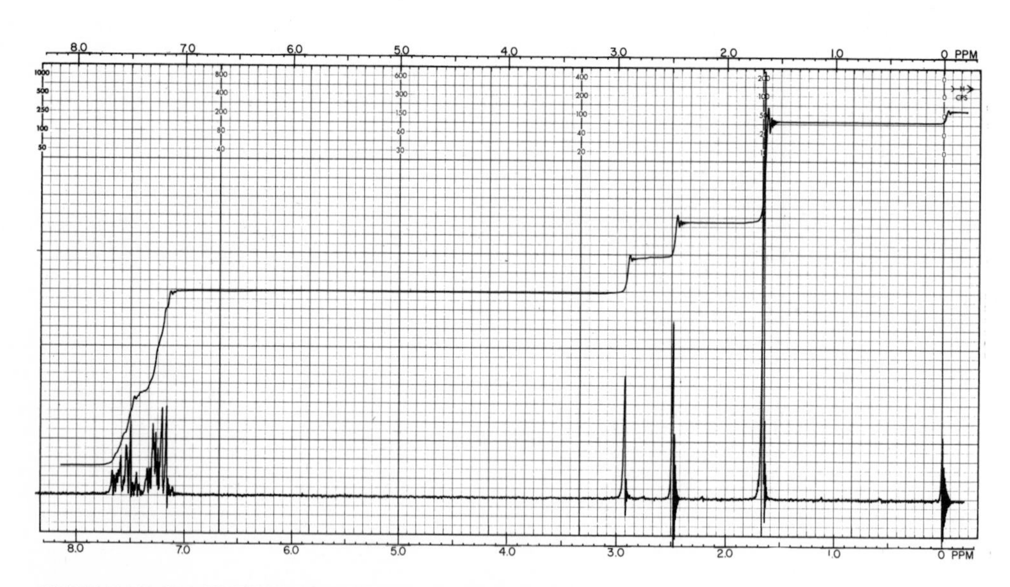

NUCLEAR MAGNETIC RESONANCE SPECTRUM:
15% in carbon tetrachloride, 500 Hz sweep width

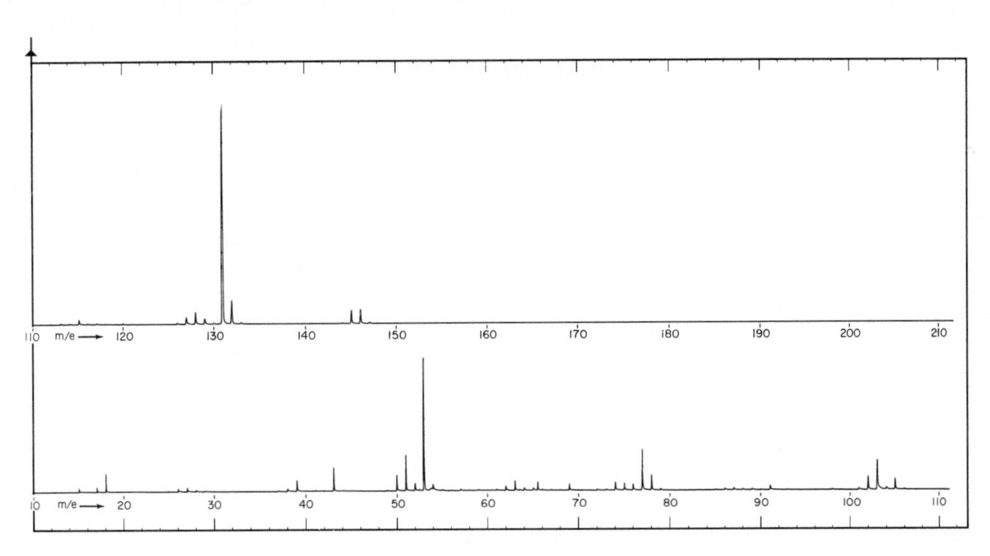

MASS SPECTRUM

MASS SPECTRAL DATA

m/e	*relative intensity*	*m/e*	*relative intensity*	*m/e*	*relative intensity*	*m/e*	*relative intensity*
15	1.3	53	64.0	75	3.2	113	0.9
17	0.8	54	2.8	76	3.1	114	0.7
18	3.1	55	0.5	77	19.3	115	2.9
26	1.4	56	0.5	78	7.4	116	1.0
27	2.2	57	0.7	79	1.0	117	1.0
28	0.7	59	0.4	85	0.4	120	1.0
29	0.4	61	0.7	86	1.1	121	0.4
37	0.5	62	2.3	87	1.2	126	1.1
38	1.4	63	4.7	88	0.6	127	4.2
39	5.4	64	1.7	89	0.9	128	7.4
40	0.5	65	1.3	91	2.3	129	3.4
41	0.8	65.5	4.2	98	0.7	130	0.7
42	0.5	66	0.7	99	0.4	131	100.0
43	7.9	67	0.5	101	1.2	132	12.8
49	0.6	68	0.4	102	5.9	133	1.0
50	9.8	69	3.2	103	13.5	144	0.4
51	18.5	72	0.9	104	1.4	145	5.9
52	3.2	73	0.4	105	5.4	146	6.6
52.5	0.8	74	3.6	106	0.7	147	0.8

COMPOUND 1-35

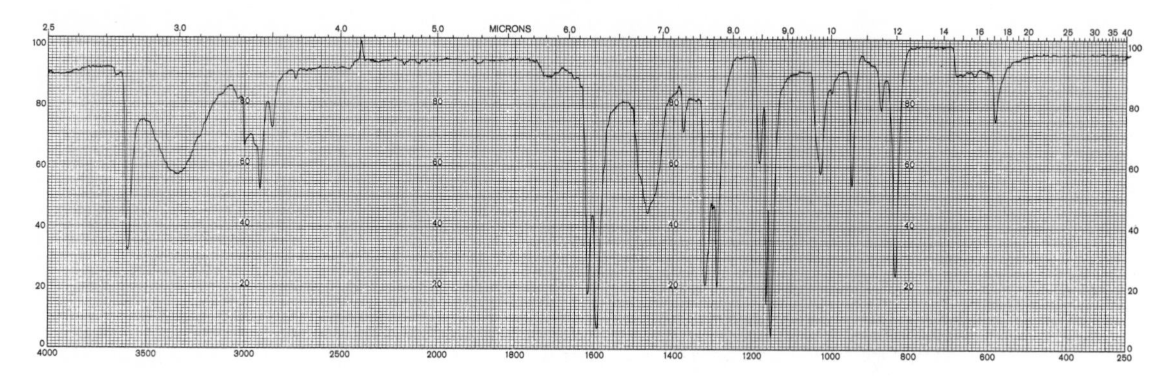

INFRARED SPECTRUM: 3.0% in chloroform

ULTRAVIOLET DATA: λ_{max}^{Hexane} 282 nm (ϵ 1,600)

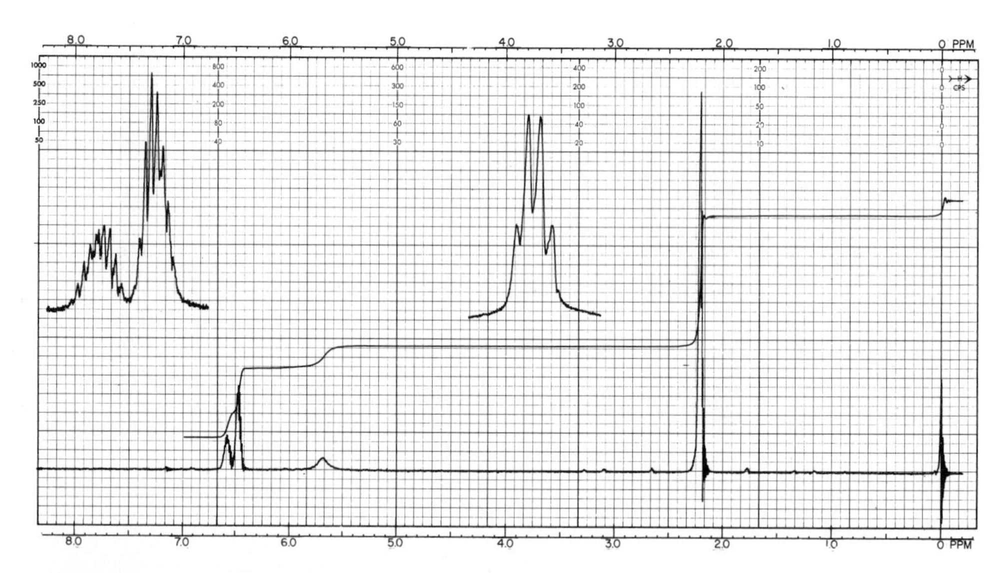

NUCLEAR MAGNETIC RESONANCE SPECTRUM:
15% in deuterochloroform, 500 Hz sweep width
Expansions: (left) 100 Hz sweep width, 300 Hz sweep offset
(right) 50 Hz sweep width, 110 Hz sweep offset

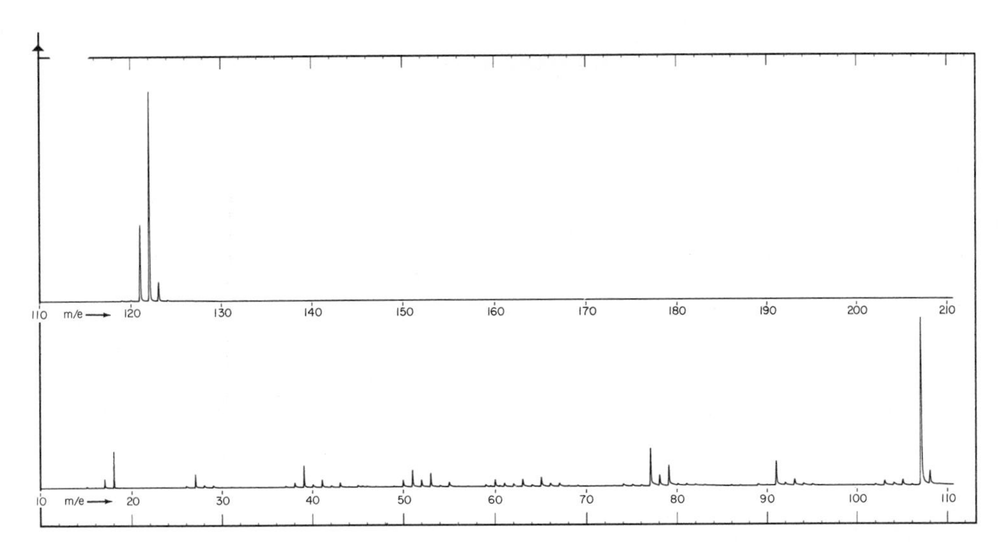

MASS SPECTRUM

MASS SPECTRAL DATA

m/e	*relative intensity*	*m/e*	*relative intensity*	*m/e*	*relative intensity*	*m/e*	*relative intensity*
15	0.4	49	0.2	67	1.4	92	1.5
17	0.7	50	2.3	68	0.3	93	2.6
18	3.1	51	5.1	69	0.3	94	1.2
26	0.8	52	2.3	74	0.8	95	0.6
27	3.5	53	3.2	75	0.6	101	0.2
28	0.9	54	0.5	76	0.5	102	0.4
29	0.8	55	1.7	77	13.5	103	2.1
31	0.2	56	0.2	78	4.5	104	1.3
37	0.4	57	0.1	79	7.8	105	2.5
38	1.4	59	0.6	80	1.3	106	0.4
39	6.8	60	2.2	81	1.0	107	60.0
40	1.0	60.5	0.5	82	0.5	108	6.6
41	2.3	61	1.3	83	0.4	109	0.8
42	0.7	62	1.1	86	0.3	120	0.9
43	1.8	63	2.7	87	0.2	121	36.0
44	0.2	64	0.8	89	0.7	122	100.0
45	0.6	65	3.5	90	0.2	123	10.3
45.5	0.4	66	1.4	91	9.1	124	0.9
46	0.3						

COMPOUND 1-36

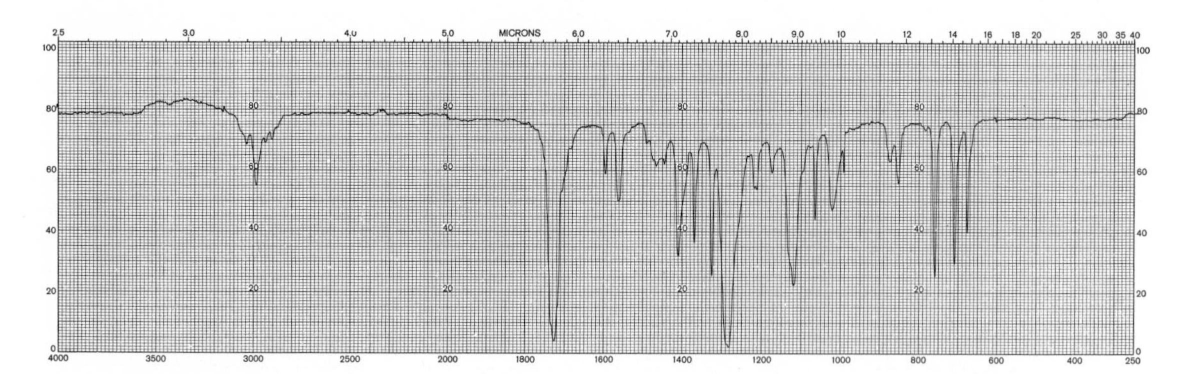

INFRARED SPECTRUM: Liquid film

ULTRAVIOLET DATA: λ_{max}^{EtOH} 211 nm (ϵ 8,550); 274 nm (ϵ 2,450).

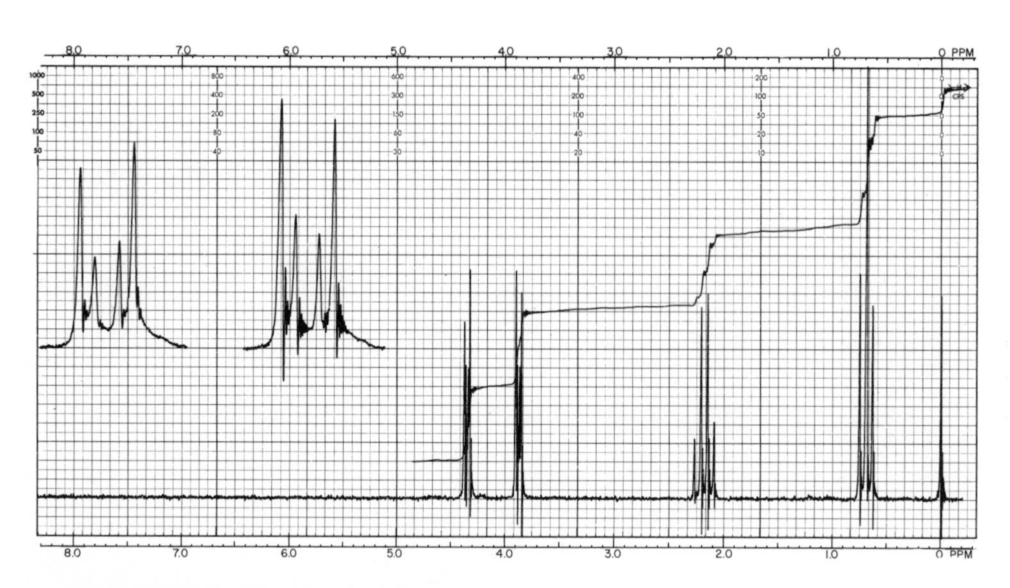

NUCLEAR MAGNETIC RESONANCE SPECTRUM:
15% in carbon tetrachloride, 1000 Hz sweep width
Expansions: (left) 100 Hz sweep width, 430 Hz sweep offset
(right) 100 Hz sweep width, 395 Hz sweep offset

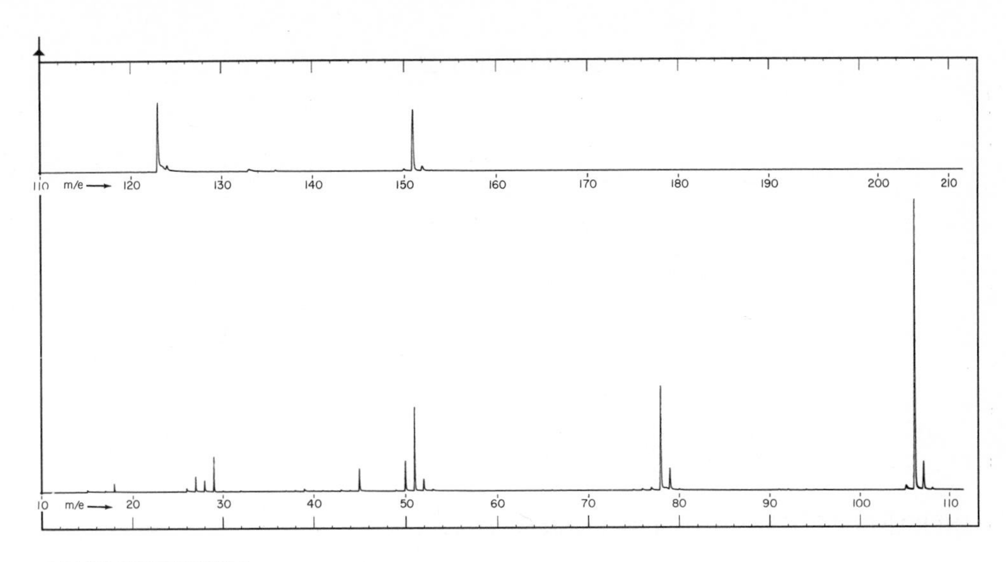

MASS SPECTRUM

MASS SPECTRAL DATA

m/e	*relative intensity*	*m/e*	*relative intensity*	*m/e*	*relative intensity*	*m/e*	*relative intensity*
13	0.2	39	1.8	67	0.5	106	100.0
14	0.4	40	0.5	68	0.3	107	10.0
15	1.4	41	0.5	71	0.3	108	1.2
17	1.0	42	0.4	75	0.7	113	0.3
18	4.8	43	1.6	76	1.4	121	0.3
25	0.3	44	1.0	77	2.3	122	0.3
26	3.9	45	8.9	78	51.0	123	41.0
27	8.9	46	0.4	79	12.6	124	5.0
28	7.5	49	1.0	80	1.0	125	0.8
29	19.3	50	14.5	91	0.7	133	1.9
30	0.7	51	43.0	92	0.8	136	1.2
31	0.4	52	7.3	93	0.3	150	1.9
32	0.8	53	1.4	94	0.5	151	40.0
37	0.3	65	0.5	95	0.3	152	4.3
38	0.9	66	0.5	105	2.5	153	0.5

COMPOUND 1-37

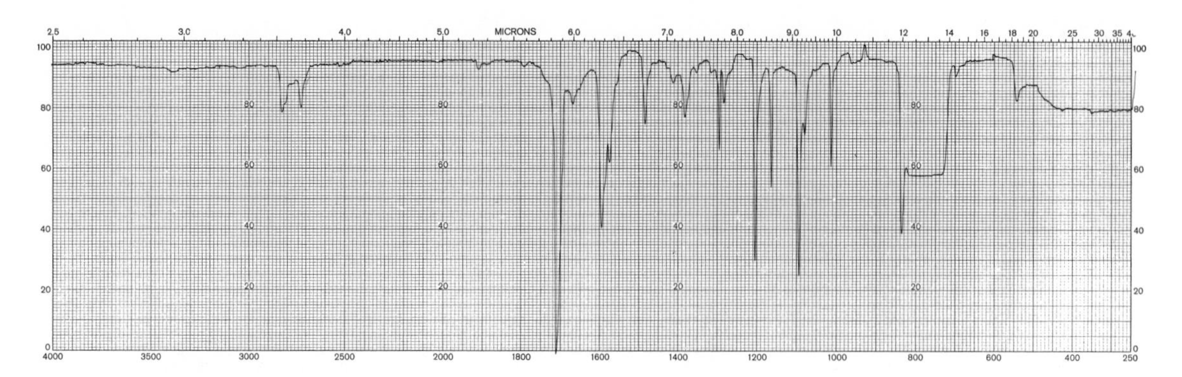

INFRARED SPECTRUM: 1.3% in carbon tetrachloride

ULTRAVIOLET DATA: λ_{max}^{EtOH} 245 nm (ϵ 16,000)

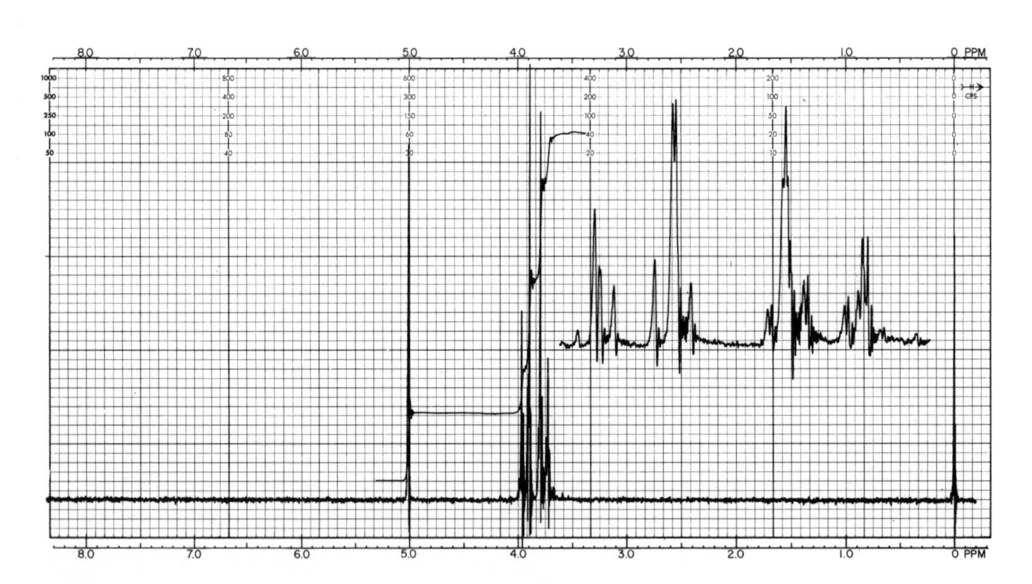

NUCLEAR MAGNETIC RESONANCE SPECTRUM:
15% in carbon tetrachloride, 1000 Hz sweep width
Expansion: 100 Hz sweep width, 435 Hz sweep offset

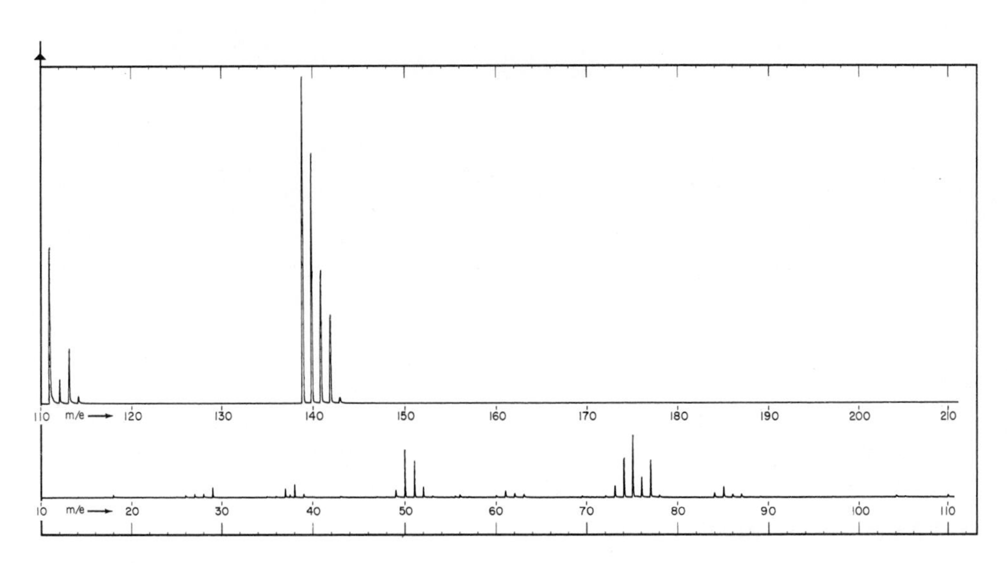

MASS SPECTRUM

MASS SPECTRAL DATA

m/e	relative intensity	m/e	relative intensity	m/e	relative intensity	m/e	relative intensity
27	0.7	51	7.7	74	9.5	110	0.8
28	0.6	52	2.4	75	16.4	111	43.0
29	2.0	56	0.8	76	5.3	112	7.7
36	0.5	60	0.5	77	9.7	113	15.6
37	1.9	61	1.8	78	0.7	114	2.5
37.5	0.7	62	1.1	84	1.3	139	100.0
38	2.4	63	1.0	85	3.0	140	74.0
39	0.9	69.5	0.6	86	1.1	141	39.0
49	1.8	72	0.5	87	1.2	142	26.1
50	11.2	73	3.3	104	0.8	143	2.2

COMPOUND 1-38

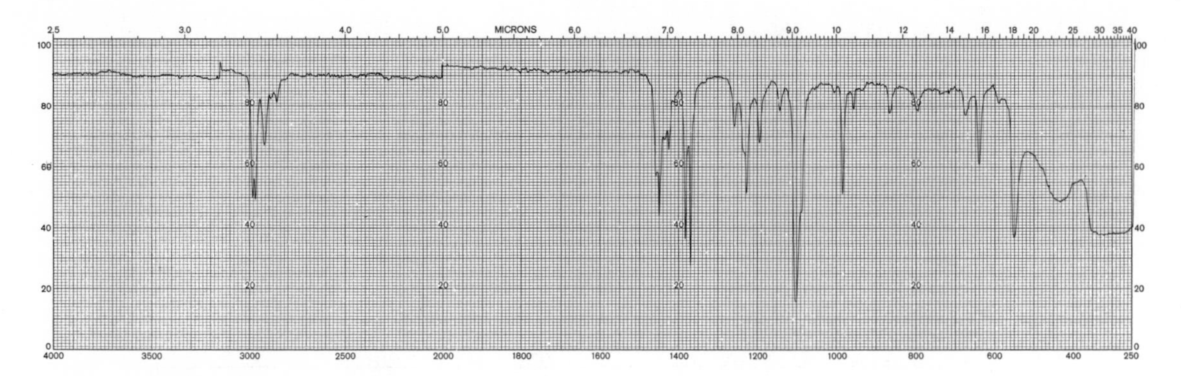

INFRARED SPECTRUM: Liquid film

ULTRAVIOLET DATA: λ_{max}^{EtOH} 212 nm (ϵ 670)

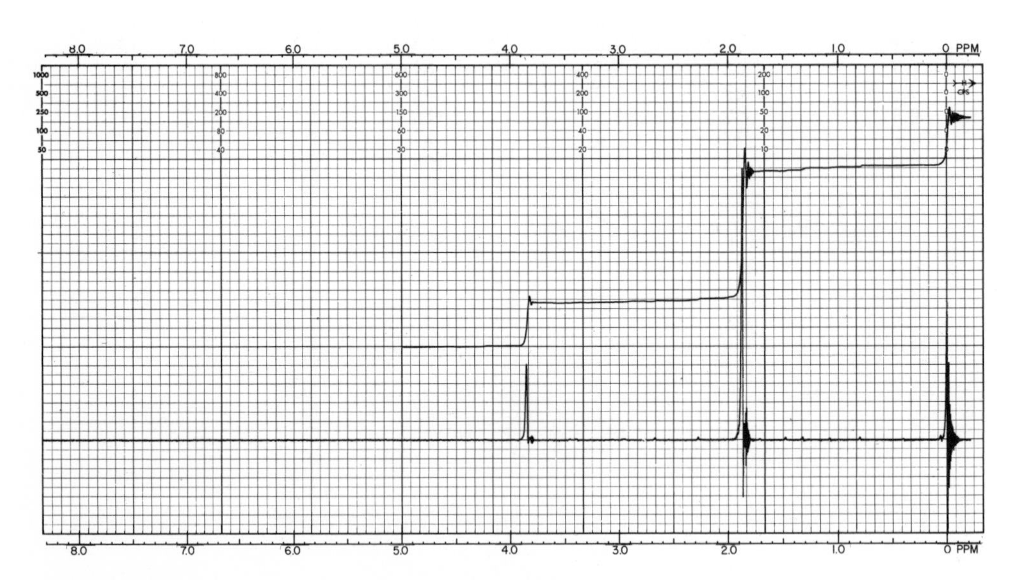

NUCLEAR MAGNETIC RESONANCE SPECTRUM:
15% in carbon tetrachloride, 500 Hz sweep width

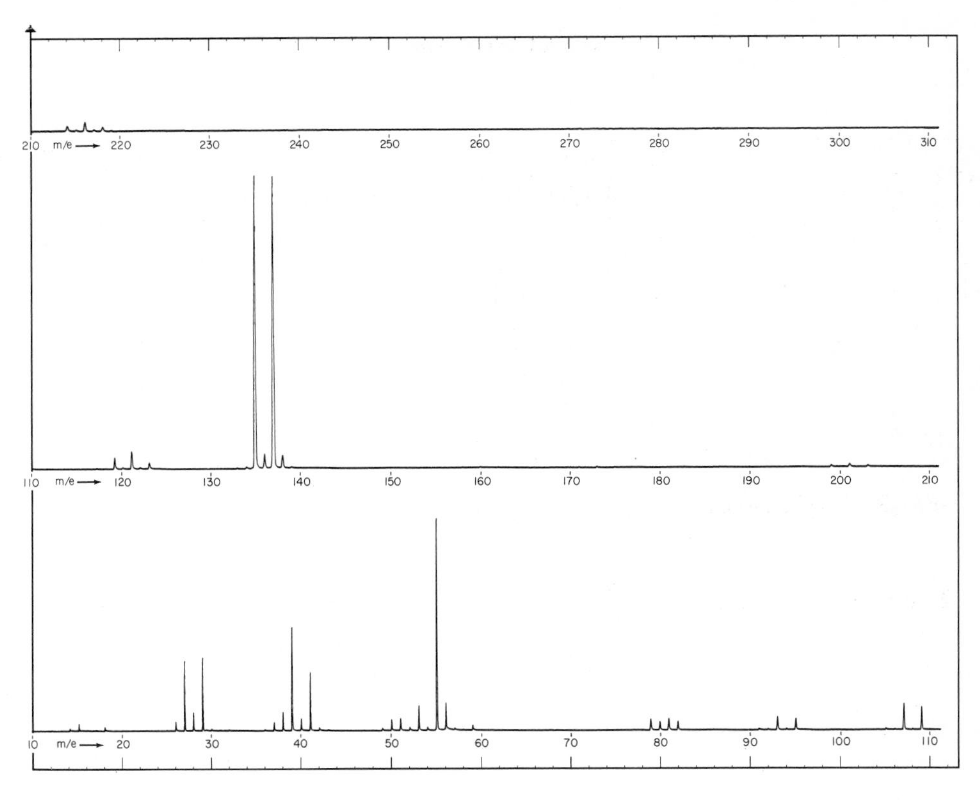

MASS SPECTRUM

MASS SPECTRAL DATA

m/e	*relative intensity*	*m/e*	*relative intensity*	*m/e*	*relative intensity*	*m/e*	*relative intensity*
15	1.2	50	2.6	92	0.6	123	2.7
17	0.7	51	2.8	93	4.6	133	0.7
18	2.8	52	1.1	94	0.6	134	0.9
26	2.0	53	7.5	95	4.4	135	100.0
27	15.0	54	0.9	105	0.6	136	8.0
28	4.0	55	79.0	107	10.0	137	99.0
29	16.0	56	8.6	108	0.5	138	7.2
37	1.9	57	1.0	109	9.3	139	1.0
38	4.2	59	1.9	110	0.2	199	1.0
39	26.0	79	3.6	117	0.6	201	2.0
40	2.9	80	3.0	119	5.0	203	1.1
41	16.0	81	3.8	120	0.7	214	1.3
42	0.9	82	3.0	121	7.4	216	2.5
49	0.9	91	0.6	122	0.8	218	1.1

COMPOUND 1-39

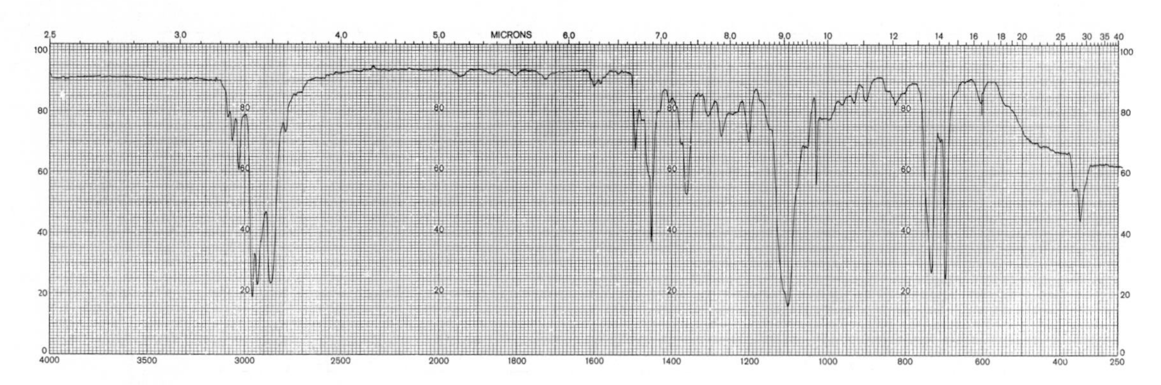

INFRARED SPECTRUM: Liquid film

ULTRAVIOLET DATA: λ_{max}^{EtOH} 206 nm (ϵ 7,530); 230 nm (ϵ 840); 247 nm (ϵ 470); 252 nm (ϵ 500); 258 nm (ϵ 490); 264 nm (ϵ 350); 267 nm (ϵ 220); 280 nm (ϵ 60).

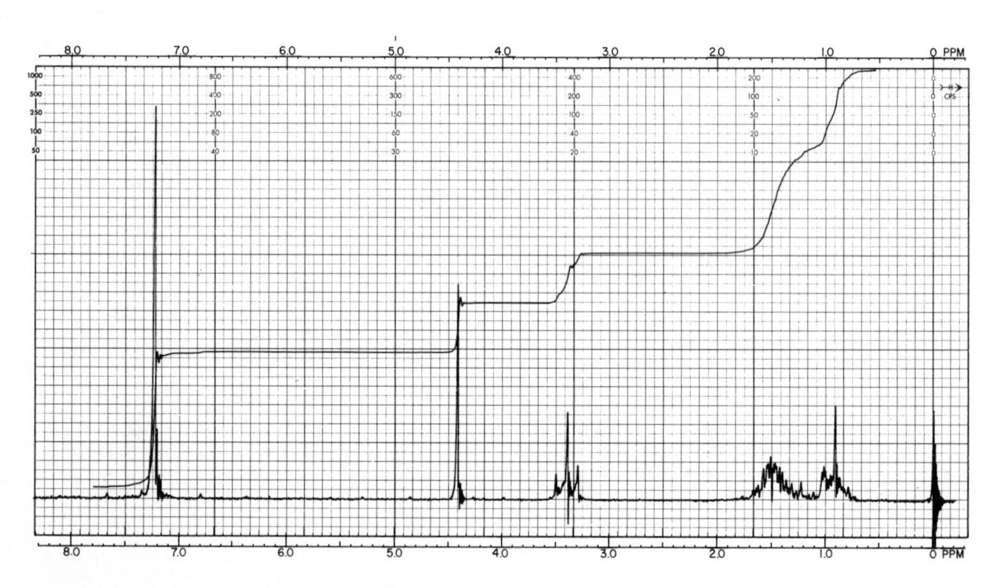

NUCLEAR MAGNETIC RESONANCE SPECTRUM:
15% in carbon tetrachloride, 500 Hz sweep width

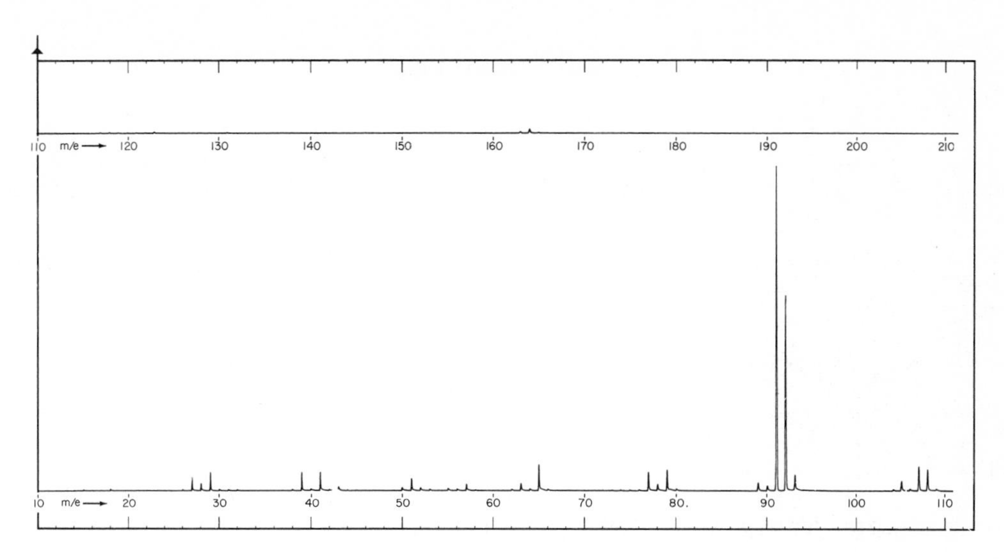

MASS SPECTRUM

MASS SPECTRAL DATA

m/e	*relative intensity*	*m/e*	*relative intensity*	*m/e*	*relative intensity*	*m/e*	*relative intensity*
26	0.3	52	0.8	77	5.2	107	7.8
27	2.6	53	0.5	78	1.8	108	7.2
28	1.3	55	0.6	79	5.8	109	0.8
29	3.3	56	0.5	80	0.7	117	0.3
30	0.3	57	1.6	87	0.2	118	0.3
31	0.2	62	0.4	89	2.7	119	0.3
38	0.4	63	1.8	90	1.7	121	0.4
39	3.9	64	0.6	91	100.0	122	0.3
40	0.4	65	6.5	92	55.0	123	0.7
41	3.8	66	0.5	93	4.9	131	0.3
42	0.4	73	0.2	94	0.3	163	0.7
43	1.0	74	0.2	104	0.6	164	1.6
50	0.8	75	0.3	105	3.2	165	0.3
51	2.7	76	0.3	106	0.7		

COMPOUND 1-40

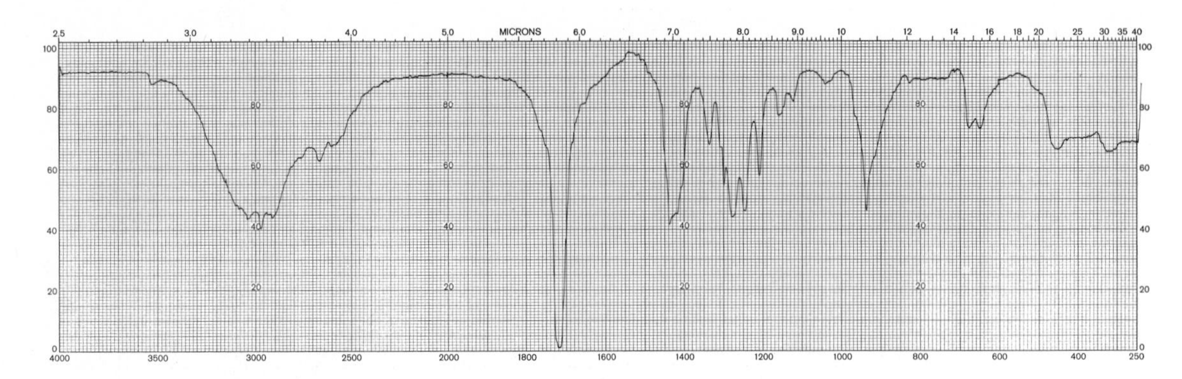

INFRARED SPECTRUM: 1.8% in carbon tetrachloride

ULTRAVIOLET DATA: λ_{max}^{EtOH} 209 nm (ϵ 37)

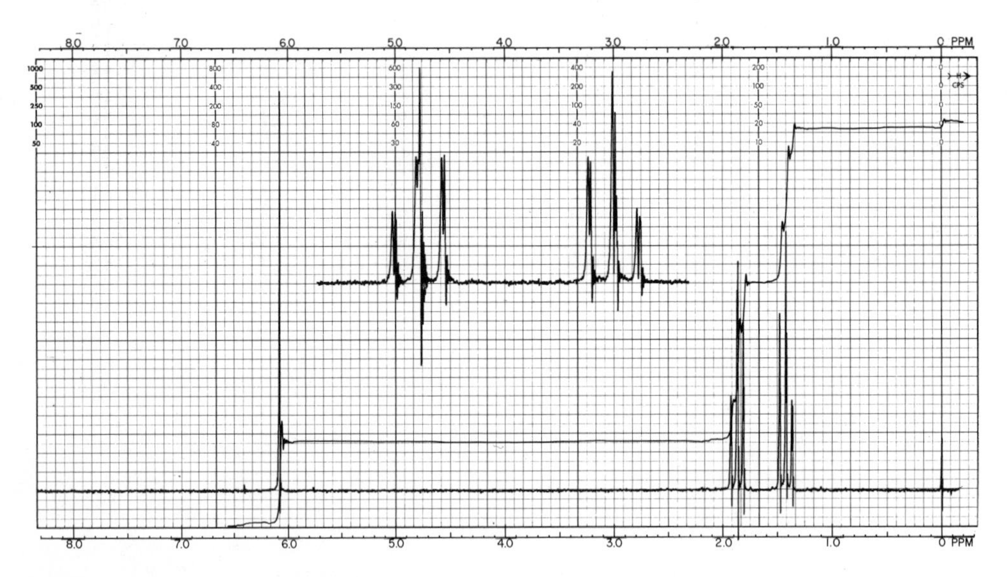

NUCLEAR MAGNETIC RESONANCE SPECTRUM:
15% in carbon tetrachloride, 1000 Hz sweep width
Expansion: 250 Hz sweep width, 80 Hz sweep offset

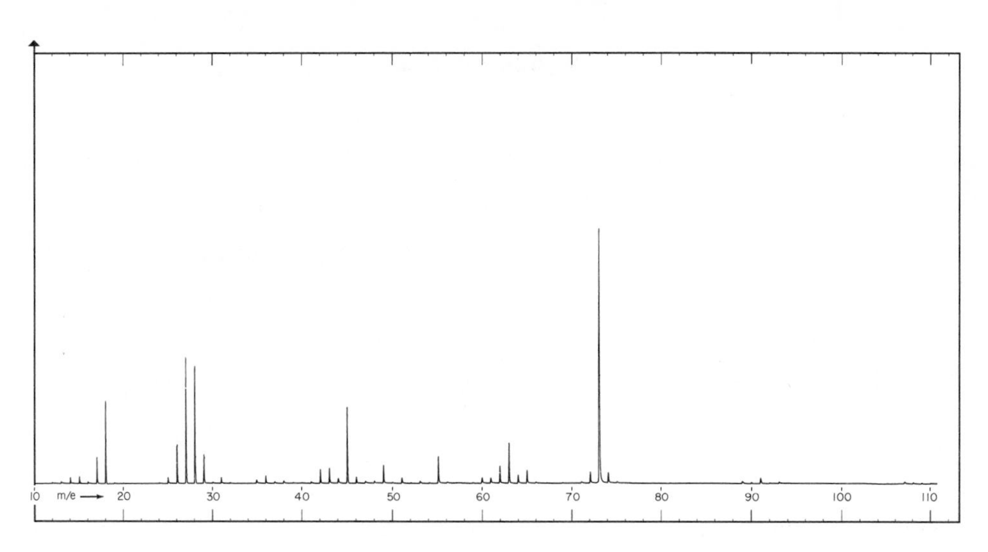

MASS SPECTRUM

MASS SPECTRAL DATA

m/e	*relative intensity*	*m/e*	*relative intensity*	*m/e*	*relative intensity*	*m/e*	*relative intensity*
12	0.2	35	1.1	51	1.9	69	0.2
13	0.4	36	2.7	52	0.2	71	0.4
14	1.2	37	0.8	53	1.0	72	4.6
15	1.2	38	1.1	54	0.3	73	100.0
16	0.4	39	0.5	55	9.7	74	4.3
17	4.0	40	0.2	56	0.5	75	0.5
18	16.5	41	0.8	57	0.3	89	1.2
19	0.3	42	4.3	59	0.1	90	0.6
24	0.3	43	4.6	60	2.2	91	2.5
25	1.7	44	2.0	61	2.1	92	0.4
26	12.2	45	24.9	62	5.8	93	0.6
27	34.0	46	2.0	63	14.2	107	1.0
28	33.0	47	0.8	64	3.0	108	0.6
29	9.1	48	0.6	65	4.6	109	0.3
30	0.5	49	5.9	66	0.4	110	0.2
31	1.8	50	0.3				

COMPOUND 1-41

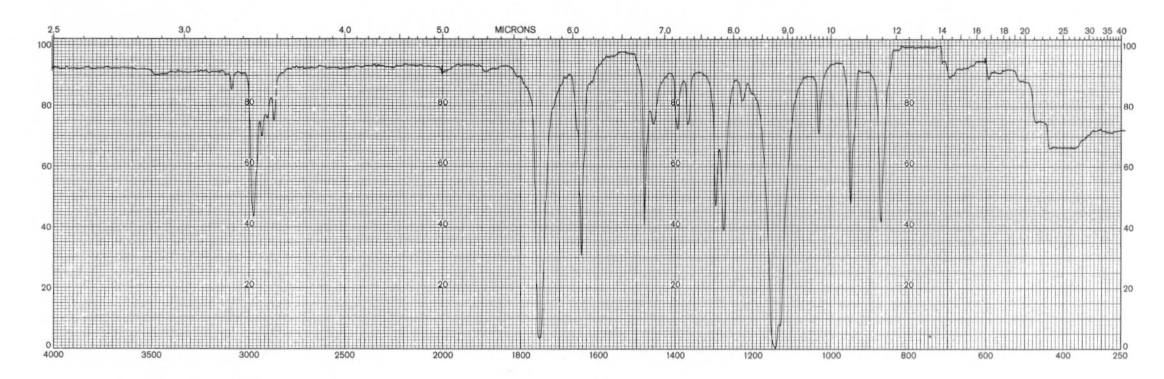

INFRARED SPECTRUM: 1.7% in carbon tetrachloride

ULTRAVIOLET DATA: no λ_{max} above 205 nm

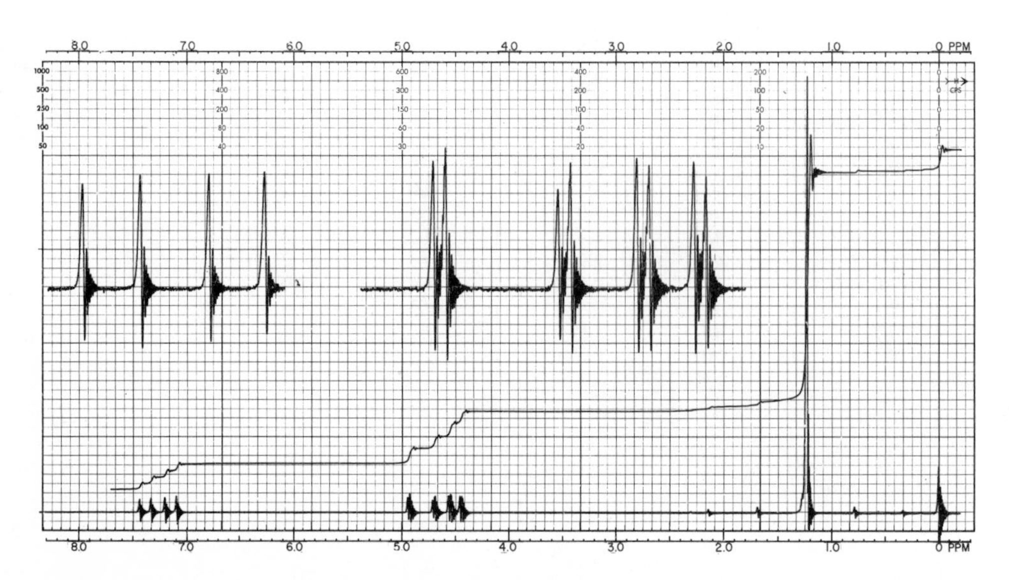

NUCLEAR MAGNETIC RESONANCE SPECTRUM:
15% in carbon tetrachloride, 500 Hz sweep width
Expansions: (left) 100 Hz sweep width, 350 Hz sweep offset
(right) 100 Hz sweep width, 240 Hz sweep offset

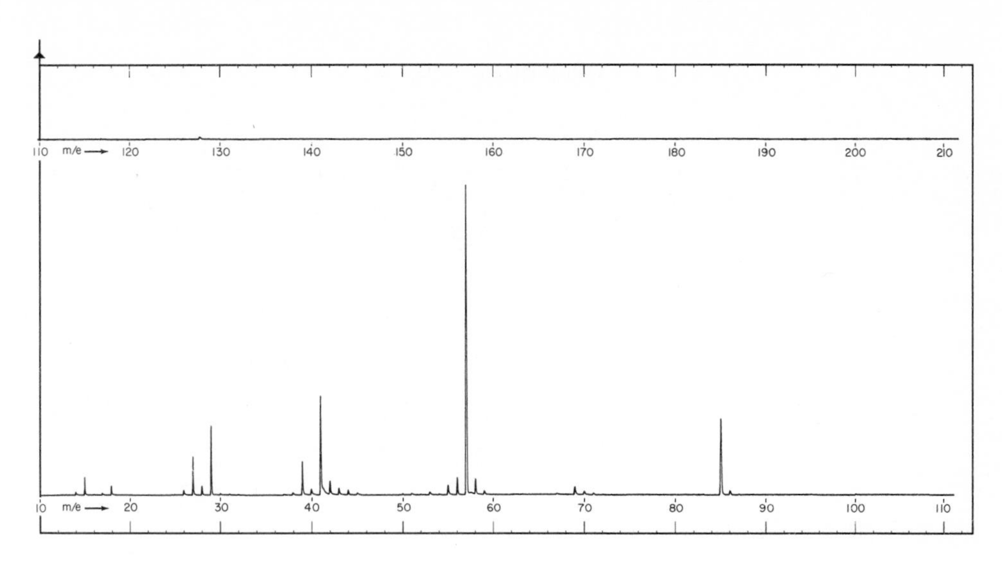

MASS SPECTRUM

MASS SPECTRAL DATA

m/e	*relative intensity*	*m/e*	*relative intensity*	*m/e*	*relative intensity*	*m/e*	*relative intensity*
14	0.6	38	0.7	51	0.6	70	1.3
15	2.9	39	9.0	53	1.0	71	0.6
18	0.7	40	1.5	54	0.4	72	0.4
26	1.4	41	25.6	55	2.9	83	0.4
27	8.3	42	4.2	56	4.9	85	24.8
28	2.2	43	2.1	57	100.0	86	1.6
29	16.4	44	1.3	58	4.9	87	0.4
30	0.4	45	0.7	59	1.3	128	0.9
32	0.3	50	0.4	69	2.5		

COMPOUND 1-42

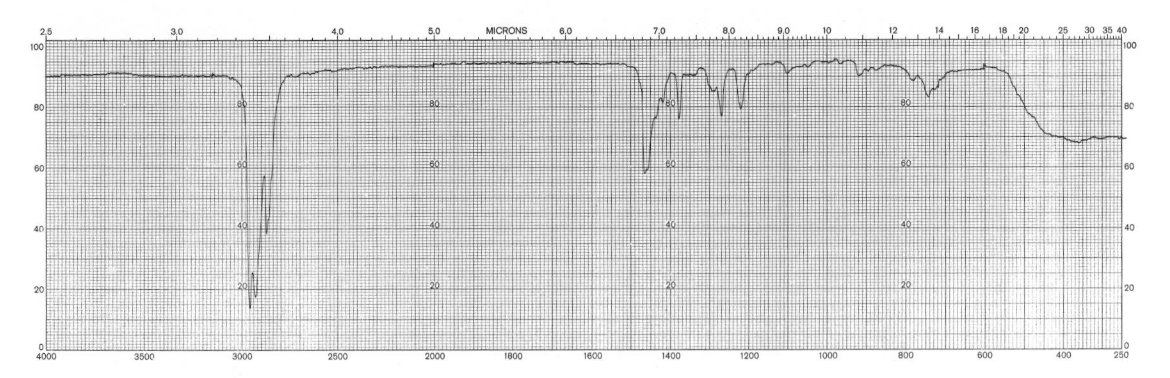

INFRARED SPECTRUM: Liquid film

ULTRAVIOLET DATA: λ_{max}^{EtOH} 210 nm (ϵ 1,200); 229(sh) nm (ϵ 145).

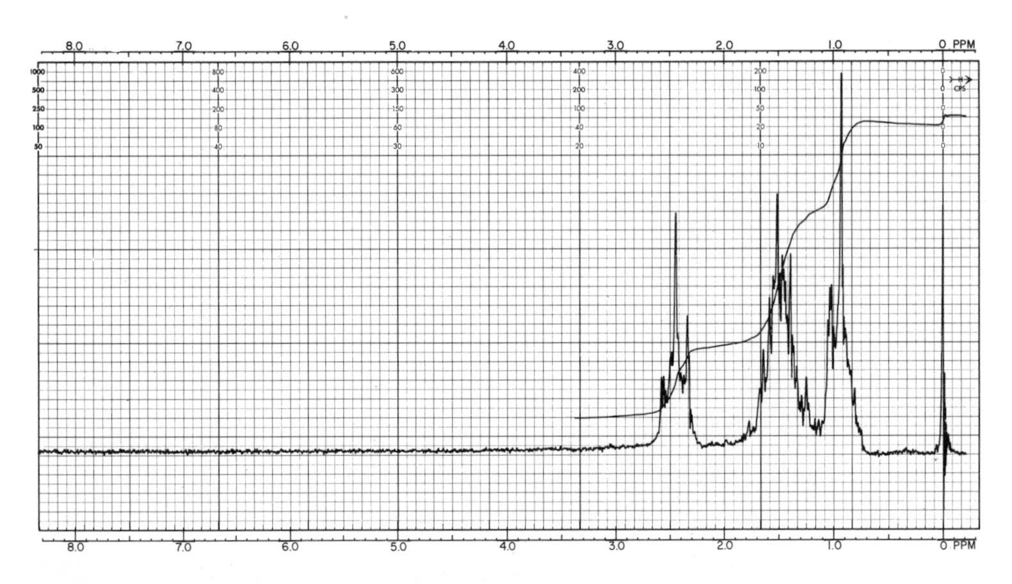

NUCLEAR MAGNETIC RESONANCE SPECTRUM:
15% in carbon tetrachloride, 500 Hz sweep width

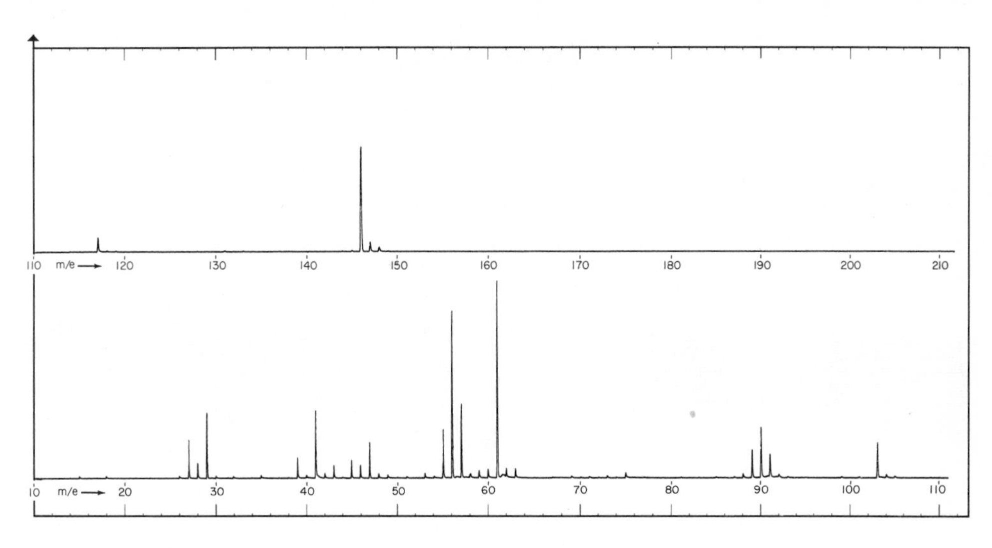

MASS SPECTRUM

MASS SPECTRAL DATA

m/e	*relative intensity*	*m/e*	*relative intensity*	*m/e*	*relative intensity*	*m/e*	*relative intensity*
26	1.5	47	14.6	59	3.4	91	12.7
27	14.6	48	1.9	60	4.3	92	2.2
28	7.6	49	1.3	61	100.0	103	18.9
29	24.8	50	0.8	62	4.9	104	2.0
31	1.1	51	0.7	63	4.5	105	1.0
35	0.8	52	0.6	73	1.2	117	8.2
39	8.6	53	1.5	75	3.2	118	0.8
40	1.1	54	1.5	85	0.9	119	0.4
41	27.5	55	20.5	87	0.9	145	1.0
42	2.0	56	86.0	88	1.9	146	69.0
43	5.5	57	38.8	89	14.6	147	7.2
45	7.2	58	2.7	90	26.0	148	3.2
46	5.1						

COMPOUND 1-43

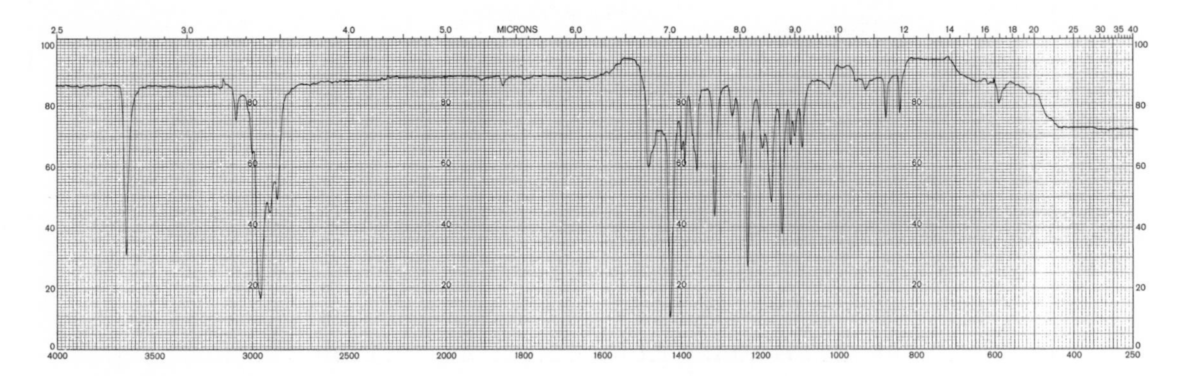

INFRARED SPECTRUM: 2.5% in carbon tetrachloride

ULTRAVIOLET DATA: λ_{max}^{EtOH} 212 nm (ϵ 7,600); 268 nm (ϵ 1,550); 274 nm (ϵ 1,500)

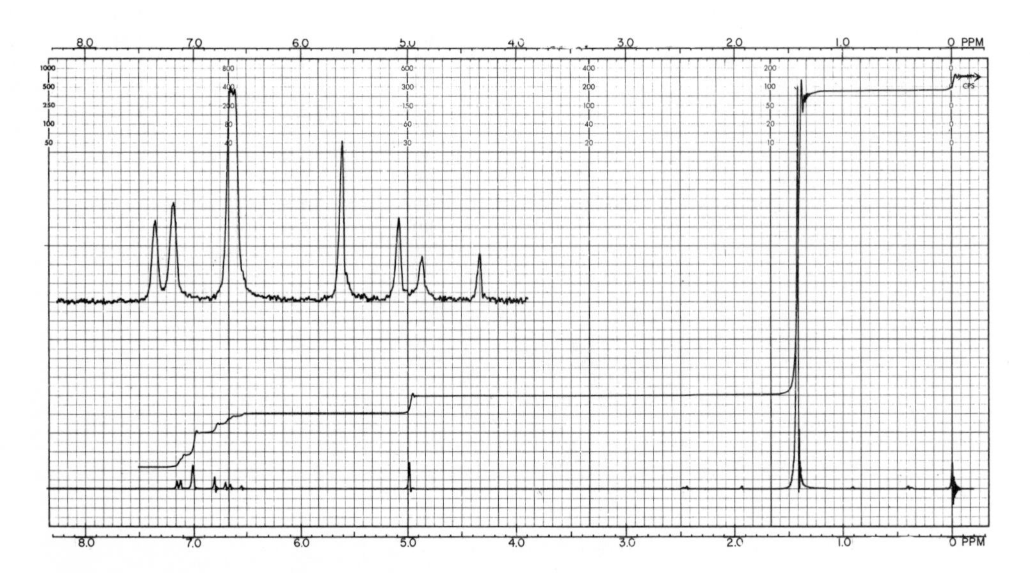

NUCLEAR MAGNETIC RESONANCE SPECTRUM:
15% in carbon tetrachloride, 500 Hz sweep width
Expansion: 100 Hz sweep width, 340 Hz sweep offset

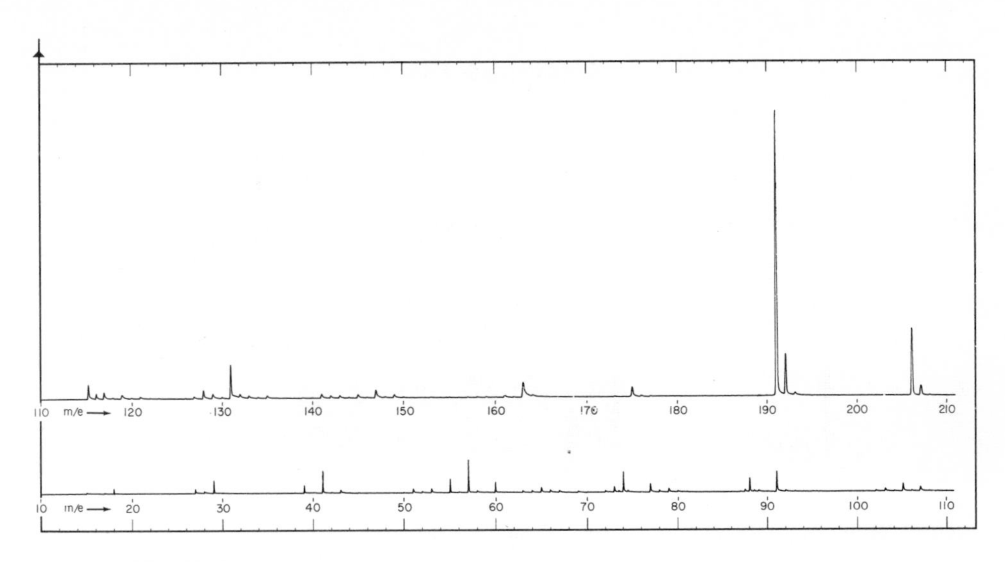

MASS SPECTRUM

MASS SPECTRAL DATA

m/e	*relative intensity*	*m/e*	*relative intensity*	*m/e*	*relative intensity*	*m/e*	*relative intensity*
27	0.9	66	0.7	115	3.6	143	0.9
28	0.5	67	0.4	116	1.5	145	1.1
29	2.1	69	0.5	117	1.9	147	2.3
39	1.5	70	0.3	118	0.5	149	1.1
41	3.9	72	0.4	119	1.3	161	0.9
43	0.8	73	1.4	121	0.8	163	4.6
51	0.9	74	4.1	127	0.6	164	1.0
52	0.6	77	2.0	128	2.5	175	2.9
53	1.0	78	0.6	129	1.5	176	0.6
55	2.7	79	1.0	131	9.2	191	100.0
57	6.4	80	0.6	132	1.4	192	16.4
58	0.6	88	2.9	133	1.0	193	1.6
60	1.9	91	4.9	134	0.4	206	24.8
63	0.6	103	0.9	135	0.7	207	3.9
64	0.5	105	2.2	141	1.4	208	0.5
65	1.2	107	1.2	142	0.9		

COMPOUND 1-44

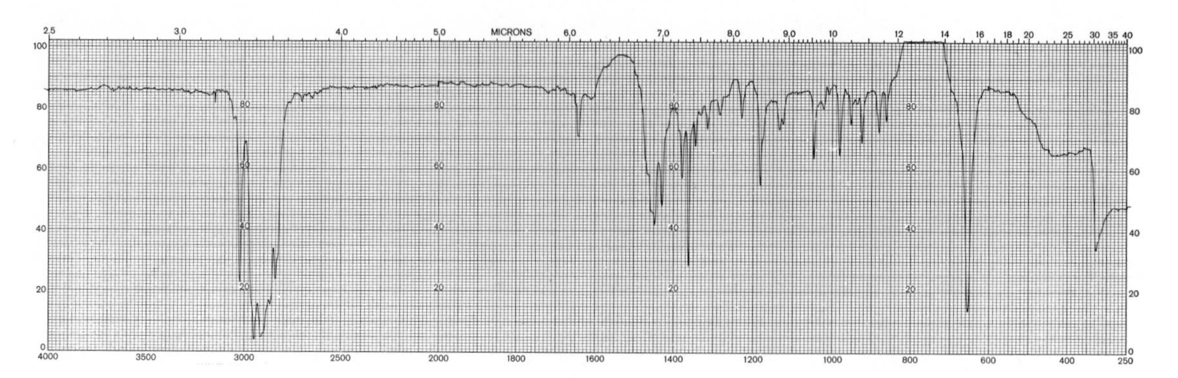

INFRARED SPECTRUM: 5.0% in carbon tetrachloride

ULTRAVIOLET DATA: no λ_{max}^{EtOH} above 205 nm

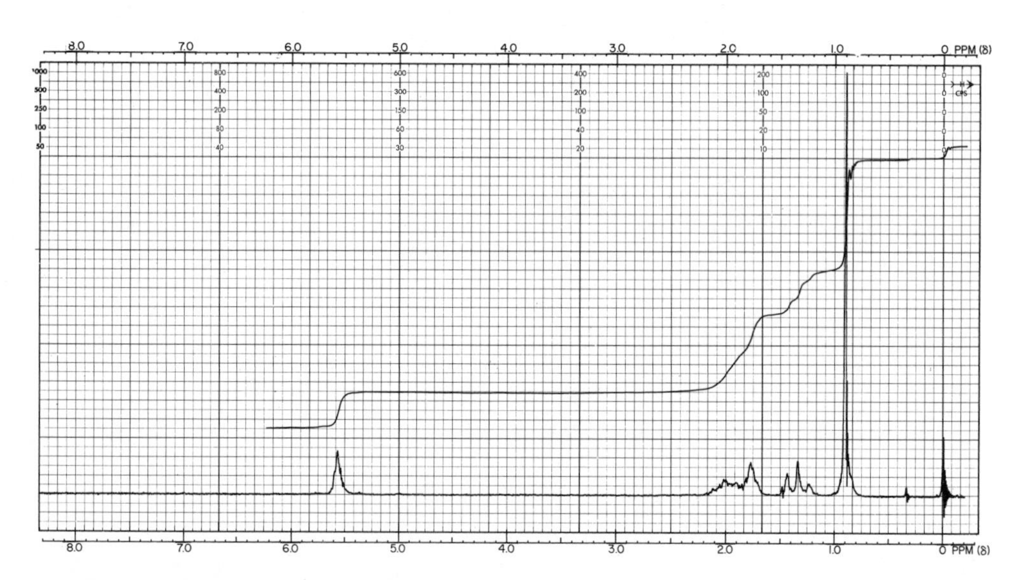

NUCLEAR MAGNETIC RESONANCE SPECTRUM:
15% in carbon tetrachloride, 500 Hz sweep width

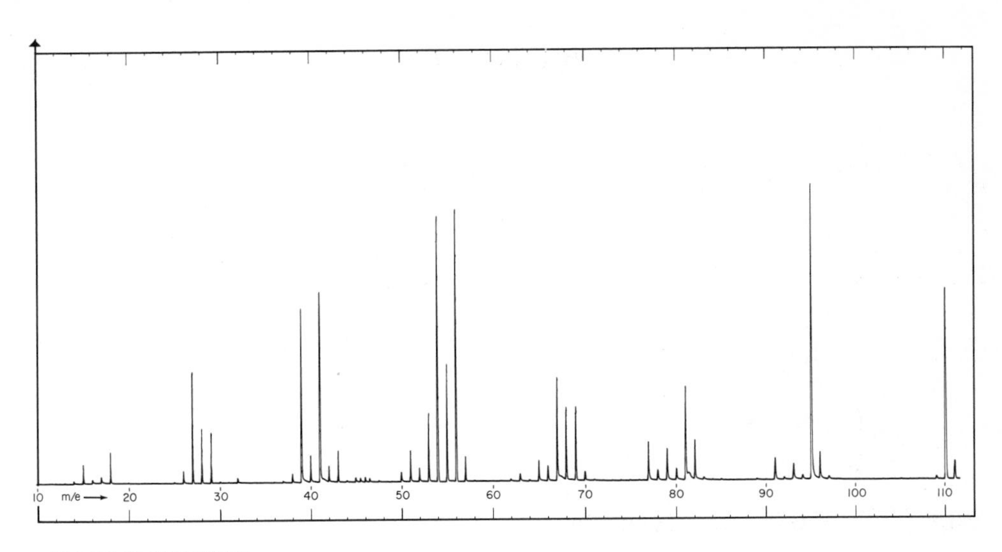

MASS SPECTRUM

MASS SPECTRAL DATA

m/e	*relative intensity*	*m/e*	*relative intensity*	*m/e*	*relative intensity*	*m/e*	*relative intensity*
14	0.7	42	3.8	62	0.6	83	0.9
15	4.1	43	8.3	63	1.9	85	0.3
16	0.8	44	0.3	64	0.4	89	0.3
17	1.2	45	1.0	65	5.3	91	4.8
18	5.0	45.5	1.1	66	3.3	92	0.8
25	0.3	46	1.3	67	25.0	93	3.8
26	3.3	46.5	1.0	68	17.0	94	1.1
27	30.0	50	2.2	69	16.6	95	81.0
28	10.4	51	7.2	70	2.1	96	6.6
29	13.2	52	3.5	71	0.3	97	1.0
32	1.2	53	16.5	77	8.9	98	0.2
37	0.5	54	94.0	78	2.2	109	1.0
38	2.1	55	29.0	79	7.0	110	46.0
39	40.0	56	100.0	80	2.7	111	4.2
40	4.7	57	5.7	81	21.2	112	0.2
41	62.0	58	0.2	82	8.7		

COMPOUND 1-45

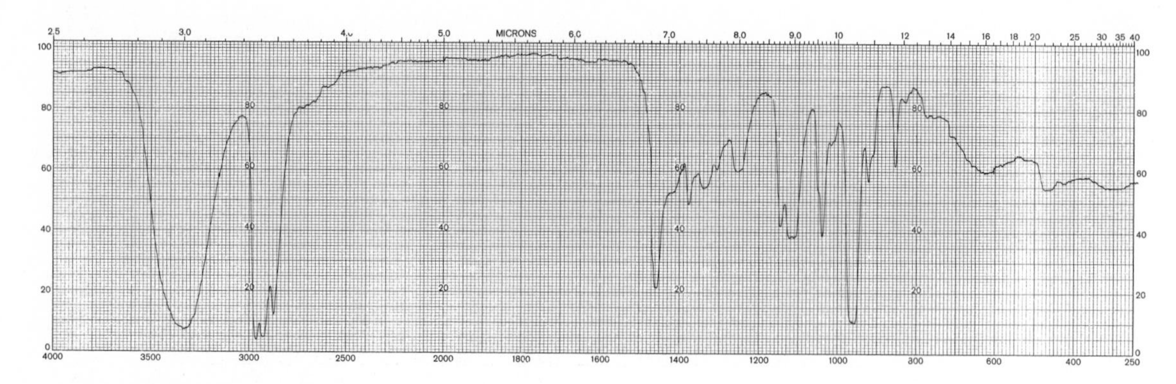

INFRARED SPECTRUM: Liquid film

ULTRAVIOLET DATA: no λ_{max} above 205 nm

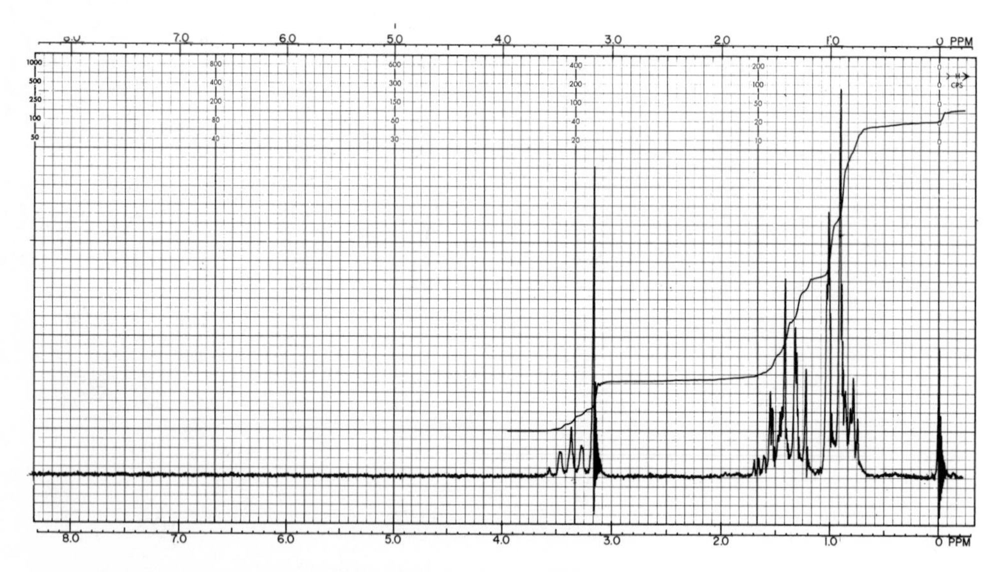

NUCLEAR MAGNETIC RESONANCE SPECTRUM:
15% in carbon tetrachloride, 500 Hz sweep width

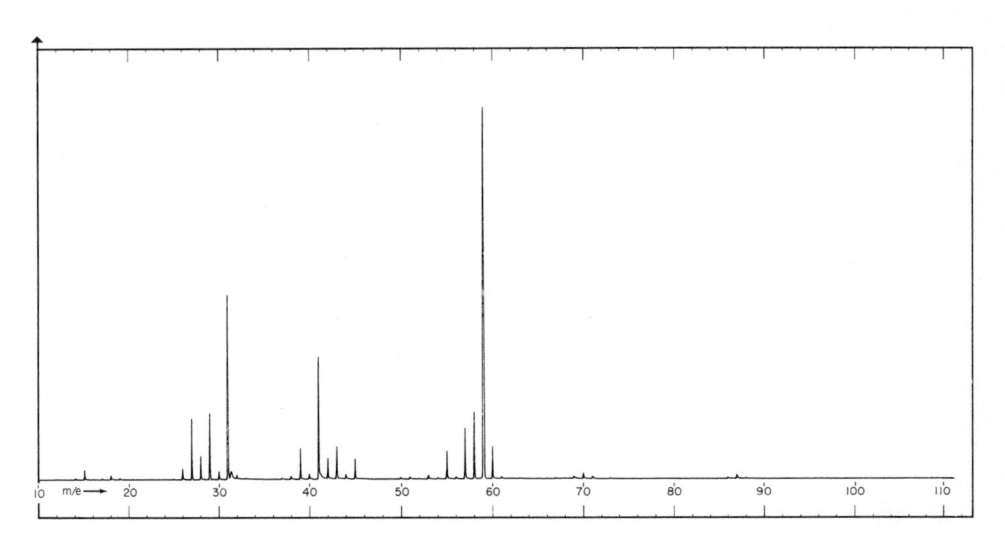

MASS SPECTRUM

MASS SPECTRAL DATA

m/e	relative intensity	m/e	relative intensity	m/e	relative intensity	m/e	relative intensity
14	0.5	32	1.9	46	0.3	59	100.0
15	2.5	37	0.5	50	0.8	60	11.4
18	1.3	38	1.0	51	1.0	61	0.9
19	0.4	39	9.9	52	0.2	69	1.4
26	3.5	40	1.9	53	1.9	70	2.6
27	20.8	41	48.0	54	0.5	71	1.4
28	8.6	42	7.9	55	11.0	86	0.7
29	25.0	43	13.1	56	1.1	87	1.8
30	2.8	44	2.1	57	18.6	88	0.3
31	89.0	45	6.9	58	25.0		

COMPOUND 2-1

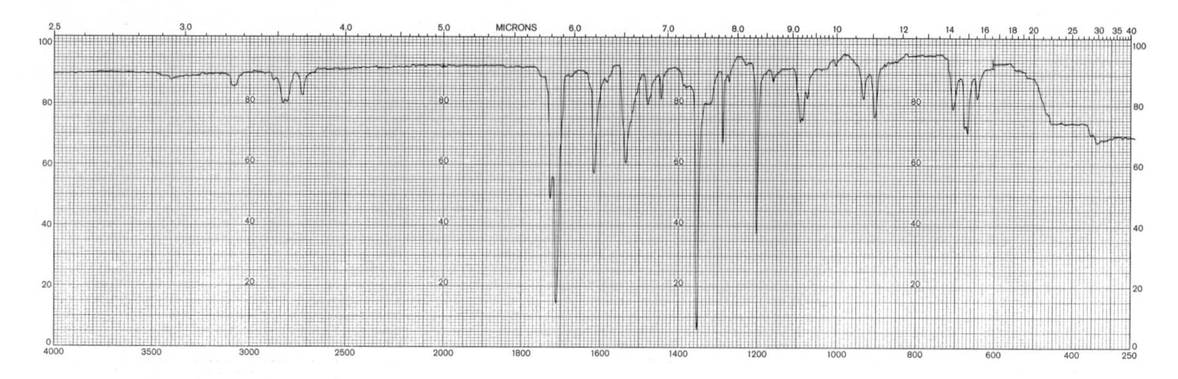

INFRARED SPECTRUM: 1.2% in carbon tetrachloride

ULTRAVIOLET DATA: λ_{max}^{EtOH} 225 nm (ϵ 13,000); 256 nm (ϵ 7,800); 335 nm (ϵ 160)

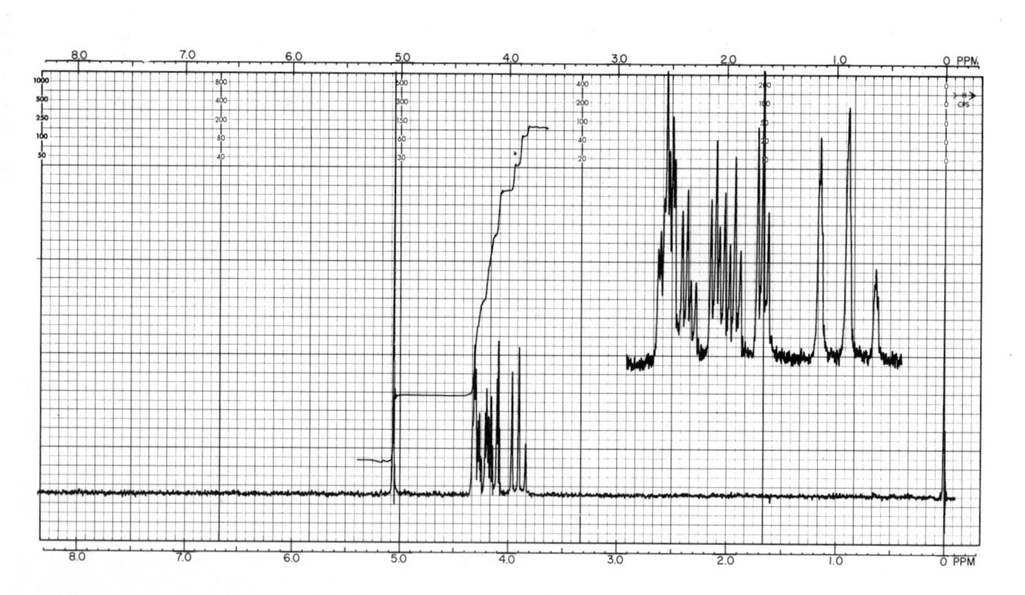

NUCLEAR MAGNETIC RESONANCE SPECTRUM:
15% in carbon tetrachloride, 1000 Hz sweep width
Expansion: 250 Hz sweep width, 440 Hz sweep offset

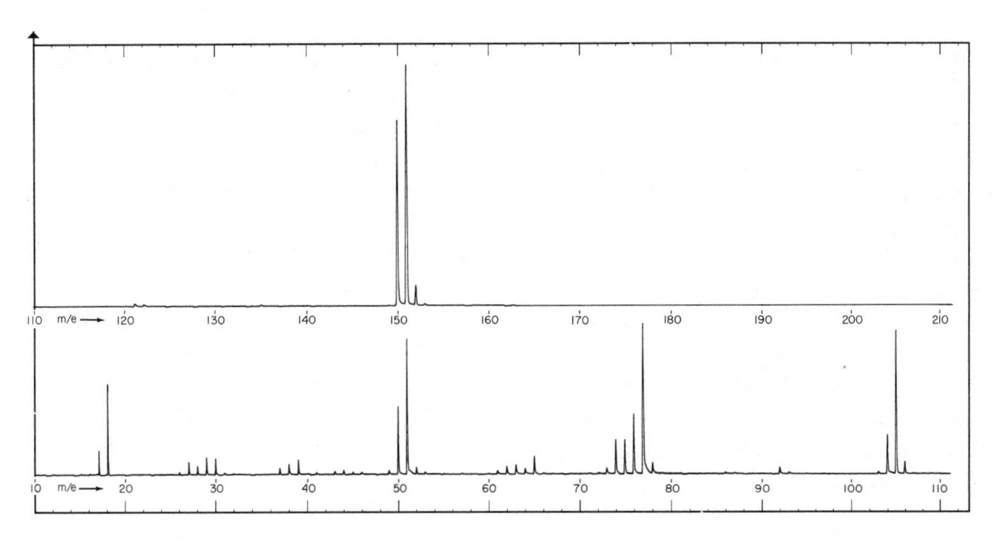

MASS SPECTRUM

MASS SPECTRAL DATA

m/e	*relative intensity*	*m/e*	*relative intensity*	*m/e*	*relative intensity*	*m/e*	*relative intensity*
15	0.1	42	0.1	66	0.5	93	0.5
16	0.2	43	0.6	72	0.3	103	0.8
17	1.3	44	0.9	73	1.4	104	10.0
18	5.3	45	0.3	74	8.5	105	35.0
26	0.5	46	0.6	75	8.4	106	3.2
27	2.3	49	0.9	76	13.7	108	0.4
28	1.4	50	13.0	77	36.0	120	0.3
29	3.3	51	29.0	78	3.4	121	1.3
30	3.2	52	1.9	85	0.3	122	1.0
31	0.3	53	0.7	86	0.5	123	0.2
32	0.3	60	0.2	87	0.4	135	0.9
37	1.3	61	0.9	88	0.2	150	77.0
38	2.1	62	1.8	89	0.3	151	100.0
39	2.9	63	2.3	90	0.2	152	9.5
40	0.2	64	1.3	91	0.3	153	0.9
41	0.3	65	4.0	92	1.9		

Solution to Problem 2-1

The IR spectrum of compound **2-1** indicates the absence of –OH or –NH absorptions. The intense absorption at 5.84 μ (1712 cm^{-1}) suggests the presence of a carbonyl group. The absorptions at 3.54 and 3.67 μ (2825 and 2725 cm^{-1}) suggests that this carbonyl group may be an aldehyde. The UV spectrum shows intense absorptions and several λ_{max} but it cannot be interpreted at this time. The NMR spectrum of the compound can be analyzed as follows (it should be noted that this NMR spectrum was recorded using the 1000 Hz sweep width):

τ	*multiplicity*	*integration squares*	*relative number of hydrogens*
–0.13	s	7.1	1
1.30-2.36	complex	28.7	4

The absorption at τ –0.13 strongly supports the presence of an aldehyde group, since no other hydrogen absorb in this region of the NMR spectrum. The presence of four hydrogens in the region τ 1.30-2.36 suggests that these hydrogens are attached to an aromatic nucleus. No other hydrogen absorptions are present in the NMR spectrum. Thus, at this stage, it might be reasonable to suggest that the compound is a monosubstituted benzaldehyde derivative.

The most abundant ion in the MS of the compound occurs at *m/e* 151, and no intense peaks are present at higher *m/e* value. The intense peak at *m/e* 150 (M^+ –1) is consistent with the hypothesis that the compound is an aldehyde because of the stability of the resulting acylium ion. If 151 is the molecular weight of the compound, the compound would probably contain one nitrogen atom. The sum of a monosubstituted benzene fragment (C_7H_5O, 105) and a nitrogen atom (14) is 119. When this mass is subtracted from 151 the result is 32 mass units. This mass could conveniently be satisfied by the presence of two oxygen atoms, possibly present as a nitro group. This is substantiated by the presence of an abundant ion at *m/e* 105 (M^+ –46, –NO_2). It is significant to recognize the abundance of the ion at *m/e* 104, which suggests that the ion at *m/e* 105 has retained the aldehyde functional group. Thus, the compound appears to be a nitrobenzaldehyde; the substitution pattern remains to be determined.

If the compound were *p*-nitrobenzaldehyde, the hydrogens attached to the aromatic nucleus would constitute an A_2B_2 system in the NMR spectrum. Such a system dictates a symmetrical pattern. Clearly this is not the case, so the compound is not *p*-nitrobenzaldehyde.

It would appear that the structure of compound **2-1** is either *o*- or *m*-nitrobenzaldehyde. In either case the NMR spectrum of the hydrogens attached to the aromatic nucleus would necessarily constitute an ABCD system. The absorption present at τ 1.30-2.36 is complex and not at all inconsistent with this assumption. If the structure is *m*-nitrobenzaldehyde, the three hydrogens *ortho* to the aldehyde and nitro functional groups would be expected to be deshielded by these functional groups, and absorb at a τ value lower than the other hydrogen *meta* to these functional groups. If the structure of the compound were *o*-nitrobenzaldehyde, the two hydrogen atoms *ortho* to the nitro and aldehyde functional groups would be expected to be deshielded by these groups and therefore absorb at a lower τ value than the two hydrogen atoms *meta* and *para* to these functional groups. If the NMR spectrum is carefully examined, it can be determined that the NMR absorption from τ 2.0-2.36 (centered at τ 2.24) is a three-line absorption containing one hydrogen atom (7.0 integration squares). The spacing of these three lines is equal (8.0 Hz), which suggests that this hydrogen is coupled to two hydrogens *ortho* to it. There is no evidence that this hydrogen is coupled to a hydrogen *meta* to it. This careful analysis eliminates *o*-nitrobenzaldehyde from consideration. Thus, the structure of the compound is *m*-nitrobenzaldehyde.

Solution to Problem 2-2

The IR spectrum of compound **2-2** shows moderate broad absorption from 2.77 to 4.10 μ (3610 to 2439 cm^{-1}), which indicates the presence of some functional group(s) other than C-H stretching. The most intense absorption in the IR spectrum occurs at 5.87 μ (1704 cm^{-1}), which suggests the presence of a carbonyl group. The UV spectrum of the compound has λ_{max} at 208 nm with a low extinction coefficient. This would indicate the presence of a simple chromophoric group, not in conjugation with any unsaturated linkage, and also the absence of any aromatic functionality. The NMR spectrum of the compound (note again that the sweep width is 1000 Hz) is simple and very informative. The spectrum may be analyzed as follows:

τ	*multiplicity*	*integration squares*	*relative number of hydrogens*
-1.60	s	6.1	1
7.51	s	11.7	2
8.83	s	17.3	3

The NMR absorption present at τ - 1.60 practically dictates the presence of a carboxylic acid functional group. The two-hydrogen singlet at τ 7.51 might suggest the presence of a methylene group adjacent to and deshielded by the carboxyl group. The three-hydrogen NMR absorption at τ 8.83 might suggest a methyl group in a relatively "saturated" environment. It is important to note that the latter two NMR absorptions are singlets; therefore, these groups cannot be adjacent to carbon atoms bonded to any hydrogen.

If the functional groups indicated from the NMR spectrum are added together ($-CO_2H$, $-CH_2-$, $-CH_3$), the sum totals 74 mass units.

The MS of the compound, which contains an ion at *m/e* 142, clearly indicates that the simple interpretation of the NMR spectrum must be expanded to include the presence of other atoms. No ion in the MS suggests the presence of chlorine, bromine, or sulfur, so these atoms can tentatively be excluded as possibilities. The base peak in the mass spectrum occurs at *m/e* 59. Remembering the NMR spectral data, this ion could very likely have the structure $^+CH_2-CO_2H$. In the high mass region of the MS, there are several interesting mass differences among the major ions. For example 142 - 127 = 15 (loss of CH_3) and 101 - 83 = 18 (loss of H_2O). Since there is no strong indication for the presence of other atoms, it might be simplest to propose that the compound contains only carbon, hydrogen, and oxygen.

If this is the case, the data derived from the NMR spectrum must be doubled. This would indicate the presence of the following functional groups: $(-CO_2H)_2$, $(-CH_3)_2$, and $(-CH_2-)_2$. The sum is 148 mass units. It is to be noted that this mass is larger than the mass of the ion having the largest mass in the MS. It would appear possible that this compound does not show a molecular ion in the mass spectrum.

If the interpretation of the NMR spectrum is correct, a high degree of symmetry is indicated for the compound; that is, the two methyl groups must be magnetically equivalent, and the two methylene groups must also be magnetically equivalent. Furthermore, the methylene and methyl groups must all be attached to an atom that carries no hydrogen atoms because those groups are all singlet NMR absorptions. A partial structural formula,

$$HO_2C{-}CH_2{-}\underset{CH_3}{\overset{CH_3}{\overset{|}{\underset{|}{\ \ }}}}{-}CH_2{-}CO_2H$$

might be suggested. The only atom that can be placed in the center of this partial structural formula is carbon. The mass of the resulting structure ($C_7H_{12}O_4$) is 160. The ion observed in the high mass region at *m/e* 142 would then correspond to the loss of water (M^+ - 18), which could be accounted for by the formation of a ketene ion.

Thus the structure of the compound is 3,3-dimethylglutaric acid. This structure, on reexamination of the spectral data, is consistent with all the data.

COMPOUND 2-2

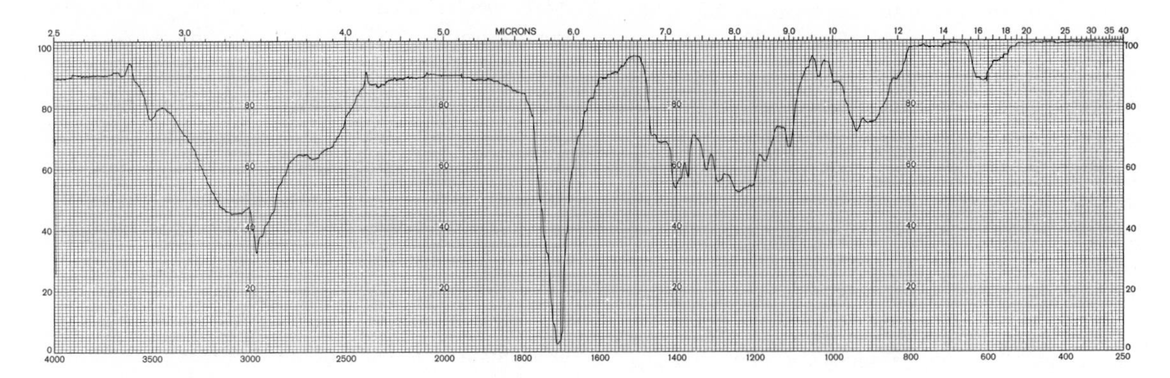

INFRARED SPECTRUM: 1.5% in chloroform

ULTRAVIOLET DATA: $\lambda_{max}^{H_2O}$ 208 nm (ϵ 100)

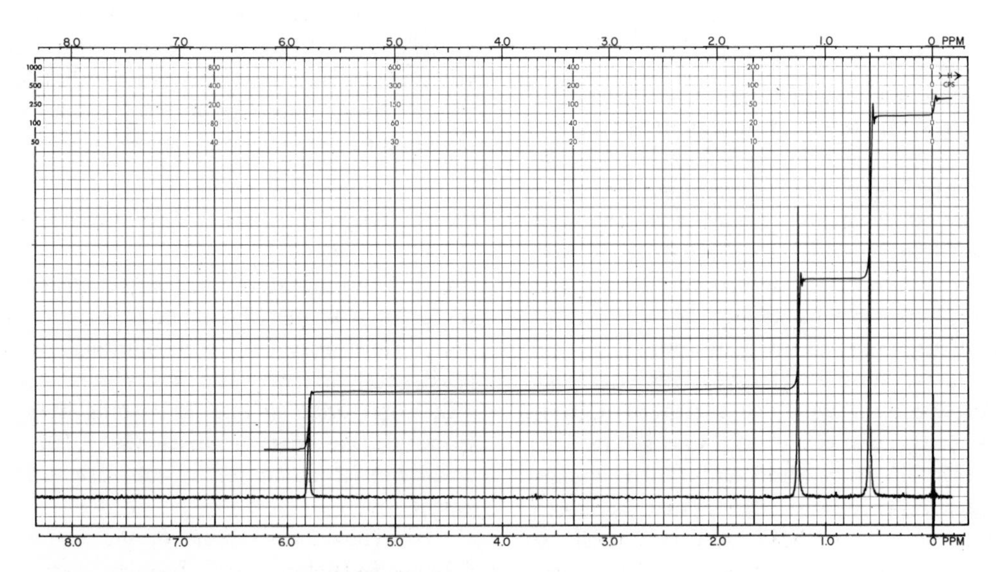

NUCLEAR MAGNETIC RESONANCE SPECTRUM:
15% in deuterochloroform, 1000 Hz sweep width

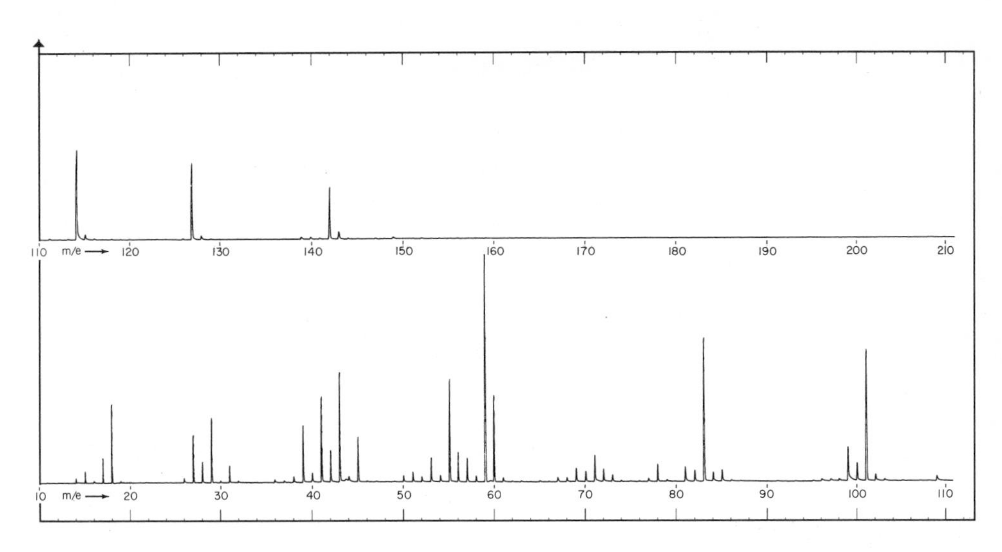

MASS SPECTRUM

MASS SPECTRAL DATA

m/e	relative intensity	m/e	relative intensity	m/e	relative intensity	m/e	relative intensity
15	1.5	50	1.8	72	4.5	102	3.0
17	3.8	51	2.8	73	2.8	103	1.6
18	16.7	52	1.4	74	1.3	109	2.8
26	1.1	53	6.9	75	1.0	111	1.5
27	10.7	54	2.2	76	1.1	113	1.2
28	7.0	55	31.0	77	1.4	114	39.0
29	13.9	56	10.2	78	4.7	115	3.0
31	3.5	57	7.7	81	4.9	116	0.2
32	1.0	58	2.1	82	4.5	127	33.0
38	1.6	59	100.0	83	56.0	128	2.7
39	15.7	60	27.4	84	3.7	129	0.2
40	2.5	61	1.6	85	4.4	139	3.3
41	22.0	65	0.9	96	1.6	140	2.7
42	8.9	67	1.9	97	1.7	141	1.3
43	30.1	68	1.7	98	1.3	142	22.6
44	2.8	69	4.9	99	14.1	143	4.4
45	12.6	70	4.2	100	7.4	145	0.2
46	0.8	71	8.4	101	55.0		

COMPOUND 2-3

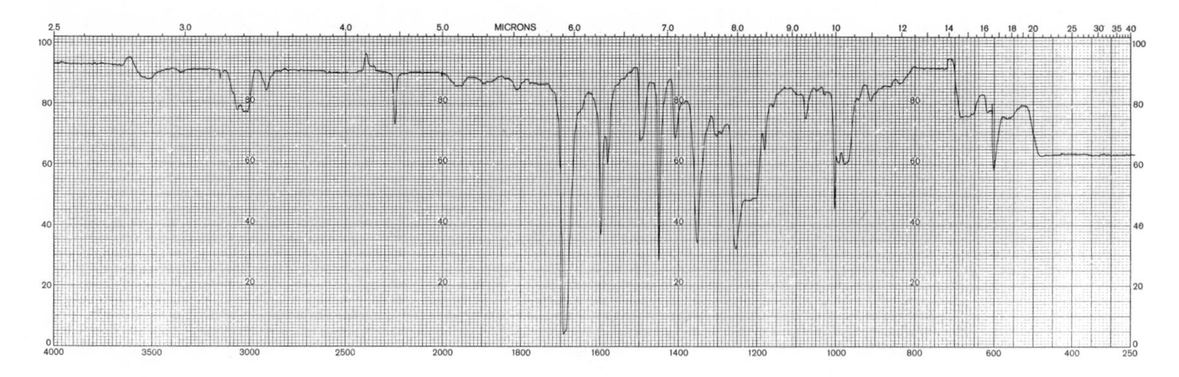

INFRARED SPECTRUM: 4.3% in chloroform

ULTRAVIOLET DATA: λ_{max}^{EtOH} 245 nm (ϵ 13,800); 280 nm (ϵ 1,080)

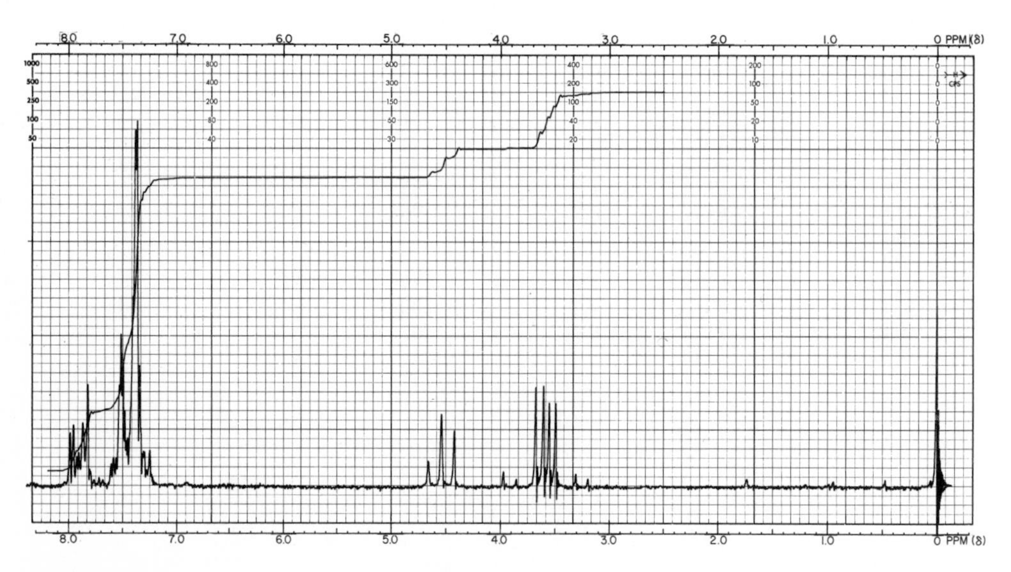

NUCLEAR MAGNETIC RESONANCE SPECTRUM:
15% in deuterochloroform, 500 Hz sweep width

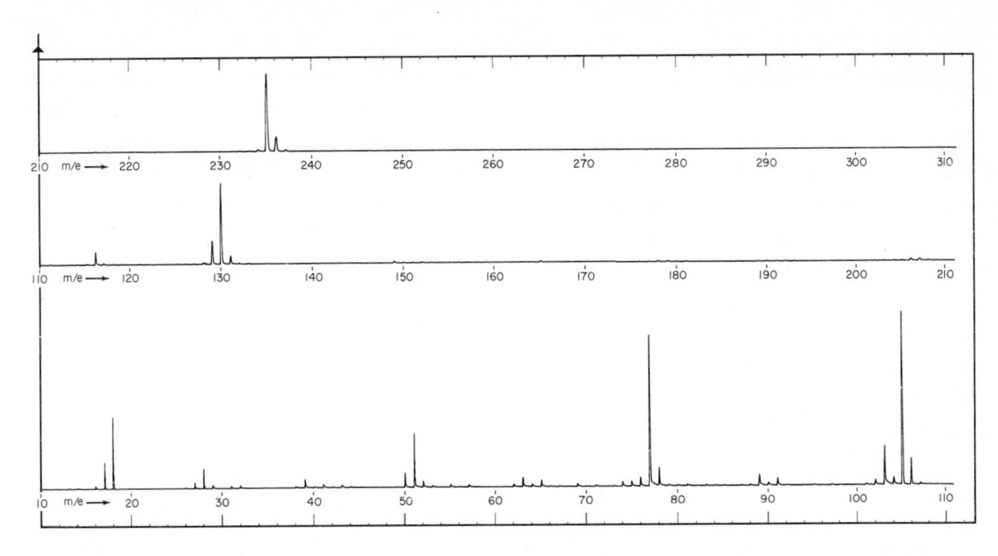

MASS SPECTRUM

MASS SPECTRAL DATA

m/e	*relative intensity*	*m/e*	*relative intensity*	*m/e*	*relative intensity*	*m/e*	*relative intensity*
16	0.4	57	0.6	90	2.0	140	0.4
17	1.2	61	0.3	91	4.4	149	0.6
18	4.1	62	1.2	92	0.6	151	0.5
26	0.4	63	3.9	95	0.5	152	0.7
27	2.1	64	1.2	101	0.9	153	0.5
28	1.9	65	3.0	102	3.0	156	0.4
29	0.5	69	0.8	103	21.9	165	0.8
31	0.8	73	0.4	104	5.1	177	0.5
32	0.3	74	2.0	105	100.0	178	0.9
38	0.6	75	2.5	106	16.0	179	0.8
39	4.3	76	4.3	107	1.3	189	0.4
40	0.4	77	72.0	114	0.6	190	0.4
41	1.0	78	9.8	115	0.8	191	0.4
42	0.4	79	0.9	116	9.8	192	0.4
43	1.0	80	0.6	117	1.1	206	1.9
46	1.5	81	0.5	119	0.4	207	1.8
50	5.9	82	0.4	127	0.4	208	0.7
51	20.8	83	0.4	128	1.3	232	0.4
52	2.9	84	0.3	129	19.1	233	0.4
53	0.5	85	0.4	130	68.0	234	1.3
54	0.2	87	0.6	131	7.1	235	64.0
55	0.6	88	1.0	132	0.8	236	12.1
56	0.2	89	6.1	133	0.5	237	1.2

Solution to Problem 2-3

The IR spectrum of compound **2-3** indicates the absence of –OH or –NH stretching absorptions and, significantly, the presence of acetylenic or nitrile stretching absorption at 4.46 μ (2242 cm^{-1}) and carbonyl stretching absorption at 5.92 μ (1689 cm^{-1}). The UV spectrum shows two strong λ_{max}; this indicates the presence of a chromophoric group. The NMR spectrum is complicated but may be analyzed as follows:

τ	*multiplicity*	*integration squares*	*relative number of hydrogens*
1.97-2.24	complex	6.4	2
2.37-2.82	complex	24.8	8
5.47	t	3.0	1
6.01-6.82	complex	6.0	2

The NMR spectrum strongly indicates the presence of at least ten hydrogen atoms attached to aromatic nuclei, and two of these hydrogen atoms (τ 1.97-2.24) would appear to be deshielded by some mechanism. The remaining hydrogen NMR absorptions, which occur at higher field, are obviously coupled to one another. These absorptions, which total three hydrogens, obviously do not constitute an AB_2 pattern. It is therefore likely that these hydrogens constitute an ABC pattern.

The MS of this compound shows an abundant ion at *m/e* 235. Because this is an odd mass, the compound may contain a nitrogen atom (or an odd number of nitrogen atoms). This would be consistent with the IR suggestion that the compound contains a nitrile functional group. The preliminary suggestion from the NMR spectrum of ten hydrogens attached to aromatic nuclei could most simply be satisfied by the presence of two monosubstituted phenyl groups. This suggestion receives support from the appearance of the ion at *m/e* 77 (benzenium ion, $C_6H_5^+$) in the MS. The hydrogens that absorb in the NMR spectrum from τ 5.47 to 6.82 were proposed to result from an ABC system. This grouping could best be satisfied by the arrangement –CH_2–CH(with bonds above and below). If this were the case, the methylene hydrogens of this group would have to be adjacent to an atom capable of asymmetry.

The IR absorption at 5.92 μ (1689 cm^{-1}) suggests the presence of a carbonyl group, but such a carbonyl group would have to be in conjugation with some other chromophoric unit. Because of the suggestion of a monosubstituted phenyl group, it is possible that this IR carbonyl absorption could be satisfied by the presence of a benzoyl (C_6H_5CO–) group. Such a proposal is supported by the MS. The abundant ions at *m/e* 105 (the base peak, which corresponds to benzoyl, $C_6H_5CO^+$) and at *m/e* 130 which corresponds to M^+ - 105, support this proposal. If the masses of the presumed functional groups are added together (two monosubstituted phenyl groups, 154; a carbonyl group, 28; a nitrile group, 26; and a hydrocarbon group, –CH_2–CH(with bonds above and below), 27) the sum is mass 235, which does in fact correspond to the presumed molecular ion in the MS. Thus the formula of the compound appears to be $C_{16}H_{13}ON$, and this formula indicates that the compound has eleven rings and/or double bonds.

Because the mass spectrum does not contain an abundant ion at *m/e* 91, which would correspond to $C_6H_5CH_2^+$ or more accurately the tropylium ion, the absence of a benzyl group is indicated. The only logical way in which the functional groups described above can be put together as a structural formula for compound 2-3 is as 1-phenyl-1-cyano-2-benzoylethane.

Solution to Problem 2-4

The strong broad absorption in the IR spectrum of compound 2-4 suggests the presence of –OH and/or –NH stretching of those functional groups. This IR spectrum is quite simple, and the absence of a number of functional groups, such as carbonyl, is indicated. The absence of a λ_{max} in the UV spectrum above 205 nm indicates the absence of any chromophoric group in the molecule. The NMR spectrum of the compound indicates that it is relatively saturated, and the spectrum may be analyzed as follows:

τ	*multiplicity*	*J, Hz*	*integration squares*	*relative number of hydrogens*
5.70-6.39	complex	–	11.2	3
5.97	s	–	6.3	2
8.32	q	5.9	6.4	2
8.78	d	6.0	9.1	3

The sharp two hydrogen absorption at τ 5.97 would indicate two equivalent hydrogens or two hydrogens in rapid equilibrium. The three hydrogen absorption centered at τ 6.05 is complex and cannot be analyzed at this time. The NMR absorptions centered at τ 8.32 and τ 8.78 appear to be amenable to first-order analysis. The absorption at τ 8.32 (a quarter) appears to contain two hydrogens adjacent to carbon(s) bearing three other hydrogens. The absorption at τ 8.78 (a doublet appears to contain three hydrogen atoms adjacent to a carbon bearing a single hydrogen.

Because the NMR indicates that the compound is completely aliphatic, the presence of aliphatic $\overset{|}{\underset{|}{NH}}$ absorption is excluded by the correlation table data. Thus the strong broad IR absorption at 3.0 μ (3333 cm^{-1}) could result only from the presence of a hydroxyl group or groups. Since the only singlet NMR absorption contains two hydrogen atoms, it could be deduced that the compound contains two hydroxyl groups. The IR and NMR spectra have suggested the presence of the following functional groups: –OH, –OH, $CH_3{-}\overset{|}{\underset{|}{C}}H$, $-CH_2-$, and $-CH_2-$. The sum of the masses of these functional groups is 90 ($C_4H_{10}O_2$, no double bonds or rings).

The MS of the compound does not show an ion at *m/e* 90. However, if 90 is the mass of the compound, there are significant ions in the MS at *m/e* 75 (M - 15, $-CH_3$) and *m/e* 72 (M -18, H_2O). The presence of these ions are consistent with the above NMR analysis. For this compound the careful interpretation of the NMR spectrum has clearly been the most useful of any.

Thus, the structure of the compound is 1,3-dihydroxybutane.

COMPOUND 2-4

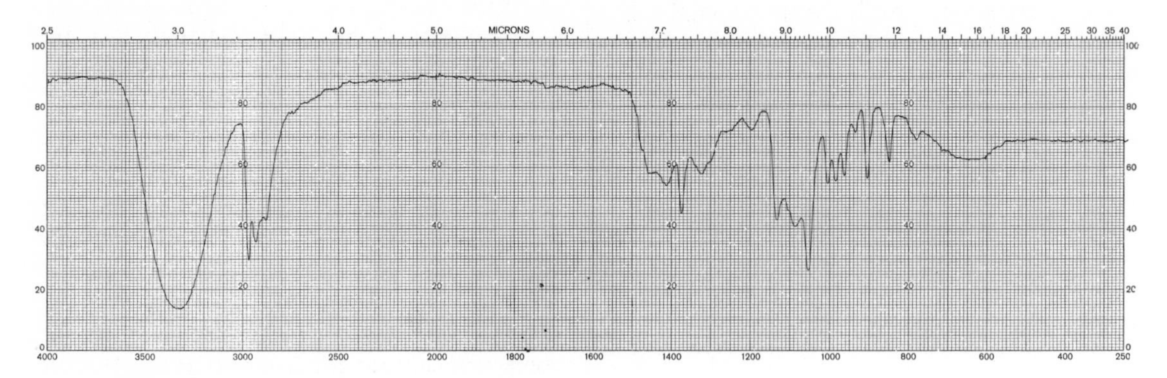

INFRARED SPECTRUM: Liquid film

ULTRAVIOLET DATA: no λ_{max} above 205 nm

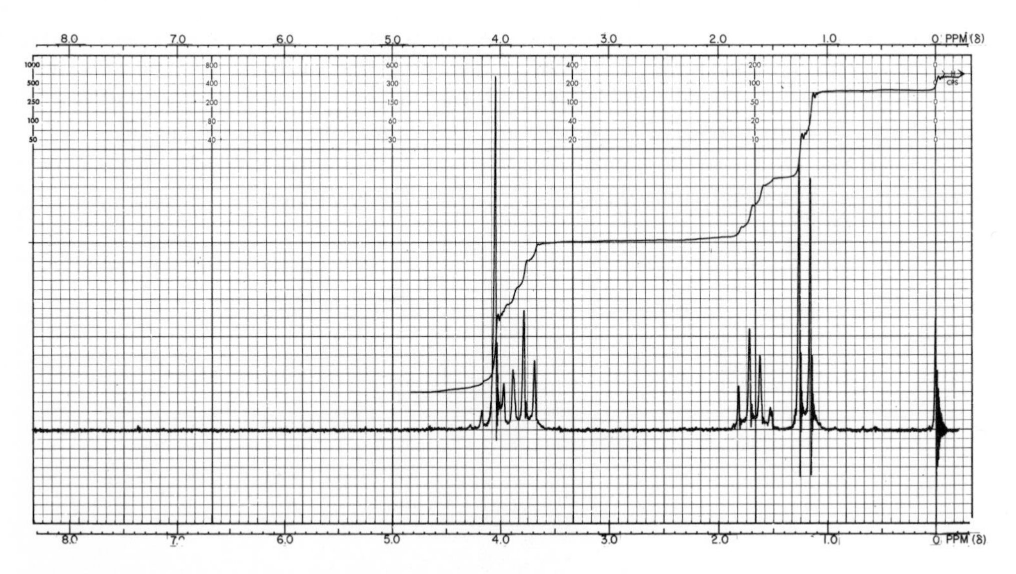

NUCLEAR MAGNETIC RESONANCE SPECTRUM:
15% in deuterochloroform, 500 Hz sweep width

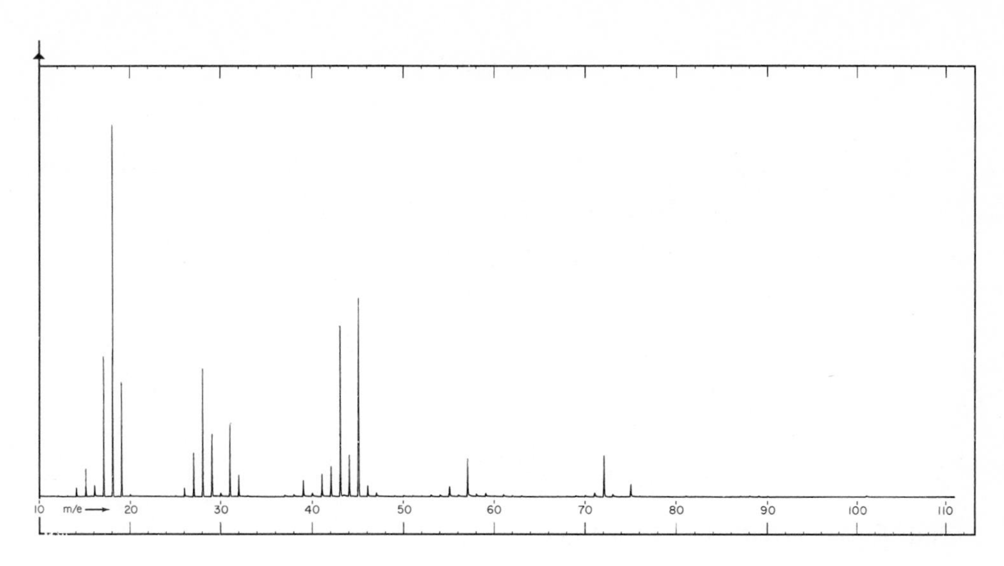

MASS SPECTRUM

MASS SPECTRAL DATA

m/e	*relative intensity*	*m/e*	*relative intensity*	*m/e*	*relative intensity*	*m/e*	*relative intensity*
14	1.4	28	23.6	42	4.5	56	0.3
15	6.8	29	12.9	43	25.3	57	5.9
16	1.3	30	0.6	44	6.3	58	0.5
17	25.3	31	12.9	45	28.0	59	0.7
18	100.0	32	2.1	46	2.0	71	0.6
19	23.0	38	0.5	47	0.6	72	6.0
20	0.3	39	2.6	53	0.3	73	0.5
26	1.6	40	0.5	54	0.3	75	2.1
27	10.9	41	3.4	55	1.6		

COMPOUND 2-5

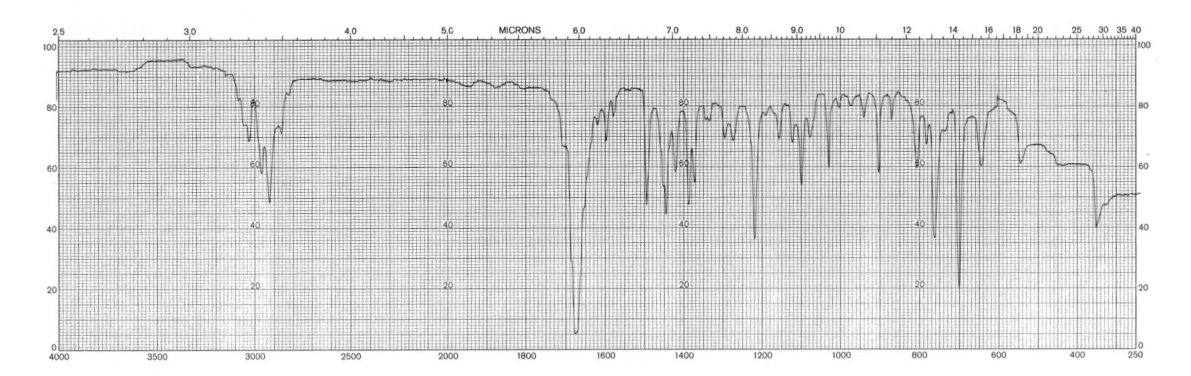

INFRARED SPECTRUM: Liquid film

ULTRAVIOLET DATA: λ_{max}^{EtOH} 209 nm (ϵ 11,300); 216(sh) nm (ϵ 10,100); 224(sh) nm (ϵ 8,400); 283(sh) nm (ϵ 274); 327 nm (ϵ 90).

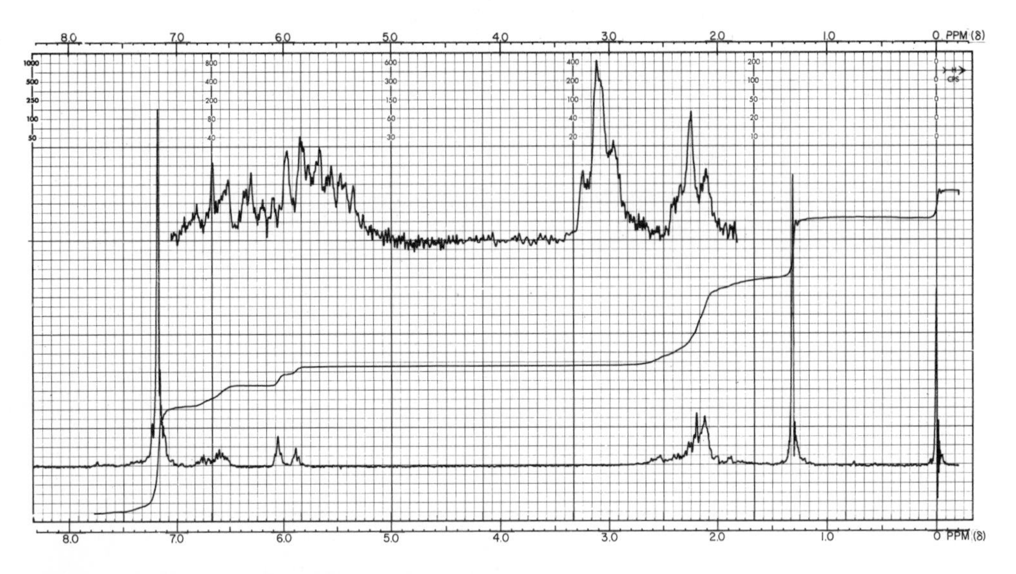

NUCLEAR MAGNETIC RESONANCE SPECTRUM:
15% in carbon tetrachloride, 500 Hz sweep width
Expansion: 100 Hz sweep width, 325 Hz sweep offset

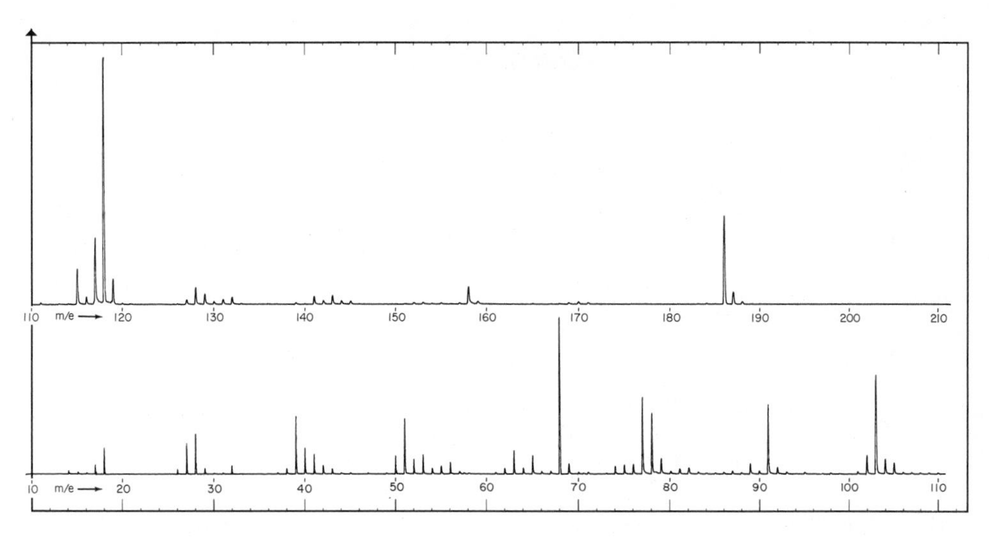

MASS SPECTRUM

MASS SPECTRAL DATA

m/e	relative intensity	m/e	relative intensity	m/e	relative intensity	m/e	relative intensity
14	0.3	61	0.3	90	0.8	139	1.0
15	0.3	62	1.1	91	12.9	141	4.1
17	0.5	63	3.9	92	1.4	142	2.2
18	1.4	64	1.0	93	0.4	143	4.3
26	0.5	65	3.1	101	0.6	144	2.1
27	3.5	66	0.6	102	3.8	145	1.5
28	3.0	67	0.5	103	19.2	146	0.4
29	0.7	68	28.0	104	3.0	152	1.0
32	0.5	69	1.8	105	2.1	153	1.1
38	0.8	70	0.3	111	1.1	154	0.4
39	8.8	71	0.4	113	0.3	155	0.8
40	3.7	74	1.5	114	0.5	156	0.3
41	2.5	75	1.7	115	14.0	157	0.7
42	1.4	76	1.9	116	4.3	158	7.9
43	0.9	77	14.0	117	30.0	159	1.5
44	0.3	78	11.9	118	100.0	165	0.4
45	0.4	79	2.8	119	11.2	167	0.4
50	2.8	80	0.6	120	0.7	168	0.3
51	7.7	81	0.9	126	0.5	169	1.0
52	2.3	82	1.3	127	2.8	170	1.8
53	3.1	83	0.4	128	7.7	171	1.1
54	1.0	85	0.4	129	5.1	172	0.3
55	1.5	86	0.5	130	1.5	184	0.8
56	1.7	87	0.6	131	2.6	186	46.0
57	0.5	88	0.3	132	3.3	187	6.8
57.5	0.3	89	2.1	133	0.4	188	1.6
58	0.2						

COMPOUND 2-6

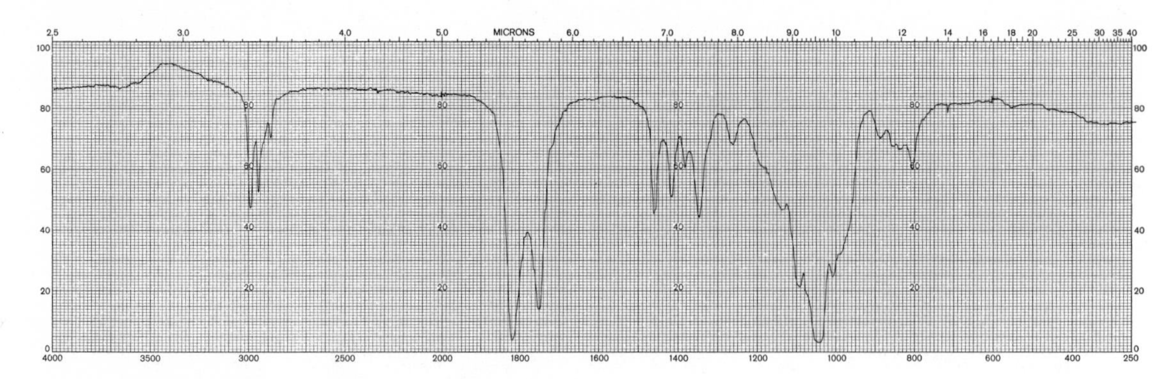

INFRARED SPECTRUM: Liquid film

ULTRAVIOLET DATA: λ_{max}^{Hexane} 225 nm (ϵ 48)

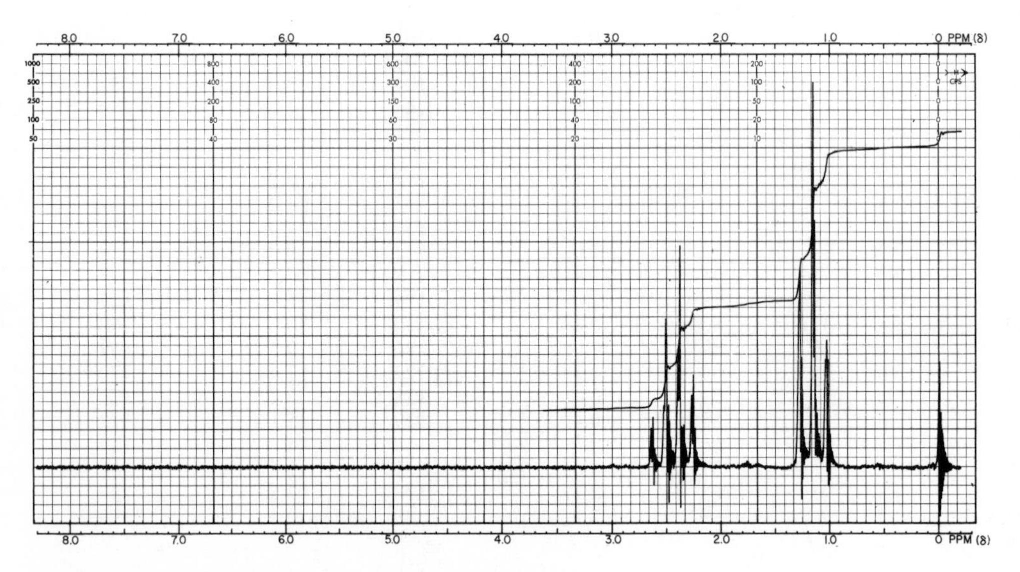

NUCLEAR MAGNETIC RESONANCE SPECTRUM:
15% in carbon tetrachloride, 500 Hz sweep width

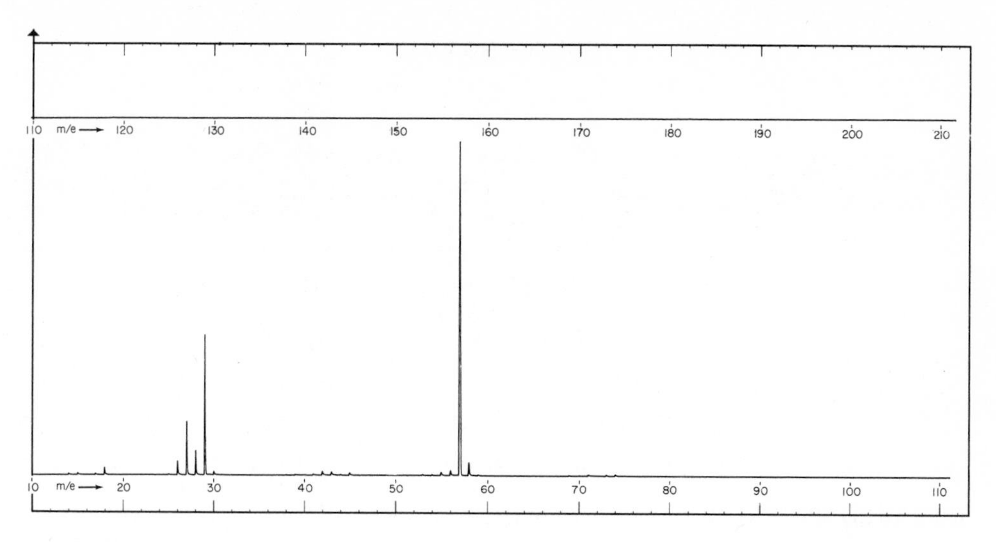

MASS SPECTRUM

MASS SPECTRAL DATA

m/e	*relative intensity*	*m/e*	*relative intensity*	*m/e*	*relative intensity*	*m/e*	*relative intensity*
14	0.5	29	38.0	44	0.3	57	100.0
15	0.6	30	1.0	45	1.0	58	4.2
17	0.3	39	0.4	53	0.4	59	0.3
18	1.4	41	0.4	54	0.4	71	0.6
26	3.8	42	1.3	55	1.4	73	0.6
27	15.4	43	1.3	56	1.8	74	0.7
28	6.5						

COMPOUND 2-7

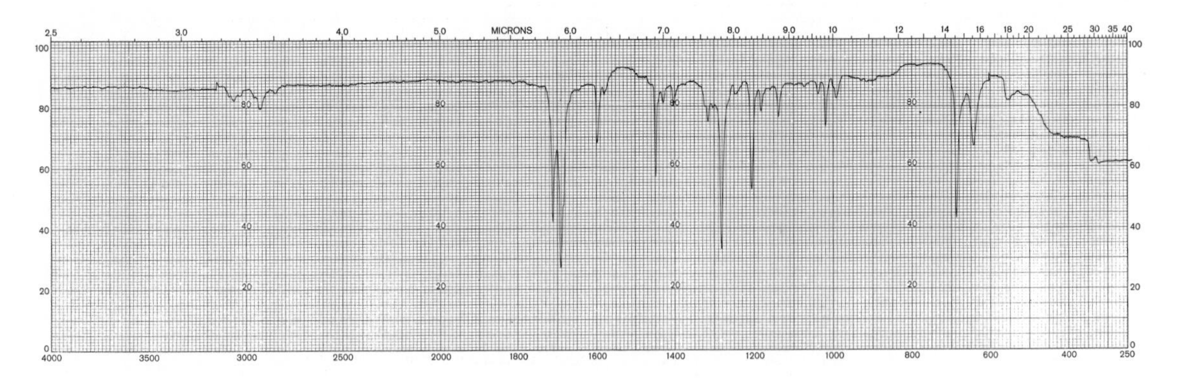

INFRARED SPECTRUM: 1.1% in carbon tetrachloride

ULTRAVIOLET DATA: λ_{max}^{Hexane} 199 nm (ϵ 19,200); 245 nm (ϵ 10,900); 280 nm (ϵ 910)

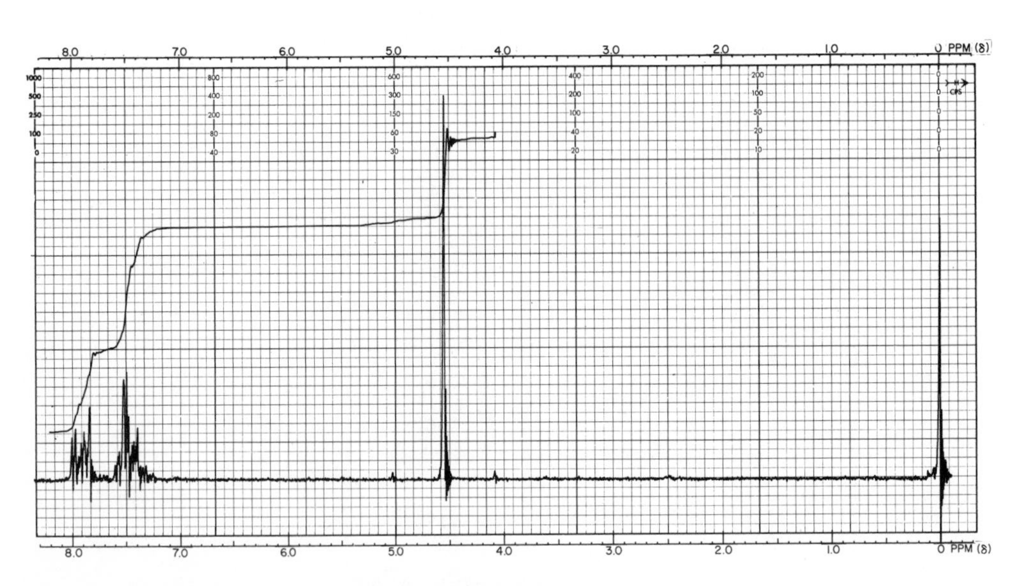

NUCLEAR MAGNETIC RESONANCE SPECTRUM:
15% in carbon tetrachloride, 500 Hz sweep width

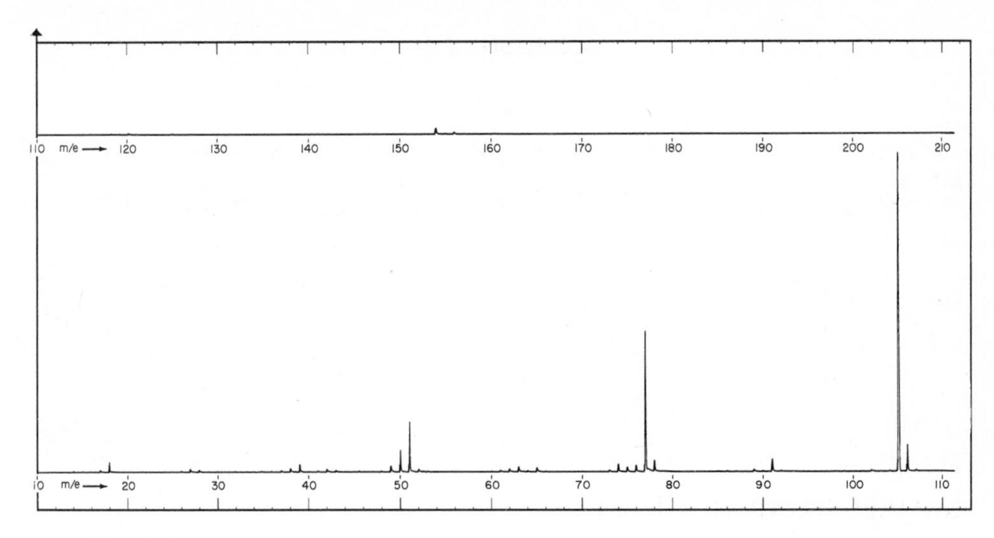

MASS SPECTRUM

MASS SPECTRAL DATA

m/e	*relative intensity*	*m/e*	*relative intensity*	*m/e*	*relative intensity*	*m/e*	*relative intensity*
14	0.4	42	1.2	65	2.1	99	0.3
17	0.5	43	0.8	73	0.8	101	0.1
18	2.4	45.5	0.3	74	3.8	102	0.7
26	0.3	47	0.2	75	2.4	104	0.4
27	1.2	48	0.3	76	3.2	105	100.0
28	0.9	49	2.9	77	61.0	106	13.4
29	0.2	50	8.3	78	5.4	107	0.9
30	0.2	51	18.9	79	0.3	118	0.3
35	0.3	52	1.4	85	0.2	119	0.2
36	0.3	52.5	0.4	86	0.4	120	0.7
37	0.7	53	0.4	87	0.3	154	3.5
38	1.5	61	0.7	89	1.4	155	0.4
39	3.1	62	1.5	90	0.4	156	1.3
40	0.2	63	2.5	91	6.3	157	0.1
41	0.5	64	0.5	92	0.6		

COMPOUND 2-8

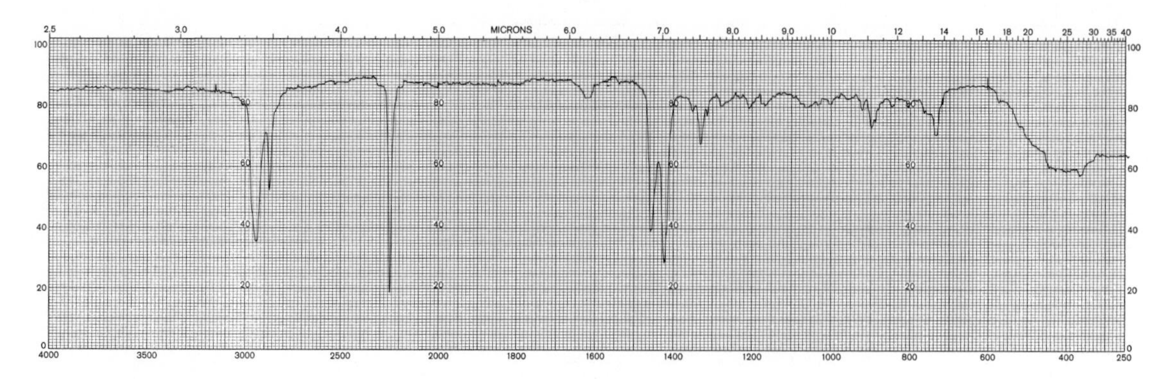

INFRARED SPECTRUM: Liquid film

ULTRAVIOLET DATA: no λ_{max}^{EtOH} above 205 nm

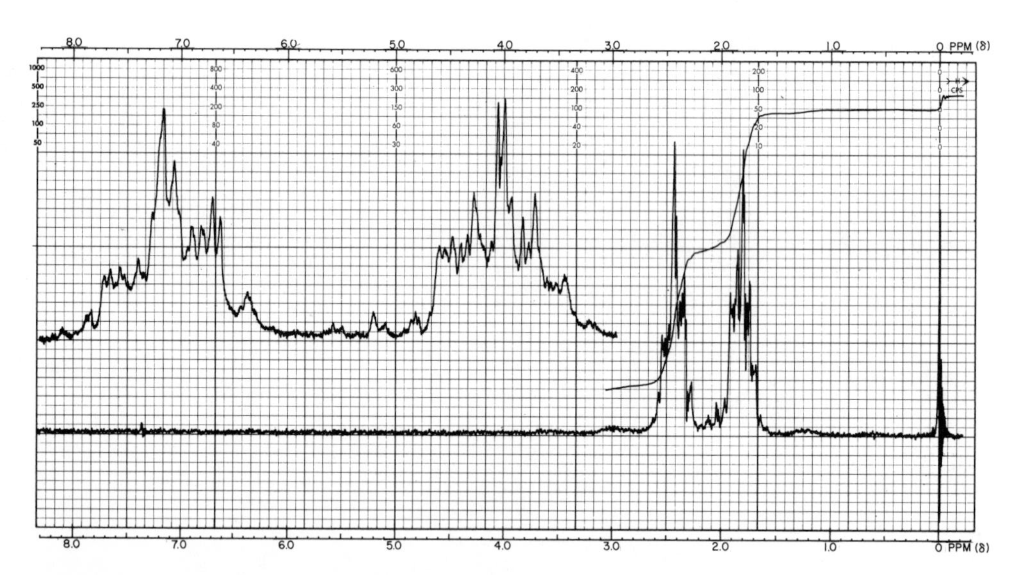

NUCLEAR MAGNETIC RESONANCE SPECTRUM:
15% in deuterochloroform, 500 Hz sweep width
Expansion: 100 Hz sweep width, 60 Hz sweep offset

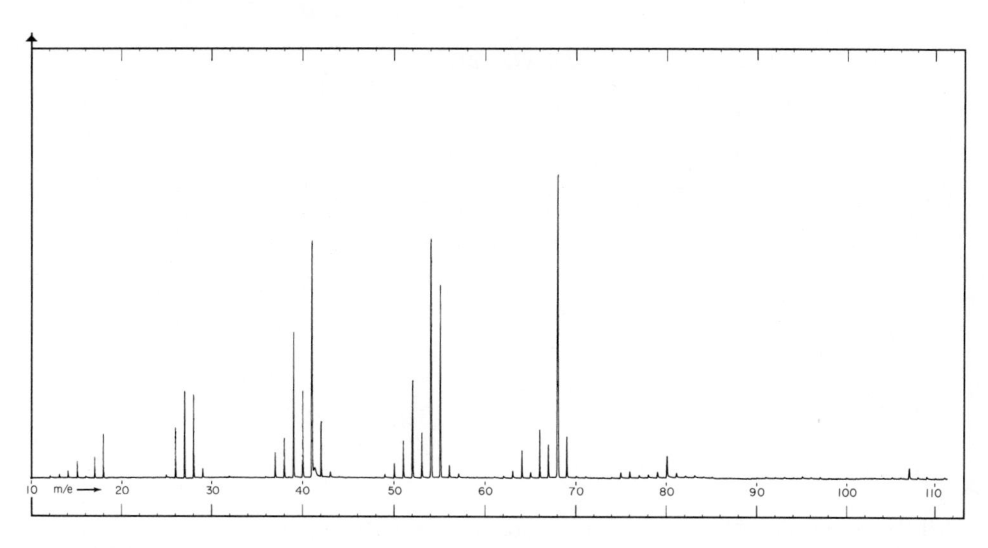

MASS SPECTRUM

MASS SPECTRAL DATA

m/e	*relative intensity*	*m/e*	*relative intensity*	*m/e*	*relative intensity*	*m/e*	*relative intensity*
13	0.6	38	6.0	54	93.0	75	1.0
14	0.7	39	25.0	55	51.0	76	1.1
15	2.4	40	14.6	56	3.4	77	0.6
17	1.5	41	100.0	62	0.6	78	0.5
18	5.9	42	10.7	63	1.3	79	1.1
25	0.6	43	1.0	64	5.3	80	4.1
26	7.7	49	0.9	65	1.3	81	1.1
27	24.0	50	2.8	66	10.7	107	2.7
28	16.0	51	7.3	67	6.2	108	0.7
29	1.3	52	1.6	68	100.0	109	0.6
37	3.4	53	8.0	69	7.9		

COMPOUND 2-9

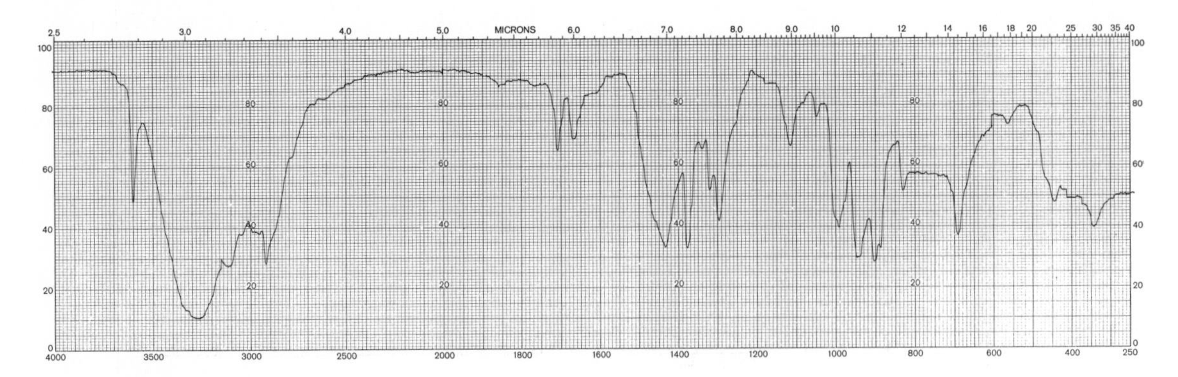

INFRARED SPECTRUM: 3.5% in carbon tetrachloride

ULTRAVIOLET DATA: no λ_{max} above 205 nm

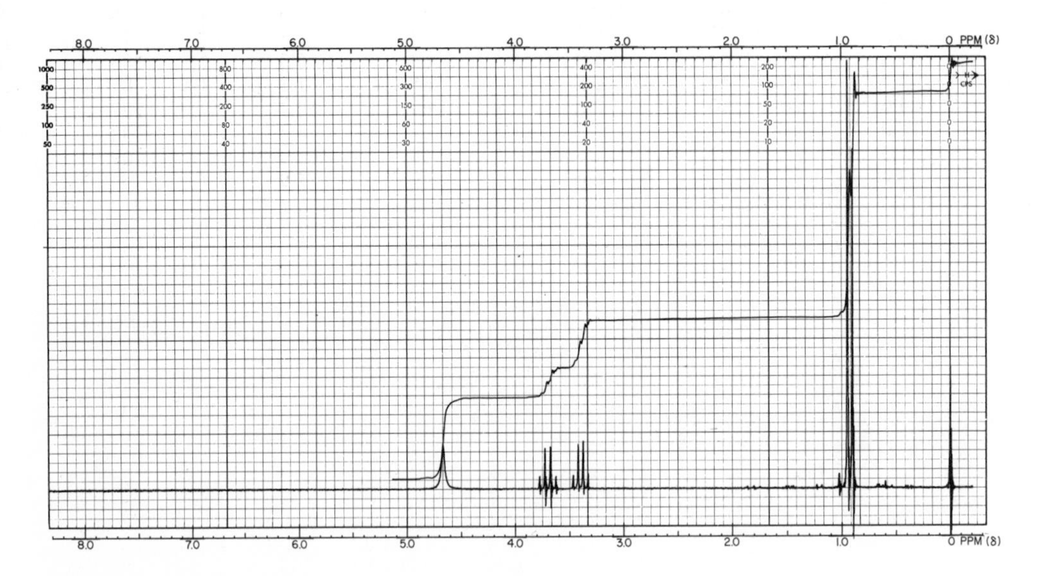

NUCLEAR MAGNETIC RESONANCE SPECTRUM:
15% in carbon tetrachloride, 1000 Hz sweep width

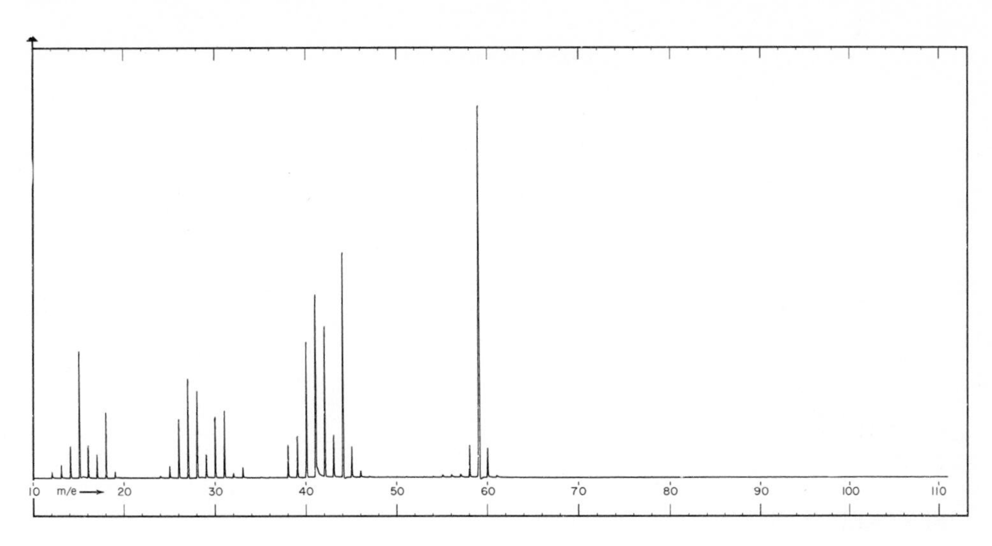

MASS SPECTRUM

MASS SPECTRAL DATA

m/e	relative intensity	m/e	relative intensity	m/e	relative intensity	m/e	relative intensity
13	0.8	26	5.0	38	3.0	44	36.0
14	2.7	27	8.8	39	4.5	45	3.2
15	20.5	28	9.1	40	14.4	58	2.9
16	2.7	29	2.0	41	24.4	59	100.0
17	1.6	30	5.8	42	15.6	60	3.2
18	3.5	31	6.0	43	4.2	61	0.2
25	0.7	33	0.7				

COMPOUND 2-10

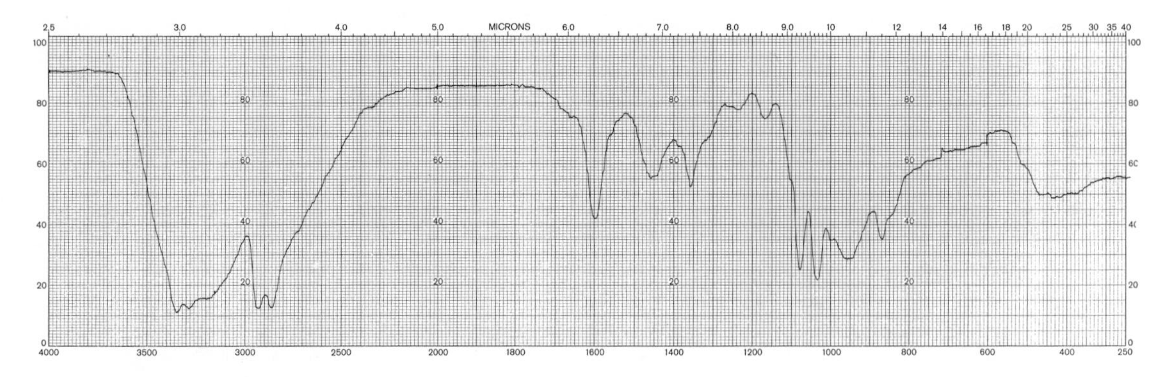

INFRARED SPECTRUM: Liquid film

ULTRAVIOLET DATA: no λ_{max} above 205 nm

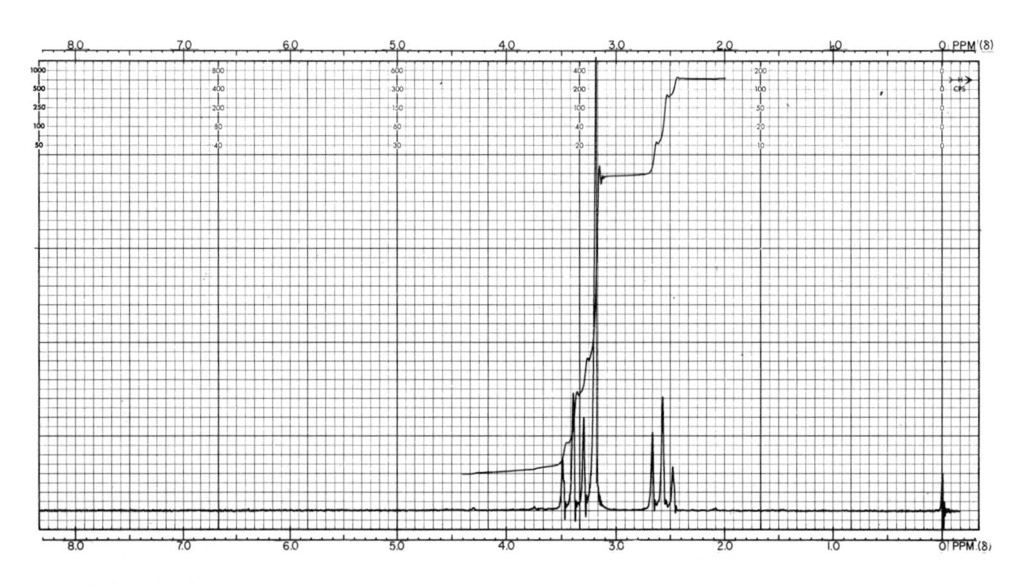

NUCLEAR MAGNETIC RESONANCE SPECTRUM:
15% in DMSO-d_6, 500 Hz sweep width

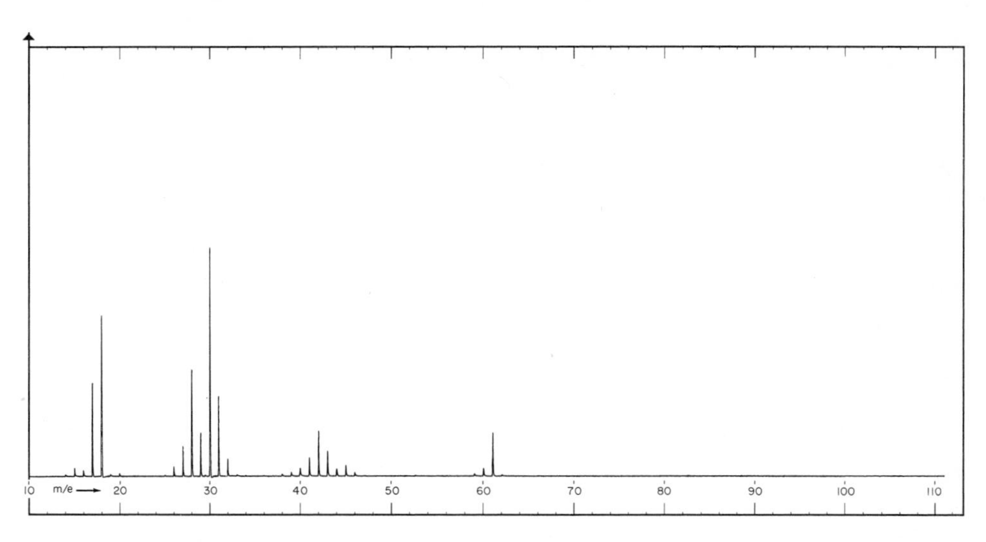

MASS SPECTRUM

MASS SPECTRAL DATA

m/e	*relative intensity*	*m/e*	*relative intensity*	*m/e*	*relative intensity*	*m/e*	*relative intensity*
14	4.2	28	21.0	39	2.6	43	4.3
17	9.2	29	6.9	40	1.9	44	2.3
18	42.9	30	100.0	41	4.1	61	13.3
20	0.1	31	8.2	42	10.7	62	0.3
27	4.4	32	0.3				

COMPOUND 2-11

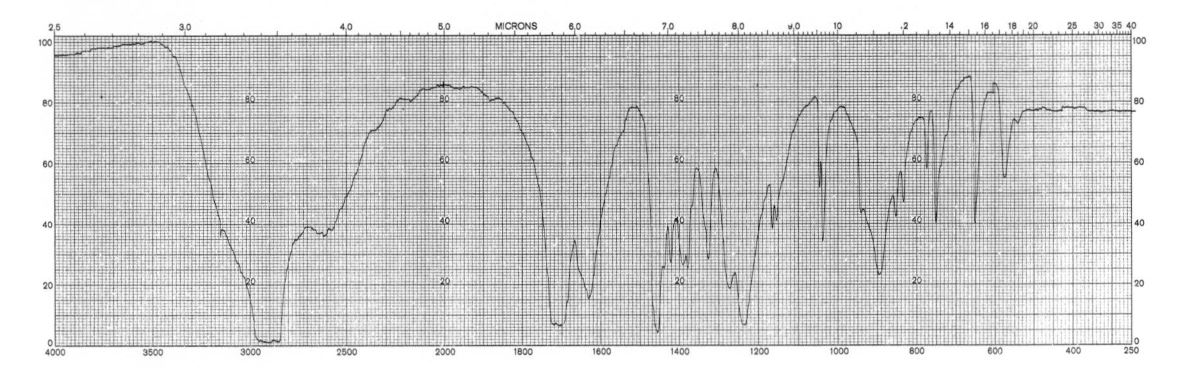

INFRARED SPECTRUM: Nujol mull

ULTRAVIOLET DATA: λ_{max}^{EtOH} 211 nm (ϵ 13,000)

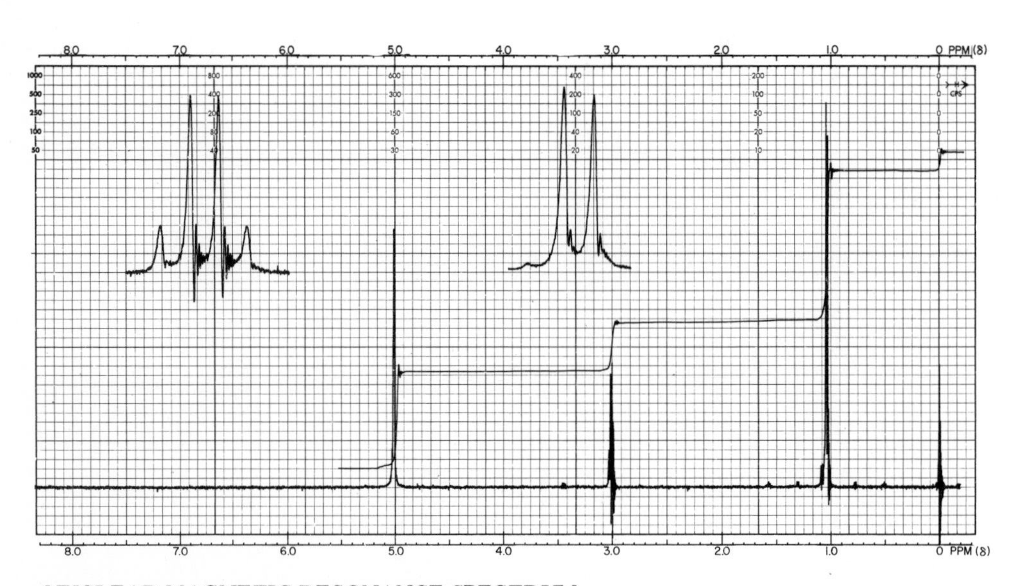

NUCLEAR MAGNETIC RESONANCE SPECTRUM:
15% in acetone-d_6, 1000 Hz sweep width
Expansions: (left) 50 Hz sweep width, 320 Hz sweep offset
(right) 50 Hz sweep width, 105 Hz sweep offset

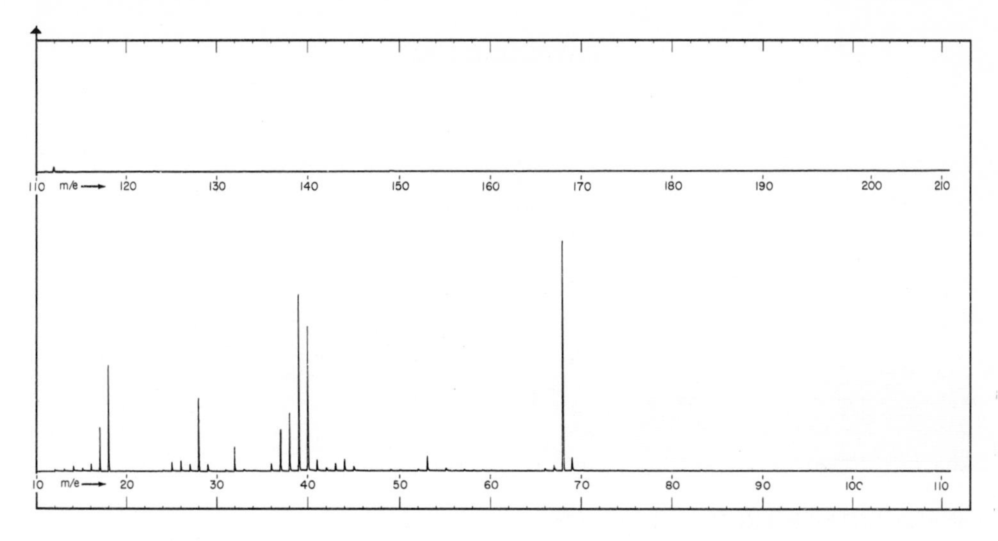

MASS SPECTRUM

MASS SPECTRAL DATA

m/e	*relative intensity*	*m/e*	*relative intensity*	*m/e*	*relative intensity*	*m/e*	*relative intensity*
12	0.4	27	1.7	39	80.0	52	1.1
13	0.4	28	9.4	40	52.0	53	6.6
14	0.8	29	2.1	41	3.8	54	0.3
15	0.5	31	0.2	42	1.1	55	0.7
16	0.9	32	2.4	43	0.9	66	1.2
17	4.6	33	0.7	44	3.4	67	2.7
18	23.4	33.5	0.3	45	1.7	68	100.0
20	0.3	34	0.7	48	0.5	69	5.9
24	0.5	36	3.2	49	0.9	70	0.3
25	2.2	37	14.3	50	0.7	112	1.7
26	2.7	38	20.1				

COMPOUND 2-12

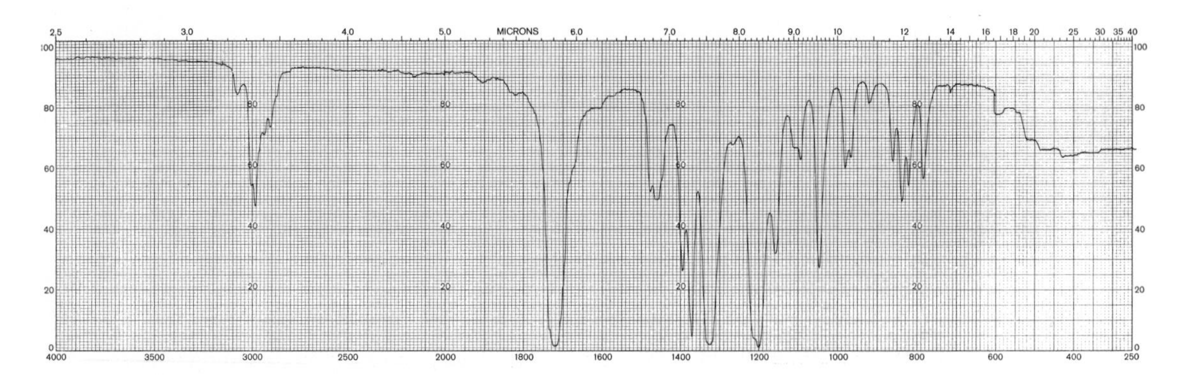

INFRARED SPECTRUM: Liquid film

ULTRAVIOLET DATA: λ_{max}^{EtOH} 242 nm (ϵ 17)

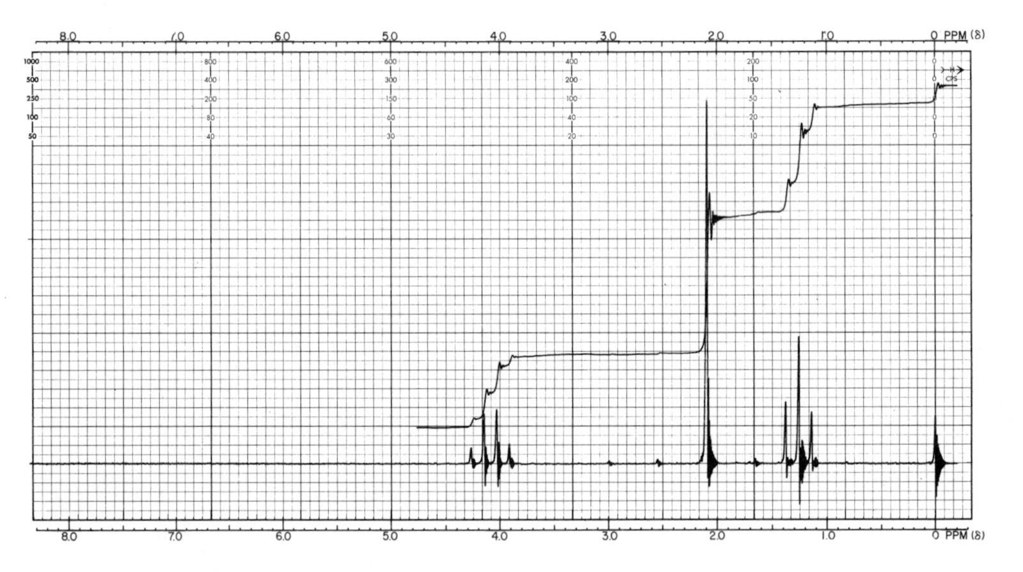

NUCLEAR MAGNETIC RESONANCE SPECTRUM:
15% in carbon tetrachloride, 500 Hz sweep width

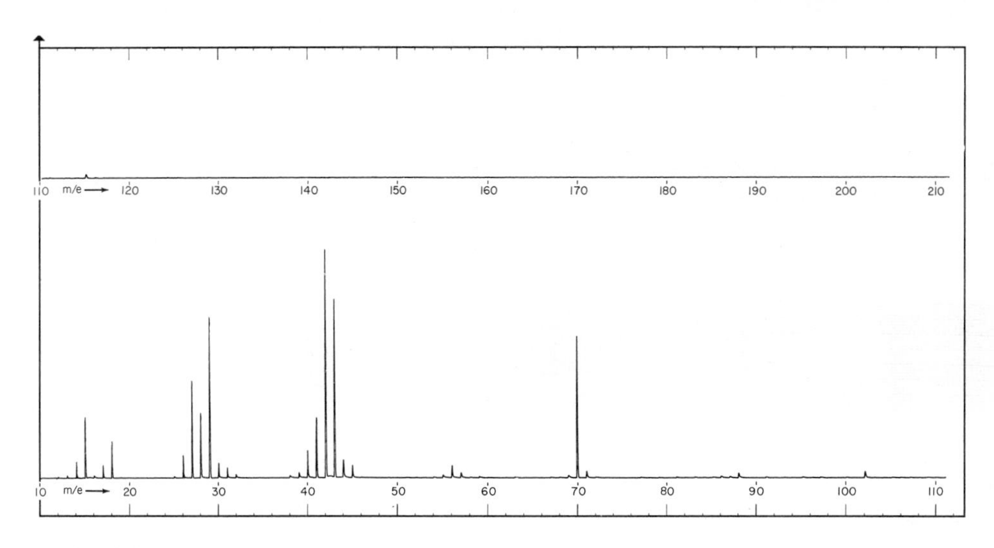

MASS SPECTRUM

MASS SPECTRAL DATA

m/e	*relative intensity*	*m/e*	*relative intensity*	*m/e*	*relative intensity*	*m/e*	*relative intensity*
14	1.9	30	4.3	43	56.0	71	2.4
15	15.2	31	2.3	44	5.9	72	0.1
17	2.0	38	1.0	45	4.3	86	0.9
18	5.2	39	1.5	56	4.1	87	0.8
26	5.2	40	0.3	57	2.1	88	2.0
27	25.8	41	16.0	69	0.8	102	2.6
28	17.2	42	100.0	70	48.0	115	1.5
29	66.0						

COMPOUND 2-13

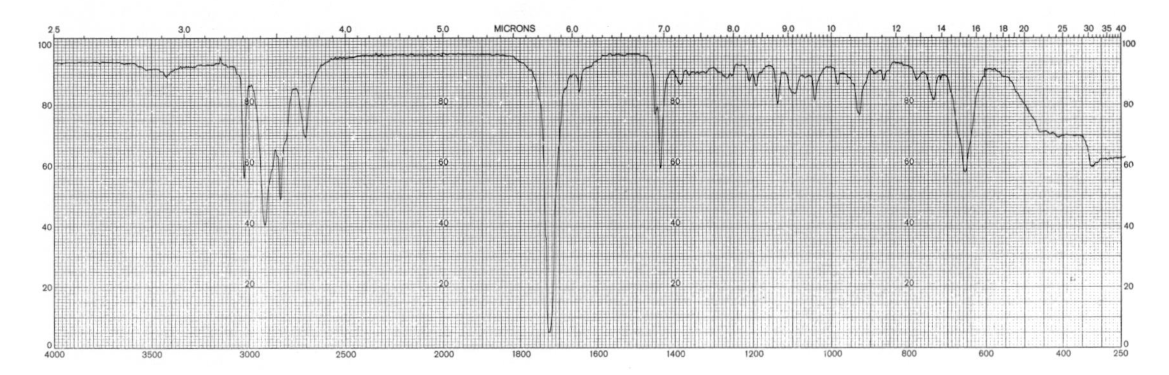

INFRARED SPECTRUM: Liquid film

ULTRAVIOLET DATA: λ_{max}^{EtOH} 255(sh) nm (ϵ 9); 285(sh) nm (ϵ 7).

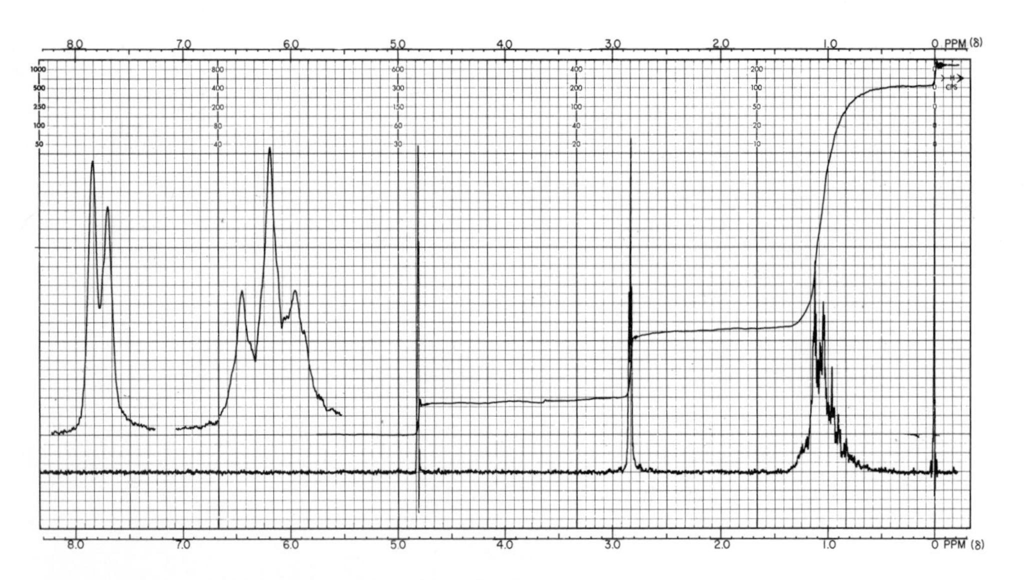

NUCLEAR MAGNETIC RESONANCE SPECTRUM:
15% in carbon tetrachloride, 1000 Hz sweep width
Expansions: (left) 50 Hz sweep width, 532 Hz sweep offset
(right) 50 Hz sweep width, 304 Hz sweep offset

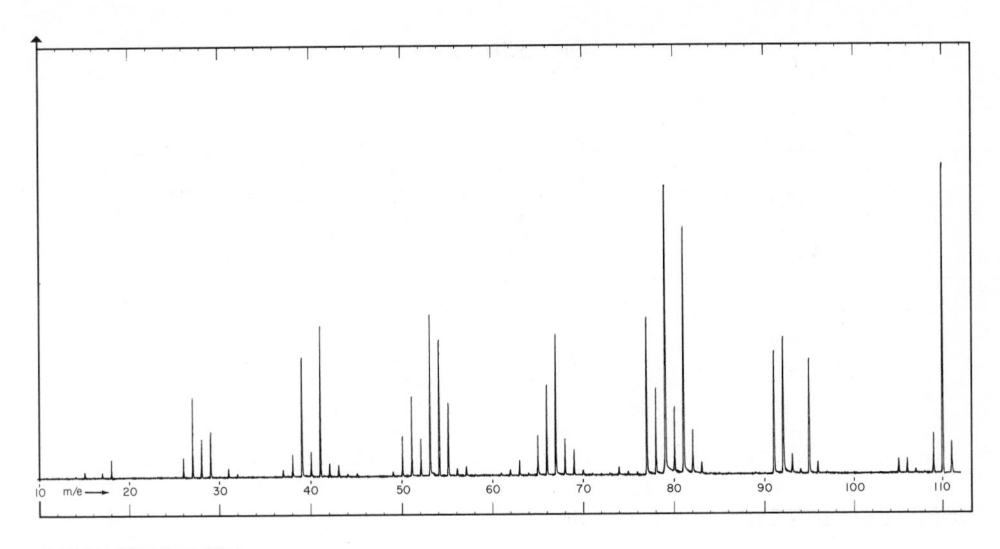

MASS SPECTRUM

MASS SPECTRAL DATA

m/e	*relative intensity*	*m/e*	*relative intensity*	*m/e*	*relative intensity*	*m/e*	*relative intensity*
15	1.0	45	0.7	67	35.0	89	0.7
17	1.0	49	1.0	68	10.5	91	30.0
18	3.2	50	9.0	69	5.9	92	37.0
26	3.6	51	16.5	70	1.1	93	5.2
27	23.0	52	7.7	71	0.6	94	1.5
28	7.8	53	44.0	73	0.5	95	34.0
29	11.0	54	35.0	74	1.6	96	3.8
31	1.5	55	18.8	75	0.8	97	0.7
32	0.6	56	1.8	76	0.6	105	4.0
37	1.5	57	2.5	77	39.0	106	3.9
38	4.3	58	0.5	78	22.9	107	0.9
39	35.0	61	0.7	79	100.0	108	0.6
40	4.6	62	1.6	80	19.7	109	11.8
41	40.0	63	3.4	81	90.0	110	99.0
42	3.1	64	0.8	82	13.1	111	9.9
43	2.8	65	9.6	83	3.4	112	0.7
44	0.6	66	20.0				

COMPOUND 2-14

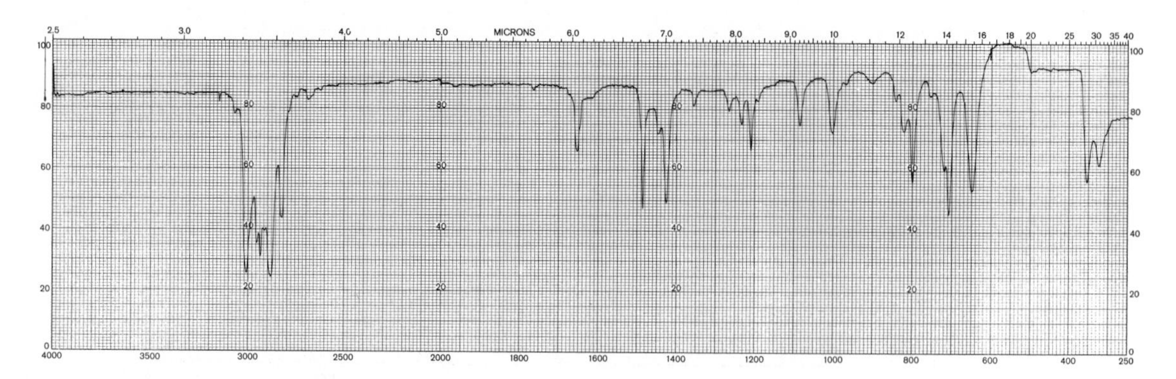

INFRARED SPECTRUM: Liquid film

ULTRAVIOLET DATA: no λ_{max}^{EtOH} above 205 nm

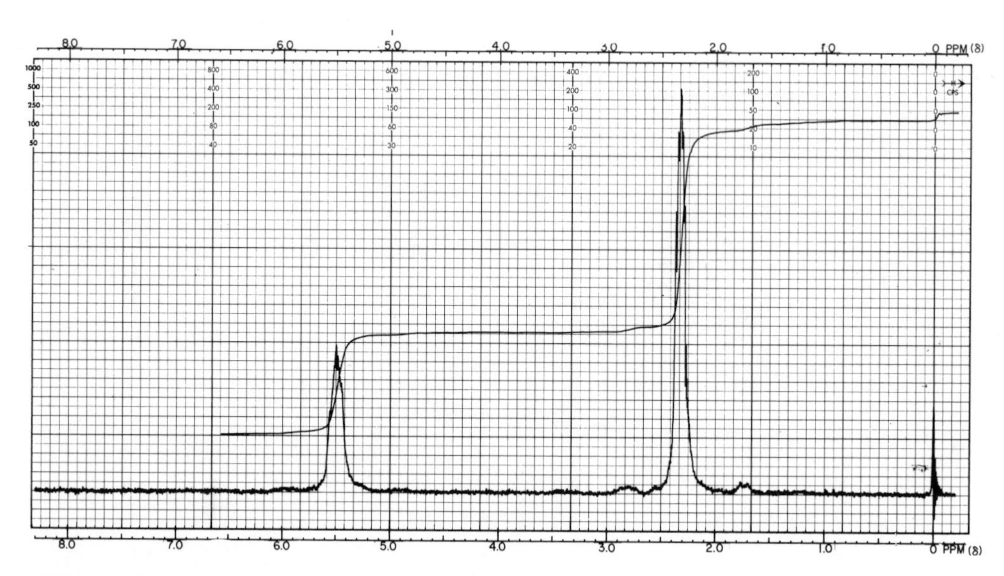

NUCLEAR MAGNETIC RESONANCE SPECTRUM:
15% in carbon tetrachloride, 500 Hz sweep width

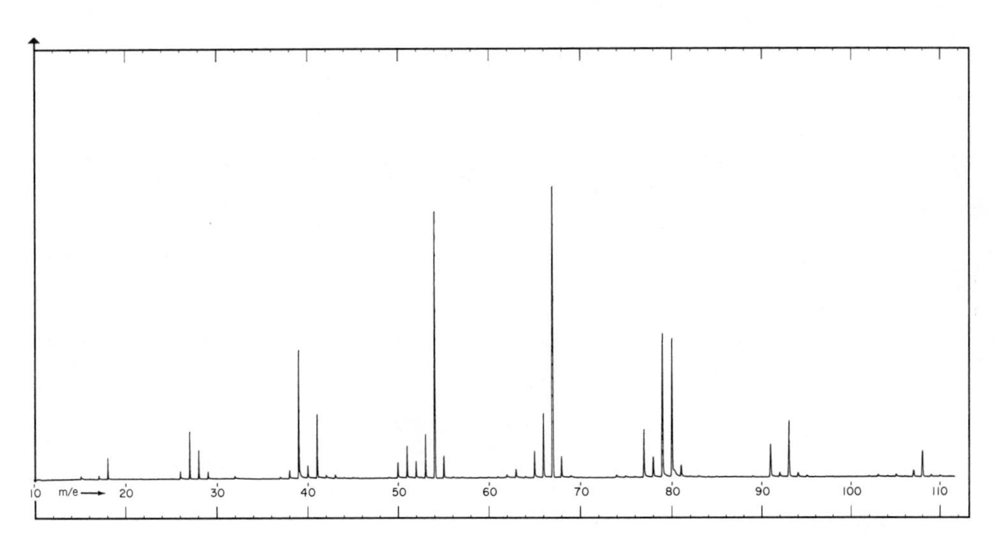

MASS SPECTRUM

MASS SPECTRAL DATA

m/e	*relative intensity*	*m/e*	*relative intensity*	*m/e*	*relative intensity*	*m/e*	*relative intensity*
15	0.8	43	1.6	67	100.0	89	1.0
17	0.9	49	0.7	68	7.3	91	12.7
18	2.0	50	4.7	69	0.9	92	1.8
26	1.7	51	9.4	74	1.3	93	21.0
27	13.7	52	6.0	75	0.7	94	2.0
28	7.7	53	14.1	76	0.8	95	1.3
29	2.2	54	98.0	77	17.6	103	1.1
32	1.0	55	7.1	78	7.5	104	0.8
37	0.9	56	0.7	79	52.0	105	1.4
38	2.8	62	1.4	80	53.0	107	2.7
39	36.0	63	3.2	81	4.6	108	9.9
40	3.8	64	0.9	82	0.9	109	1.3
41	18.7	65	10.3	83	0.9	110	0.8
42	1.3	66	21.0				

COMPOUND 2-15

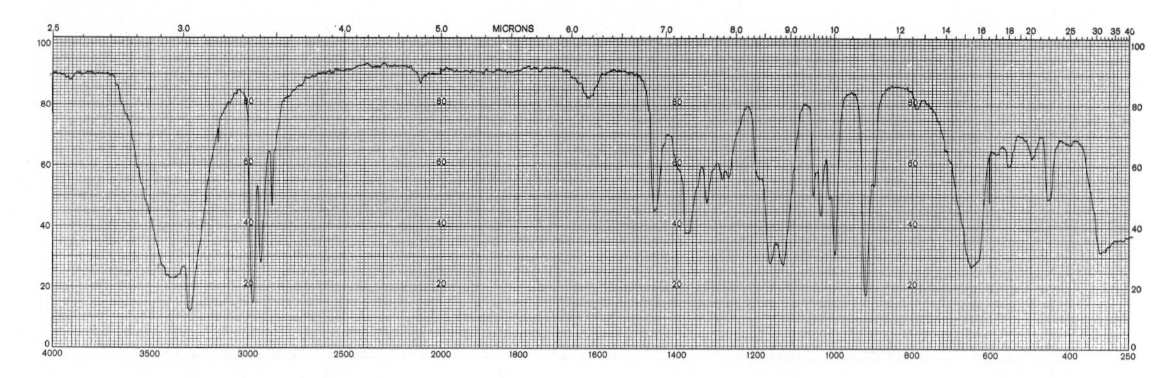

INFRARED SPECTRUM: Liquid film

ULTRAVIOLET DATA: no λ max above 205 nm

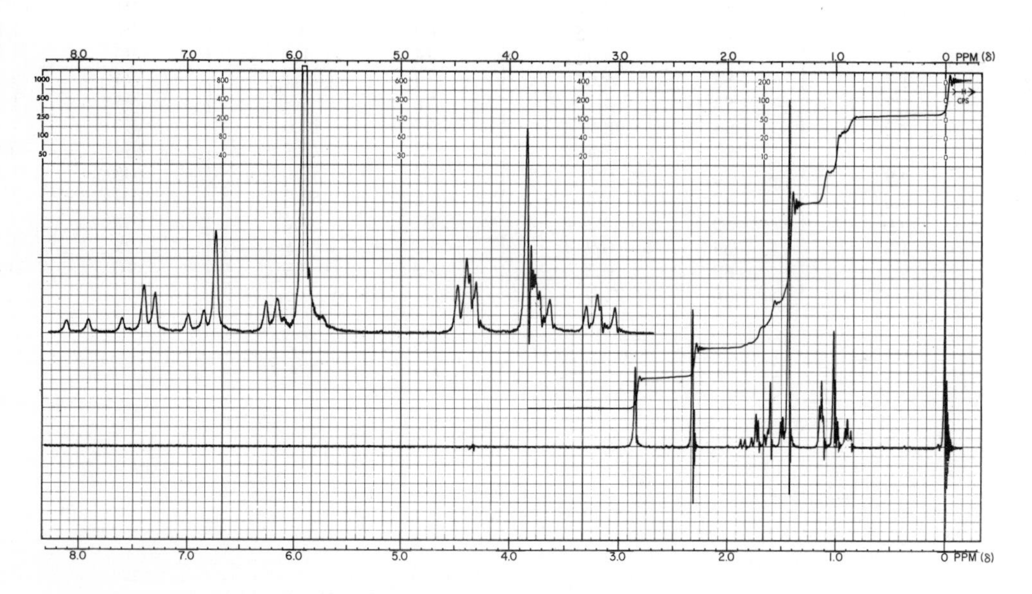

NUCLEAR MAGNETIC RESONANCE SPECTRUM:
15% in carbon tetrachloride, 500 Hz sweep width
Expansion: 100 Hz sweep width, 15 Hz sweep offset

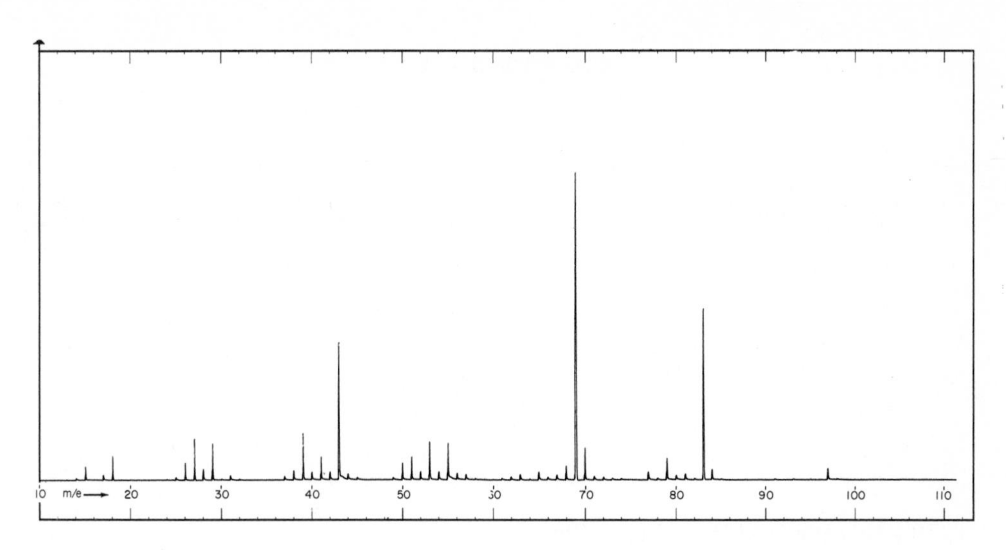

MASS SPECTRUM

MASS SPECTRAL DATA

m/e	*relative intensity*	*m/e*	*relative intensity*	*m/e*	*relative intensity*	*m/e*	*relative intensity*
15	1.9	41	4.1	57	1.2	73	0.6
17	0.6	42	1.5	61	0.4	74	0.5
18	2.2	43	35.0	62	0.7	77	1.8
25	0.6	44	1.3	63	1.3	78	0.5
26	2.6	45	0.6	65	1.7	79	4.7
27	7.6	49	0.5	66	0.6	80	1.1
28	2.1	50	3.1	67	1.2	81	1.4
29	6.5	51	4.5	68	3.0	82	0.5
31	1.0	52	1.7	69	100.0	83	38.0
37	0.8	53	9.0	70	6.5	84	2.3
38	1.7	54	1.7	71	0.9	85	0.3
39	8.9	55	7.6	72	0.8	97	2.3
40	1.4	56	1.4				

COMPOUND 2-16

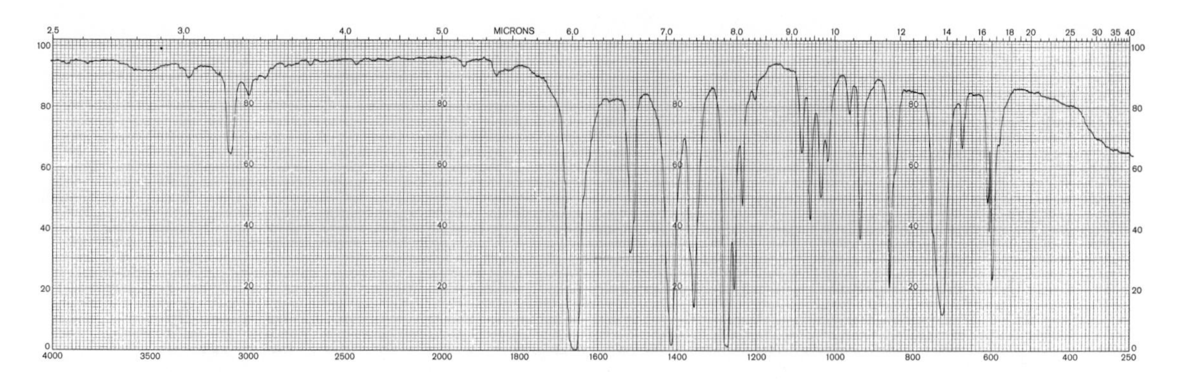

INFRARED SPECTRUM: Liquid film

ULTRAVIOLET DATA: λ_{max}^{EtOH} 260 nm (ϵ 10,000); 282 nm (ϵ 6,900).

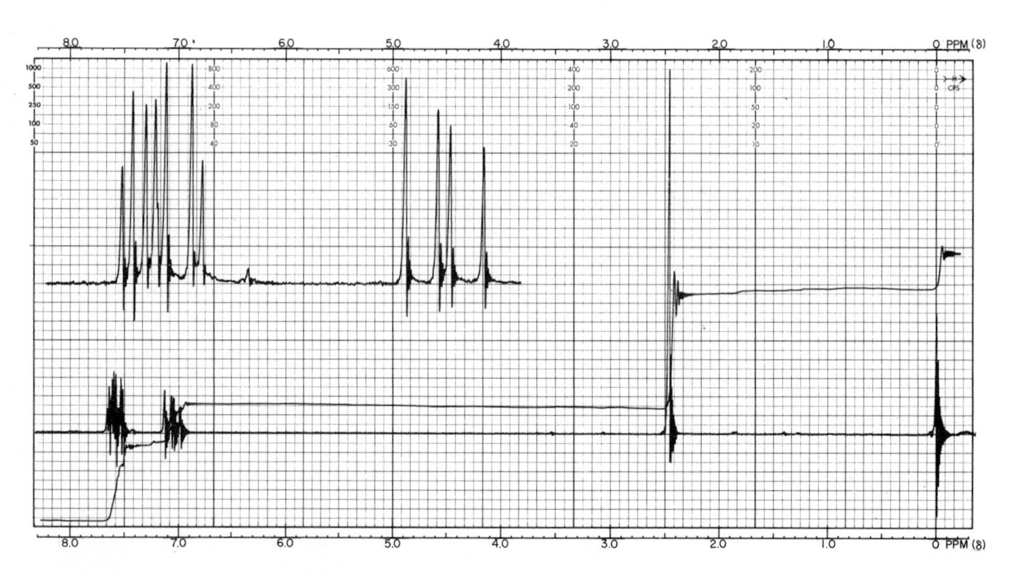

NUCLEAR MAGNETIC RESONANCE SPECTRUM:
15% in carbon tetrachloride, 500 Hz sweep width
Expansion: 100 Hz sweep width, 370 Hz sweep offset

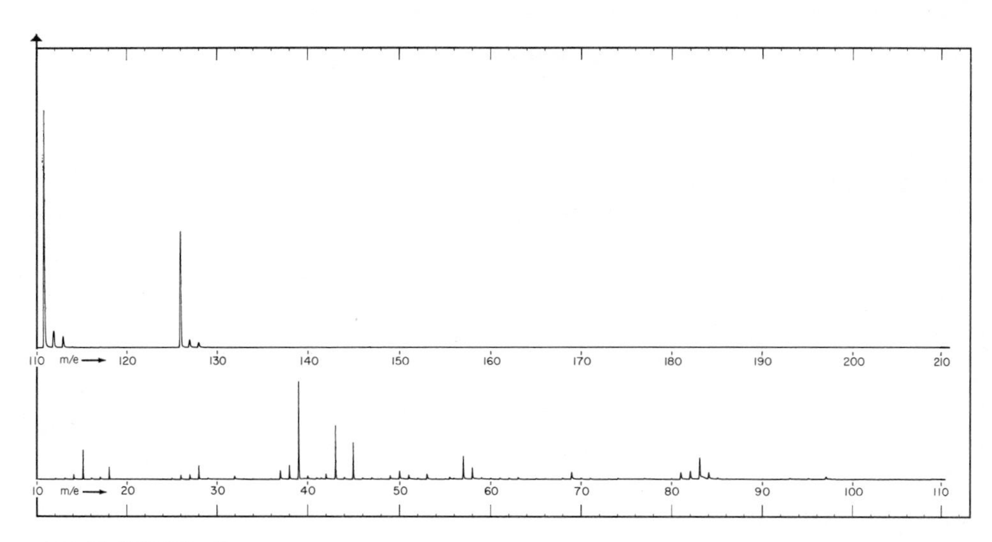

MASS SPECTRUM

MASS SPECTRAL DATA

m/e	*relative intensity*	*m/e*	*relative intensity*	*m/e*	*relative intensity*	*m/e*	*relative intensity*
14	2.2	38	5.4	53	2.4	93	0.5
15	10.6	39	34.0	57	8.1	95	0.3
16	0.6	40	1.5	58	4.4	97	1.1
17	1.1	41	0.7	69	2.5	111	100.0
18	4.3	42	1.9	70	0.3	112	6.7
25	0.5	43	20.8	71	0.4	113	5.2
26	1.7	44	1.0	80	0.3	114	0.5
27	1.9	45	12.3	81	2.8	126	49.0
28	4.6	49	2.0	82	3.2	127	3.7
29	0.7	50	2.7	83	8.1	128	2.5
32	1.3	51	1.7	84	2.7	129	0.2
37	2.8	52	1.0	85	0.5		

COMPOUND 2-17

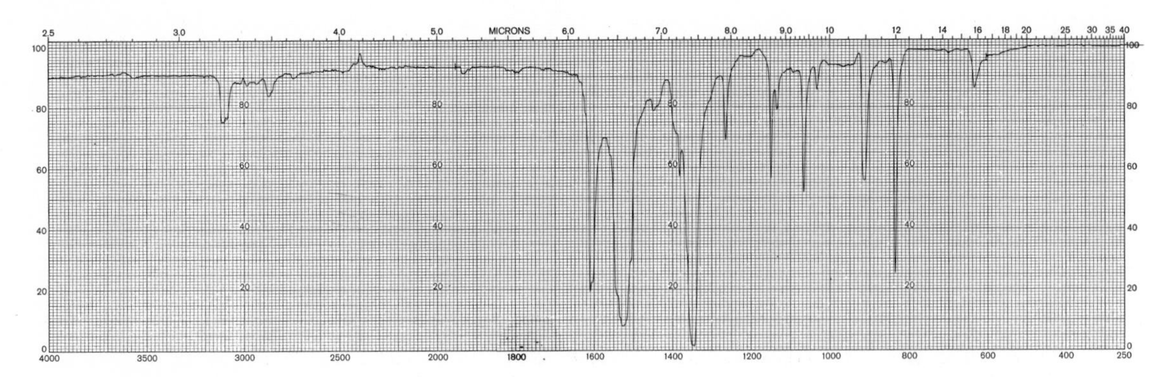

INFRARED SPECTRUM: 3.0% in chloroform

ULTRAVIOLET DATA: λ_{max}^{EtOH} 252 nm (ϵ 14,000).

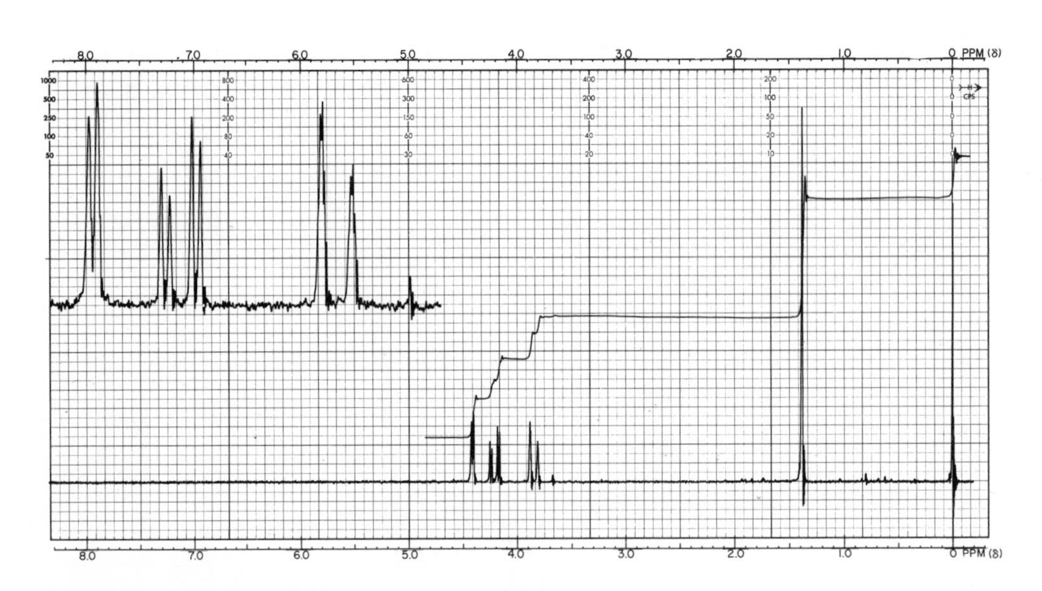

NUCLEAR MAGNETIC RESONANCE SPECTRUM:
15% in deuterochloroform, 1000 Hz sweep width
Expansion: 250 Hz sweep width, 290 Hz sweep offset

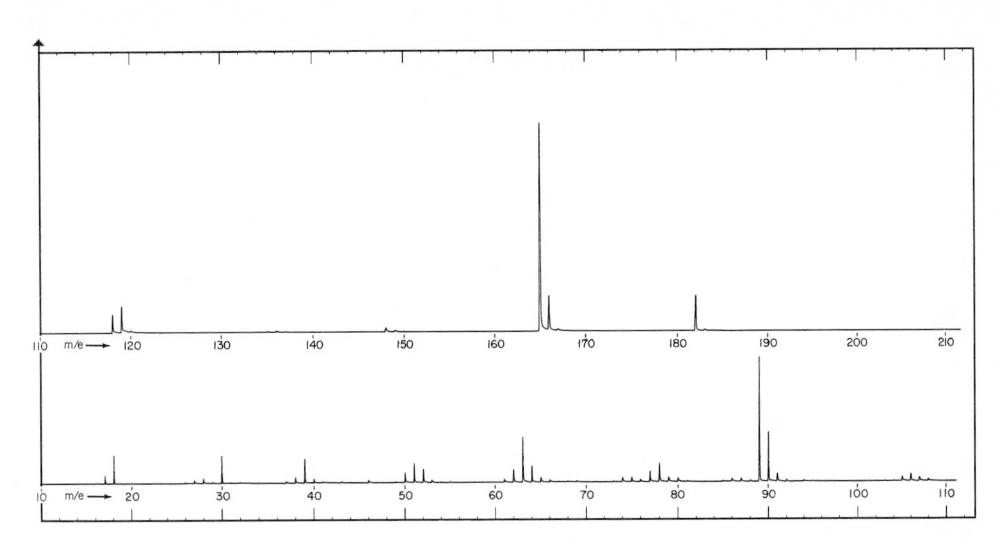

MASS SPECTRUM

MASS SPECTRAL DATA

m/e	relative intensity	m/e	relative intensity	m/e	relative intensity	m/e	relative intensity
17	1.5	52	6.7	78	12.1	107	3.3
18	6.1	53	1.8	79	3.0	108	2.1
26	0.8	54	0.5	80	2.6	118	11.8
27	1.6	55	0.8	81	0.9	119	19.0
28	2.0	61	1.9	85	0.8	120	2.2
29	0.9	62	7.3	86	2.3	121	0.8
30	12.8	63	25.0	87	2.5	135	0.9
37	0.8	64	10.8	88	1.1	136	1.8
38	2.4	65	2.6	89	66.0	137	0.8
39	11.0	66	2.1	90	24.0	148	3.3
40	2.1	67	0.7	91	6.6	149	1.8
41	0.7	68	0.7	92	1.5	165	100.0
43	0.8	69	0.8	93	0.7	166	16.2
44	0.5	73	0.5	94	1.2	167	2.1
46	1.7	74	2.6	95	0.7	182	17.0
49	0.5	75	3.0	104	0.5	183	1.5
50	4.6	76	2.1	105	4.0	184	0.2
51	9.5	77	7.7	106	5.6		

COMPOUND 2-18

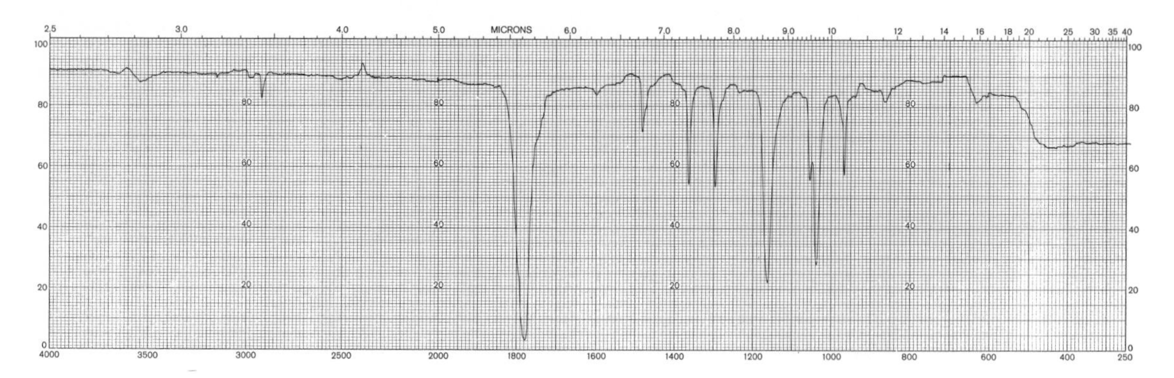

INFRARED SPECTRUM: 1.1% in chloroform

ULTRAVIOLET DATA: λ_{max}^{EtOH} 209 nm (ϵ 84)

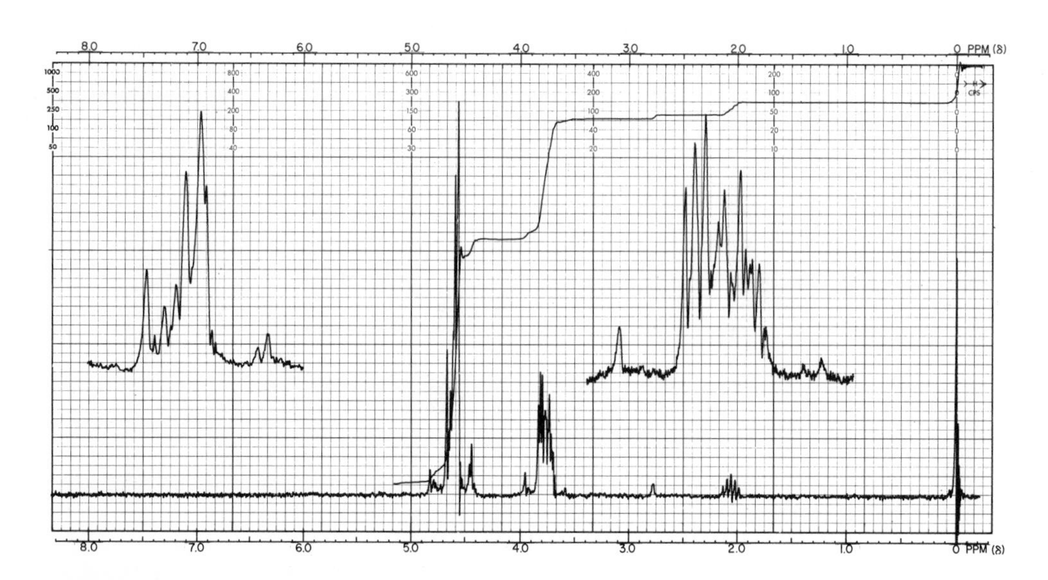

NUCLEAR MAGNETIC RESONANCE SPECTRUM:
15% in acetone-d_6, 500 Hz sweep width
Expansions: (left) 100 Hz sweep width, 190 Hz sweep offset
(right) 100 Hz sweep width, 200 Hz sweep offset

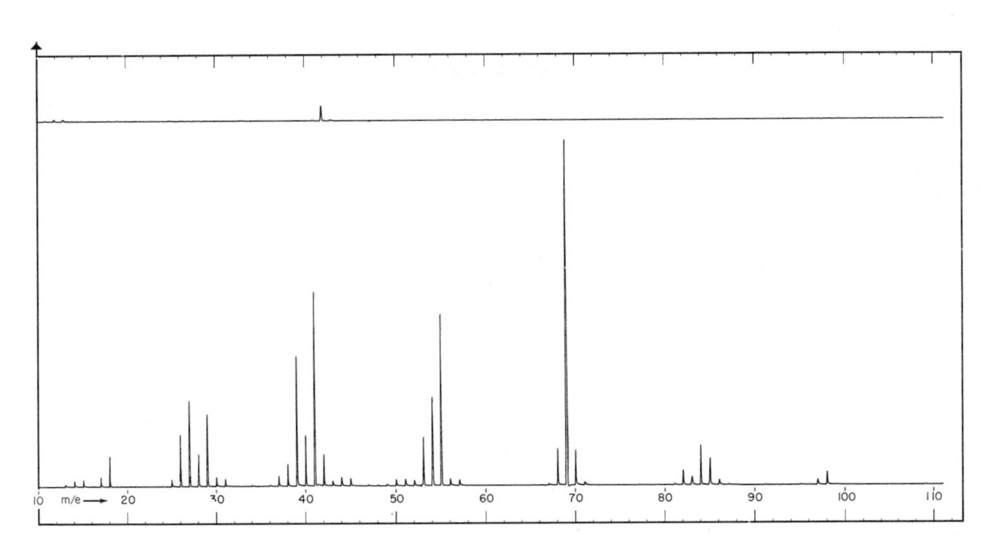

MASS SPECTRUM

MASS SPECTRAL DATA

m/e	*relative intensity*	*m/e*	*relative intensity*	*m/e*	*relative intensity*	*m/e*	*relative intensity*
14	1.1	38	3.6	53	10.2	81	1.5
15	1.1	39	24.3	54	18.4	82	2.8
17	1.6	40	8.0	55	34.1	83	2.2
18	6.5	41	46.1	56	2.0	84	8.5
25	1.2	42	5.6	57	2.1	85	6.1
26	8.7	43	1.7	67	1.0	86	1.6
27	15.0	44	2.7	68	8.0	95	1.3
28	6.8	45	1.6	69	100.0	96	0.9
29	14.1	50	1.3	70	7.3	97	2.0
30	1.6	51	1.6	71	1.3	98	3.4
31	1.5	52	1.3	79	0.5	142	3.1
37	2.0						

COMPOUND 2-19

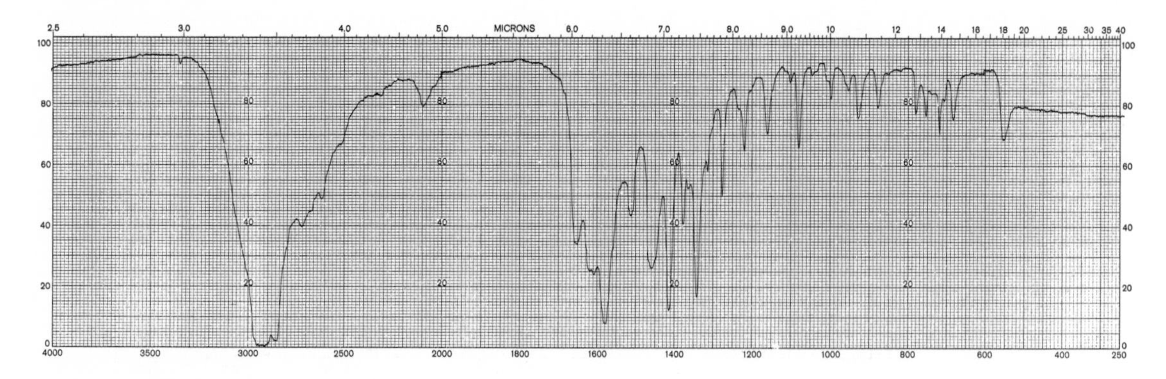

INFRARED SPECTRUM: Nujol mull

ULTRAVIOLET DATA: no λ_{max} above 205 nm

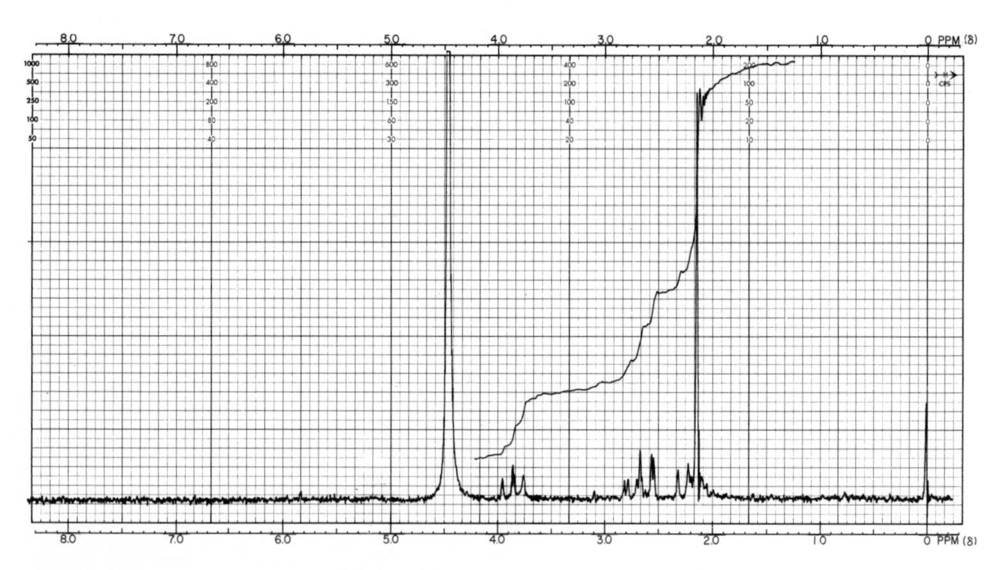

NUCLEAR MAGNETIC RESONANCE SPECTRUM:
15% in deuterium oxide, 80°, 500 Hz sweep width

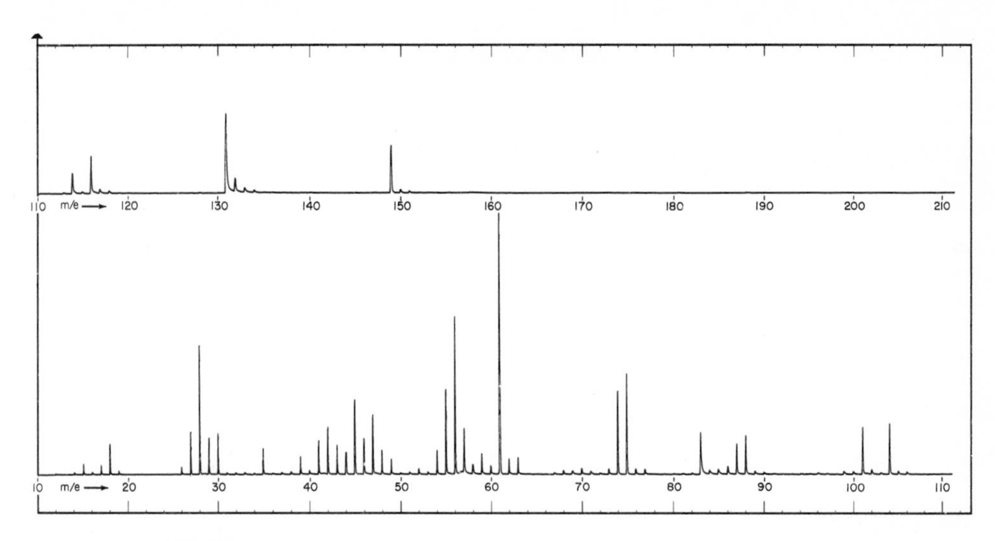

MASS SPECTRUM

MASS SPECTRAL DATA

m/e	*relative intensity*	*m/e*	*relative intensity*	*m/e*	*relative intensity*	*m/e*	*relative intensity*
15	1.7	43	7.1	64	0.4	99	1.1
16	0.4	44	6.0	67	0.4	100	0.9
17	1.4	45	20.9	68	1.1	101	17.0
18	4.6	46	12.4	69	1.3	102	1.8
19	0.6	47	16.2	70	2.0	103	0.5
26	1.3	48	6.0	71	0.9	104	17.7
27	10.0	49	4.4	72	0.4	105	1.4
28	32.7	50	0.4	73	1.7	106	1.0
29	9.5	51	0.4	74	26.5	113	0.3
30	9.7	52	1.6	75	31.7	114	6.9
31	0.4	53	0.7	76	1.8	115	0.7
32	0.6	54	6.6	77	1.6	116	13.3
33	0.5	55	23.2	82	0.4	117	1.6
34	0.2	56	47.0	83	13.1	118	0.8
35	6.0	57	15.2	84	1.6	131	29.7
37	0.4	58	3.4	85	1.9	132	6.0
38	0.8	59	6.2	86	2.8	133	2.2
39	4.4	60	2.4	87	9.7	134	1.3
40	0.8	61	100.0	88	12.6	149	18.3
41	8.0	62	4.9	89	1.4	150	1.4
42	11.9	63	4.7	90	0.8	151	1.1

COMPOUND 2-20

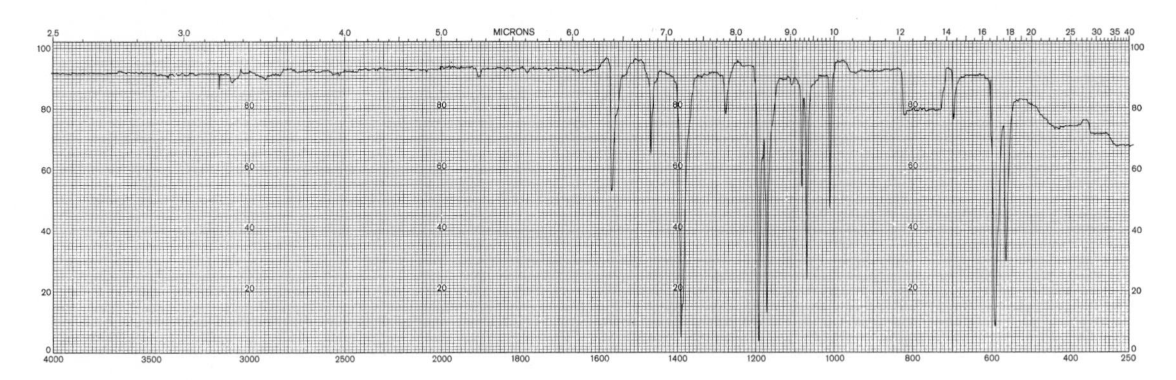

INFRARED SPECTRUM: 2.0% in carbon tetrachloride

ULTRAVIOLET DATA: λ_{max}^{Hexane} 248 nm (ϵ 16,900); 280 nm (ϵ 780)

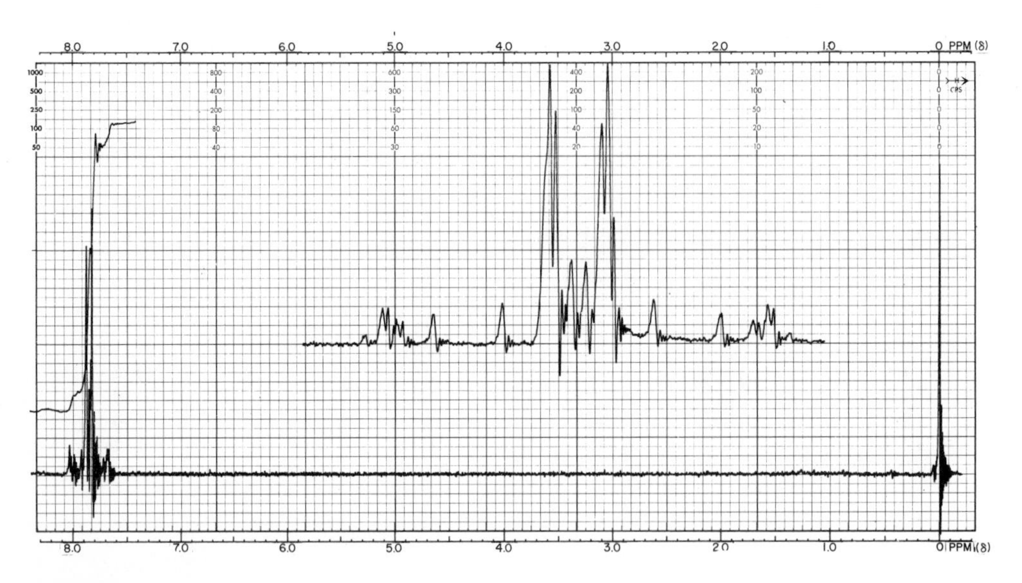

NUCLEAR MAGNETIC RESONANCE SPECTRUM:
15% in carbon tetrachloride, 500 Hz sweep width
Expansion: 50 Hz sweep width, 450 Hz sweep offset

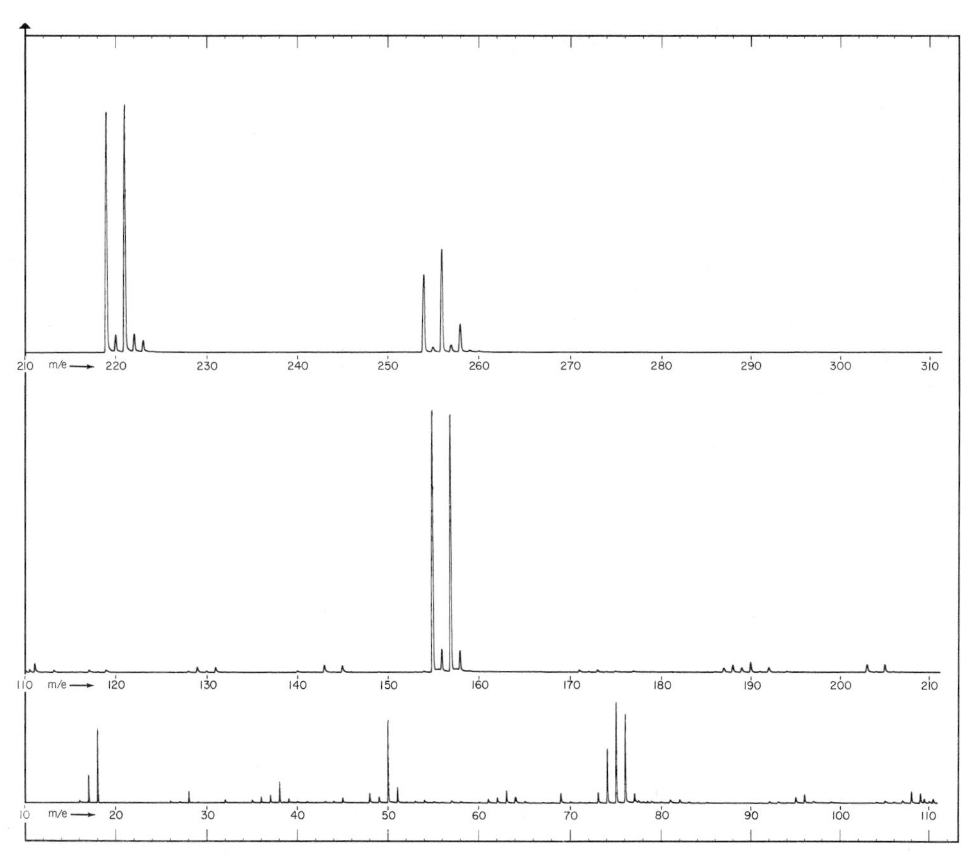

MASS SPECTRUM

MASS SPECTRAL DATA

m/e	*relative intensity*	*m/e*	*relative intensity*	*m/e*	*relative intensity*	*m/e*	*relative intensity*
17	1.8	73	2.6	110	0.8	188	2.8
18	7.7	74	14.3	110.5	1.4	189	2.1
28	2.5	75	25.9	111	3.1	190	4.0
32	0.7	76	23.8	113	1.0	191	0.5
35	0.6	77	2.8	117	1.2	192	2.1
36	1.5	77.5	1.1	119	1.1	194	0.6
37	1.5	78	0.6	128	0.7	203	3.3
38	3.7	78.5	0.7	129	2.1	204	0.5
39	0.9	79	0.8	130	0.9	205	3.6
44	0.7	80	0.6	131	1.9	206	0.4
45	1.3	81	1.0	140	0.8	207	0.3
48	2.3	82	1.3	143	2.7	219	95.0
49	1.6	85	0.5	145	2.6	220	7.6
50	18.4	92	0.6	154	0.5	221	100.0
51	3.1	93	0.4	155	94.0	222	7.9
54	0.8	95	2.0	156	8.7	223	5.0
57	0.7	96	2.9	157	91.0	224	0.4
58	0.4	97	0.6	158	7.9	254	31.0
61	1.0	99	0.6	159	0.8	255	2.3
62	1.3	105	1.0	171	1.0	256	42.0
63	3.0	106	0.5	173	1.0	257	3.2
64	2.0	107	0.9	175	0.6	258	11.6
65	0.7	108	3.3	177	0.7	259	0.8
69	2.3	109	2.8	187	1.6	260	0.7
70	0.5	109.5	1.3				

COMPOUND 2-21

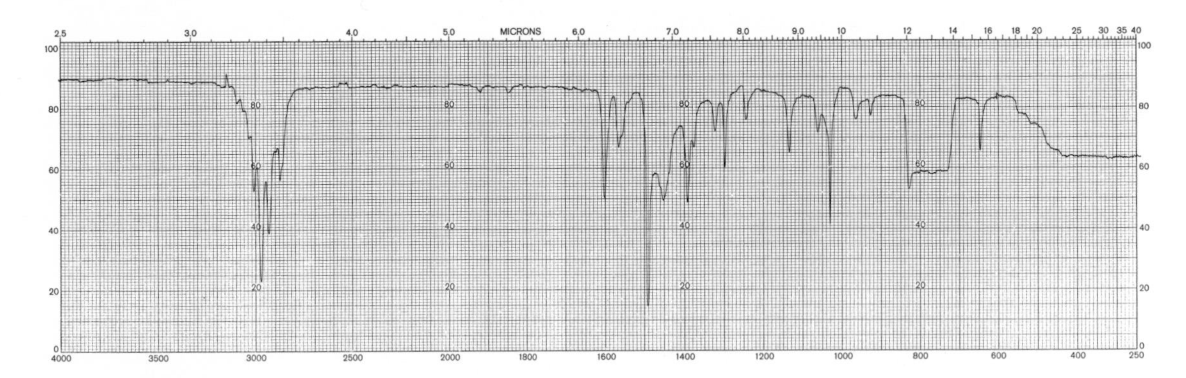

INFRARED SPECTRUM: 3.4% in carbon tetrachloride

ULTRAVIOLET DATA: λ_{max}^{EtOH} 270 nm (ϵ 4,000); 280 nm (ϵ 3,150)

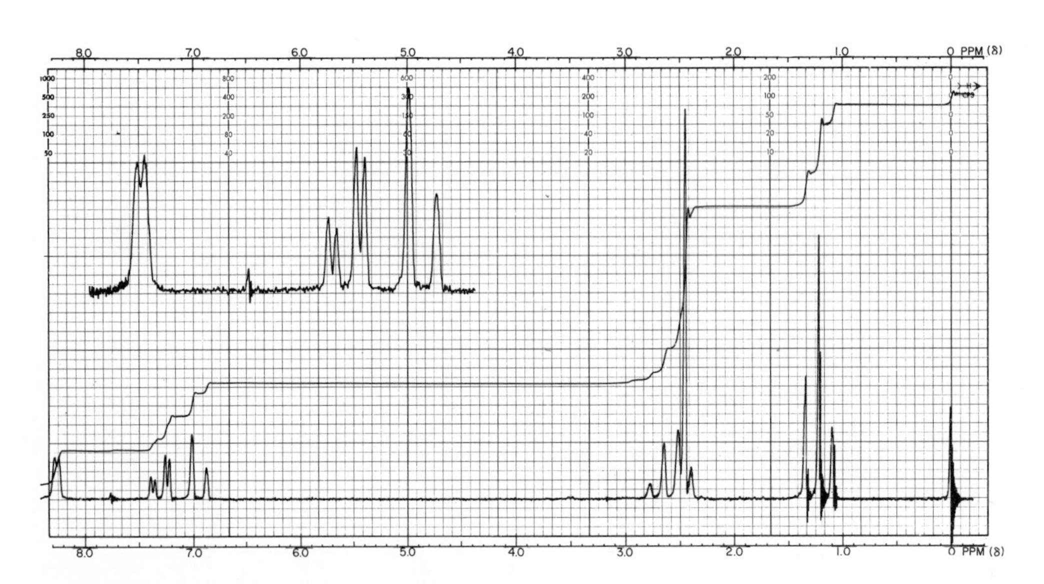

NUCLEAR MAGNETIC RESONANCE SPECTRUM:
15% in carbon tetrachloride, 500 Hz sweep width
Expansion: 250 Hz sweep width, 270 Hz sweep offset

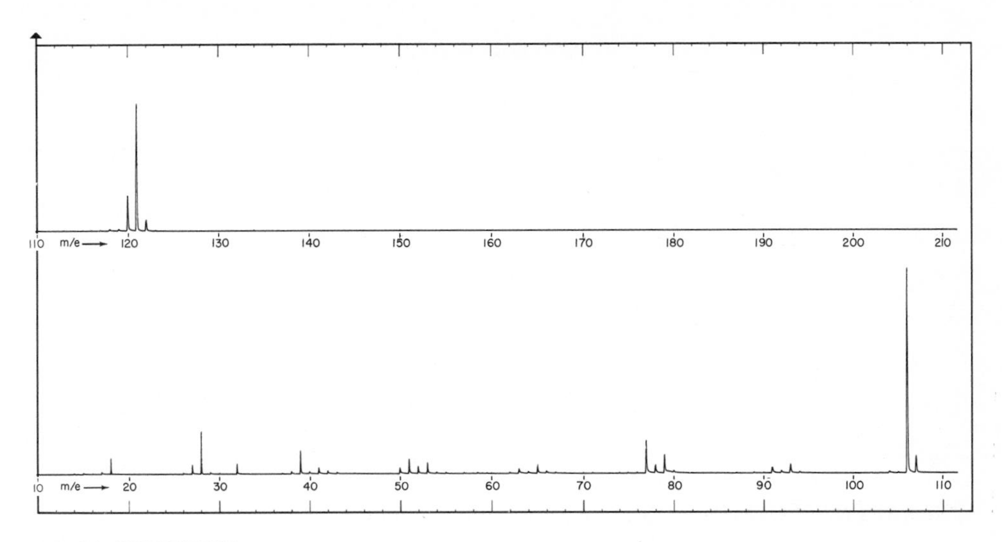

MASS SPECTRUM

MASS SPECTRAL DATA

m/e	*relative intensity*	*m/e*	*relative intensity*	*m/e*	*relative intensity*	*m/e*	*relative intensity*
14	0.5	41	1.2	64	1.2	90	0.2
15	0.8	42	0.3	65	4.8	91	3.2
16	0.3	43	0.3	66	1.6	92	1.4
17	1.4	50	2.8	67	0.7	93	4.3
18	6.3	51	6.1	68	0.3	94	0.8
26	0.8	52	4.2	70	0.3	104	1.1
27	5.0	53	5.9	71	0.5	105	0.7
28	20.2	54	1.0	74	0.3	106	100.0
29	1.2	55	0.9	75	0.3	107	8.5
30	0.4	56	0.2	76	0.5	108	0.3
32	4.1	57	0.8	77	13.7	117	0.4
35	0.3	58	0.2	78	4.1	118	1.0
36	1.5	59	0.2	79	8.1	119	1.1
37	9.7	60	0.3	80	1.7	120	18.8
38	1.3	61	0.3	87	0.2	121	61.0
39	3.0	62	0.8	88	0.1	122	5.6
40	1.8	63	2.6	89	0.5	123	0.1

COMPOUND 2-22

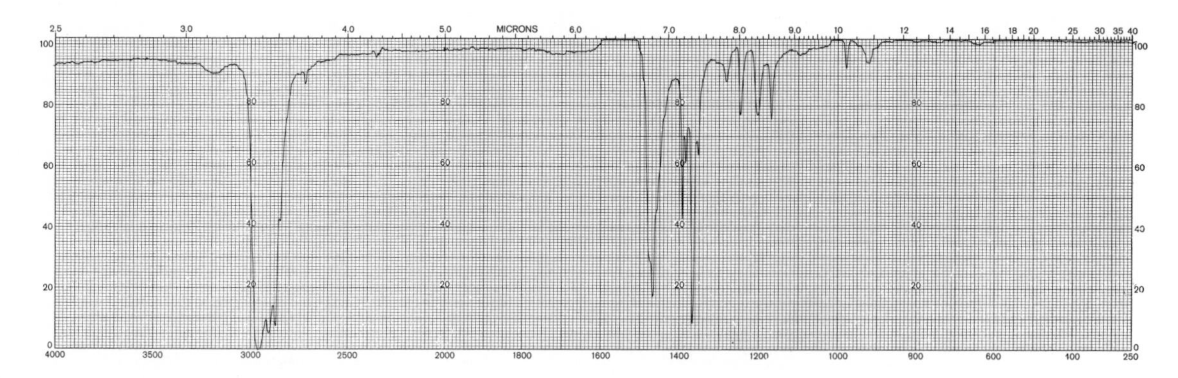

INFRARED SPECTRUM: 5.1% in carbon tetrachloride

ULTRAVIOLET DATA: λ_{max}^{EtOH} above 205 nm

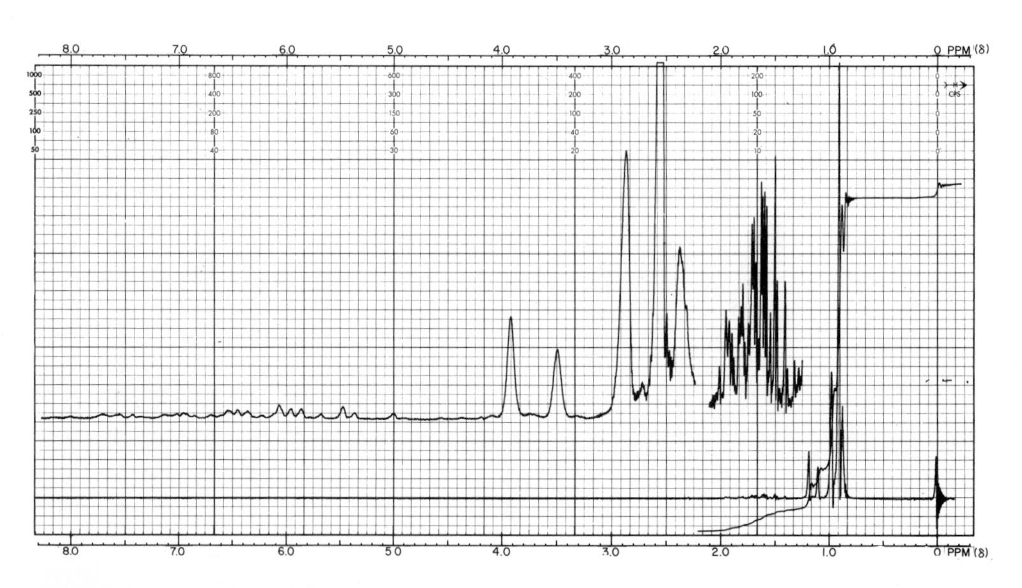

NUCLEAR MAGNETIC RESONANCE SPECTRUM:
15% in carbon tetrachloride, 500 Hz sweep width
Expansions: (left) 100 Hz sweep width, 23 Hz sweep offset
(right) amplitude X 50

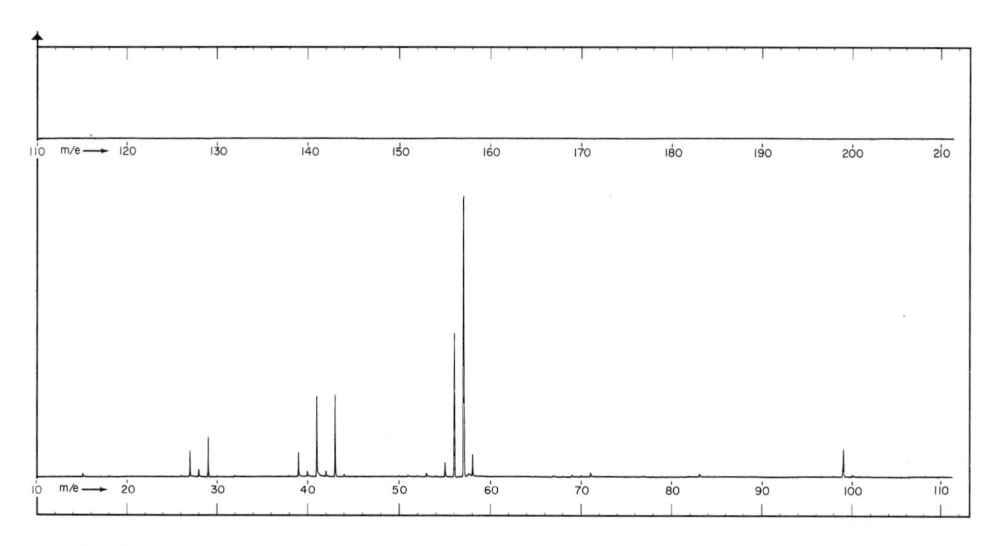

MASS SPECTRUM

MASS SPECTRAL DATA

m/e	*relative intensity*	*m/e*	*relative intensity*	*m/e*	*relative intensity*	*m/e*	*relative intensity*
15	0.8	41	19.9	56	40.0	70	0.4
27	5.6	42	2.0	57	100.0	71	1.2
28	4.5	43	19.5	58	6.1	81	0.4
29	8.5	44	1.0	65	0.4	83	1.2
32	1.5	51	0.6	67	0.8	98	0.4
39	6.6	53	1.3	68	0.4	99	9.8
40	1.7	55	4.1	69	0.7	100	0.8

COMPOUND 2-23

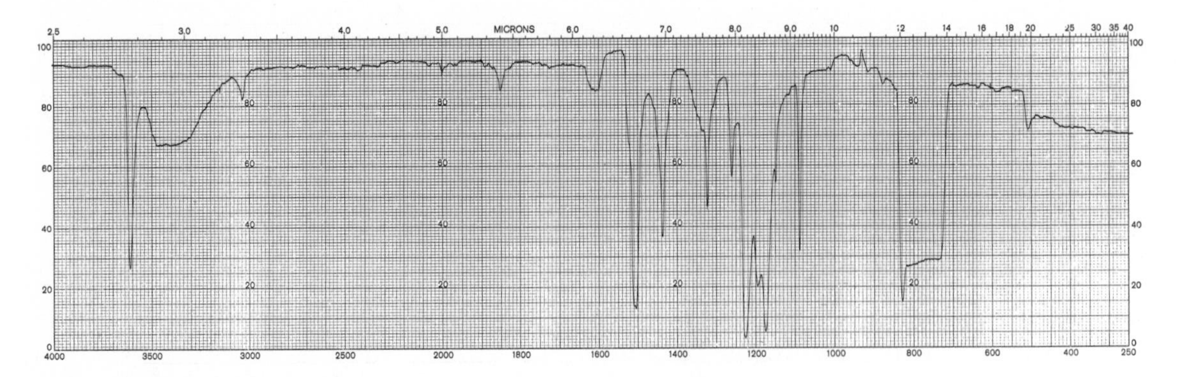

INFRARED SPECTRUM: 2.9% in carbon tetrachloride

ULTRAVIOLET DATA: λ_{max}^{Hexane} 211 nm (ϵ 4,200); 276 nm (ϵ 2,800); 279 nm (ϵ 3,100)

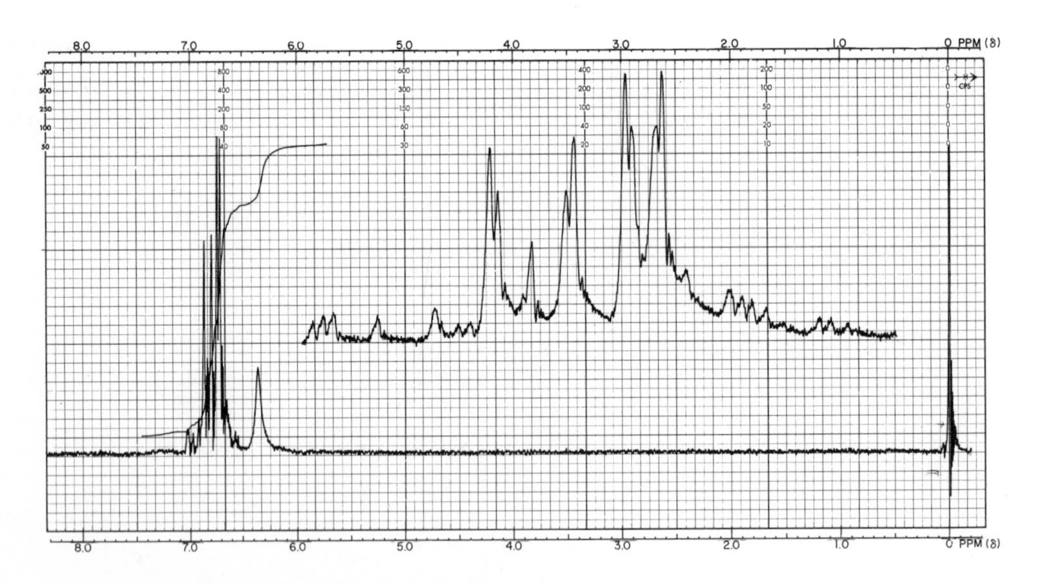

NUCLEAR MAGNETIC RESONANCE SPECTRUM:
15% in carbon tetrachloride, 500 Hz sweep width
Expansion: 50 Hz sweep width, 385 Hz sweep offset

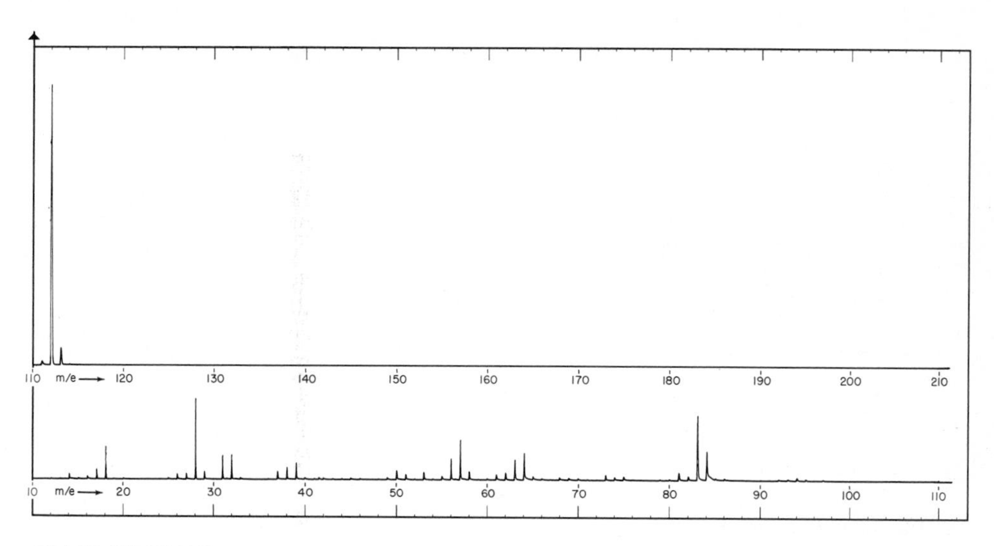

MASS SPECTRUM

MASS SPECTRAL DATA

m/e	*relative intensity*	*m/e*	*relative intensity*	*m/e*	*relative intensity*	*m/e*	*relative intensity*
14	1.7	40	0.7	58	3.7	80	0.4
15	0.3	41	0.4	59	0.2	81	3.3
16	0.6	41.5	0.7	60	0.4	82	1.7
17	3.0	42	0.6	61	2.3	83	28.0
18	13.9	43	0.4	62	2.8	84	11.9
20	0.5	44	0.4	63	8.0	85	0.9
25	0.5	45	0.6	64	10.8	86	0.9
26	2.0	46	0.4	65	1.8	93	0.4
27	2.3	49	1.1	66	0.6	94	0.5
28	40.0	50	3.6	67	0.3	95	1.3
29	3.0	51	2.3	68	1.1	96	0.8
31	7.2	52	0.4	69	1.0	98	0.4
32	8.3	53	3.2	70	0.7	110	0.2
33	0.9	54	0.4	71	0.3	111	1.7
36	0.3	55	1.7	73	2.6	112	100.0
37	2.8	56	8.8	74	1.4	113	6.6
38	4.4	56.5	0.4	75	1.7	114	0.4
39	5.9	57	15.6	79	0.3		

COMPOUND 2-24

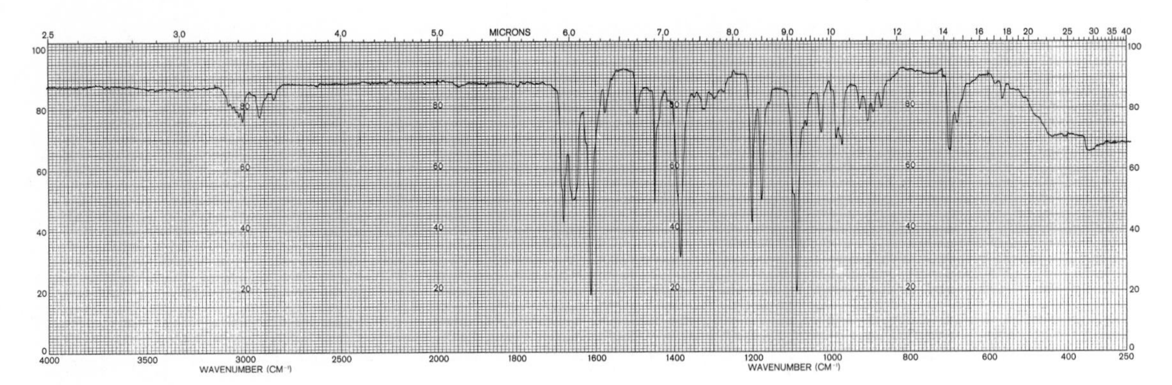

INFRARED SPECTRUM: 1.3% in carbon tetrachloride

ULTRAVIOLET DATA: λ_{max}^{EtOH} 222 nm (ϵ 13,200); 291 nm (ϵ 24,000)

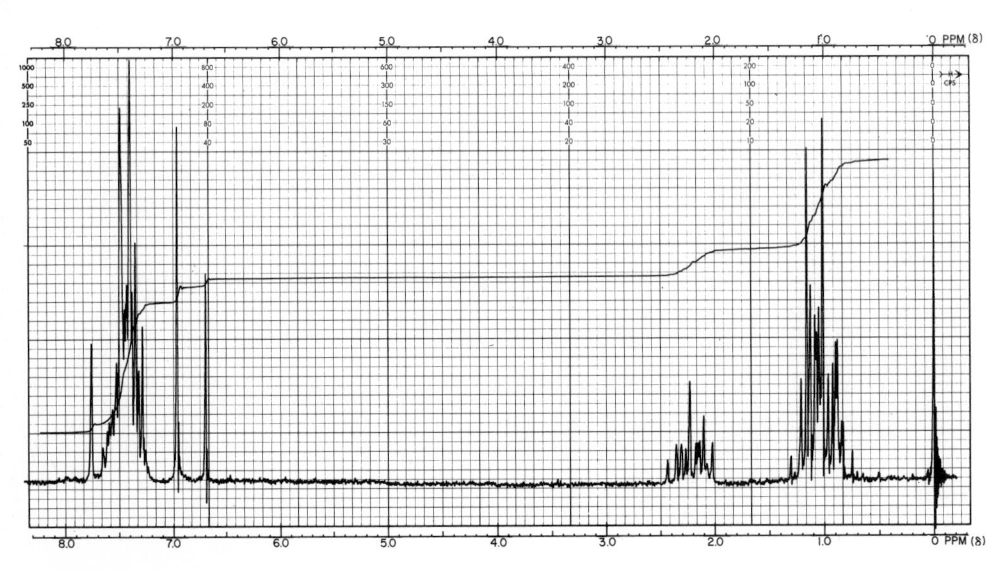

NUCLEAR MAGNETIC RESONANCE SPECTRUM:
15% in deuterochloroform, 500 Hz sweep width

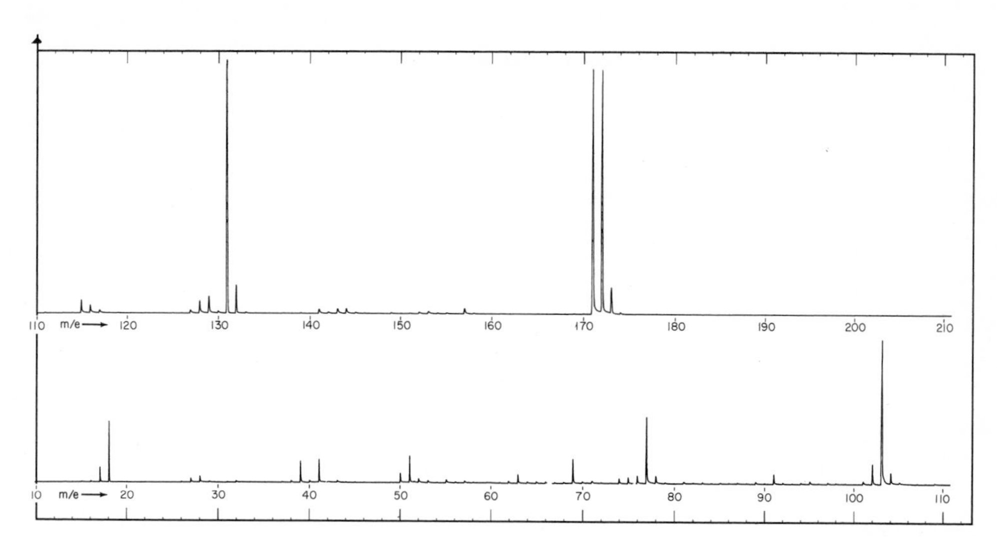

MASS SPECTRUM

MASS SPECTRAL DATA

m/e	relative intensity	m/e	relative intensity	m/e	relative intensity	m/e	relative intensity
15	0.5	58	0.4	85	0.5	129	11.7
16	0.3	61	0.5	86	0.7	130	2.2
17	1.1	62	2.3	87	1.2	131	100.0
18	4.2	63	6.9	88	0.5	132	20.4
27	0.7	64	1.4	89	2.0	133	1.3
28	2.8	65	2.0	90	0.6	139	0.8
29	1.1	65.5	0.5	91	7.9	141	4.0
30	0.6	66	1.1	92	1.0	142	1.7
32	0.6	67	0.6	94	0.9	143	3.7
37	0.4	68	1.1	95	1.5	144	4.7
38	1.5	69	16.7	98	0.8	145	1.6
39	13.3	69.5	0.5	101	2.4	149	0.7
40	1.7	70	1.2	102	16.8	151	0.6
41	14.9	70.5	0.5	103	91.0	152	2.3
42	1.0	71	1.5	104	10.0	153	3.4
43	0.6	71.5	0.3	105	1.4	154	1.7
45	0.4	72	0.7	107	0.5	155	1.2
46	0.3	73	0.5	113	0.6	156	0.5
50	5.8	74	4.3	114	0.6	157	4.4
51	18.8	75	5.2	115	11.3	158	0.7
51.5	1.2	76	6.7	116	7.7	169	0.7
52	2.9	77	55.0	117	2.7	170	0.6
53	2.0	78	5.9	118	0.7	171	98.0
55	1.7	79	0.7	126	0.5	172	98.0
57	0.5	81	0.9	127	2.8	173	20.2
57.5	0.5	83	0.3	128	9.1	174	1.6

COMPOUND 2-25

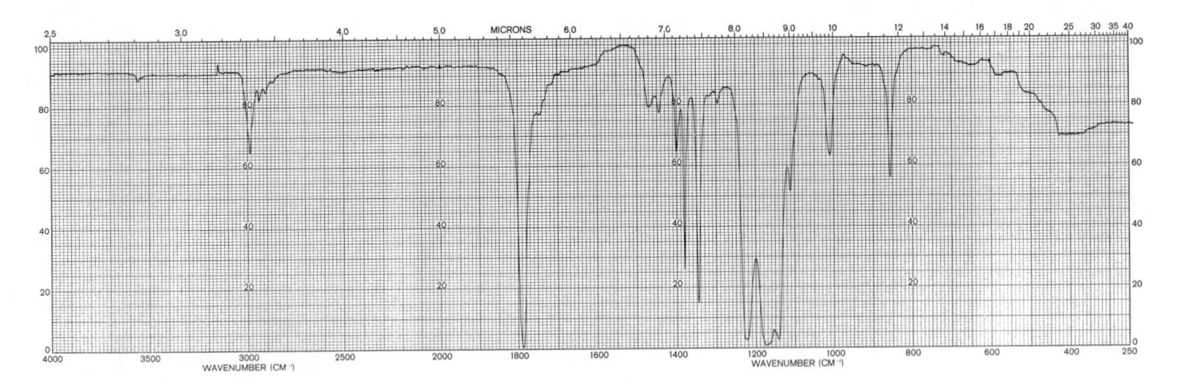

INFRARED SPECTRUM: 3.3% in carbon tetrachloride

ULTRAVIOLET DATA: no λ_{max} above 205 nm

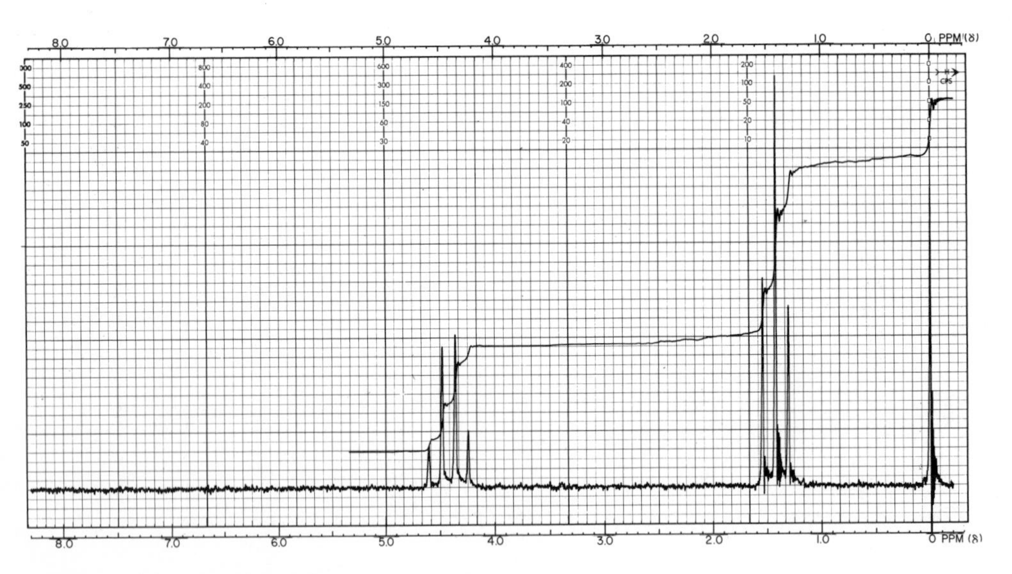

NUCLEAR MAGNETIC RESONANCE SPECTRUM:
15% in carbon tetrachloride, 500 Hz sweep width

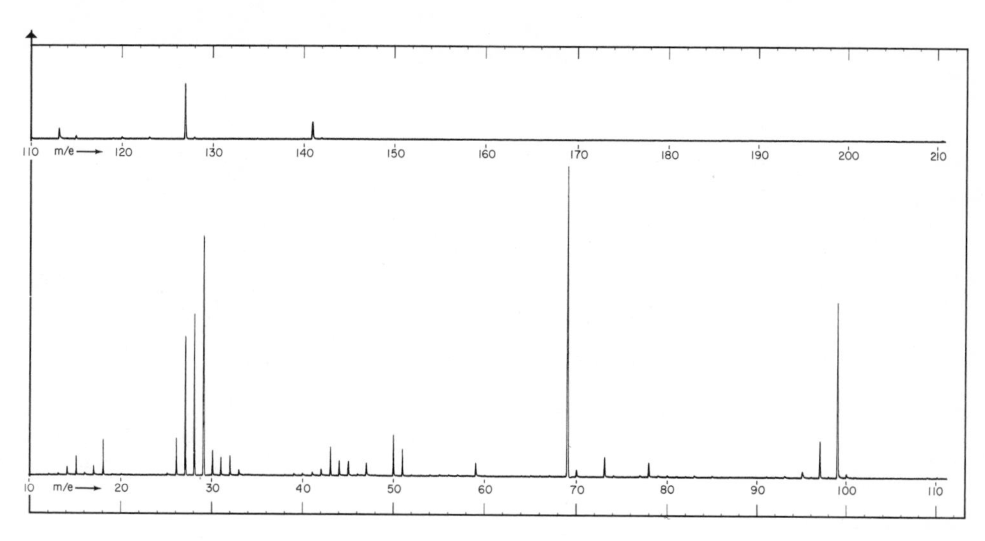

MASS SPECTRUM

MASS SPECTRAL DATA

m/e	*relative intensity*	*m/e*	*relative intensity*	*m/e*	*relative intensity*	*m/e*	*relative intensity*
14	1.1	31	2.1	51	0.4	99	36.0
15	2.1	32	2.0	59	2.0	100	0.9
16	0.4	33	0.9	69	100.0	113	2.4
17	1.0	41	0.8	70	1.3	115	0.7
18	4.0	42	1.2	73	3.3	120	0.5
26	3.5	43	3.8	78	2.5	123	0.5
27	29.0	44	2.1	83	0.5	127	11.2
28	33.0	45	2.1	93	0.6	128	0.5
29	95.0	47	1.8	95	1.1	141	3.7
30	3.1	50	5.4	97	6.6	142	0.4

COMPOUND 3-1

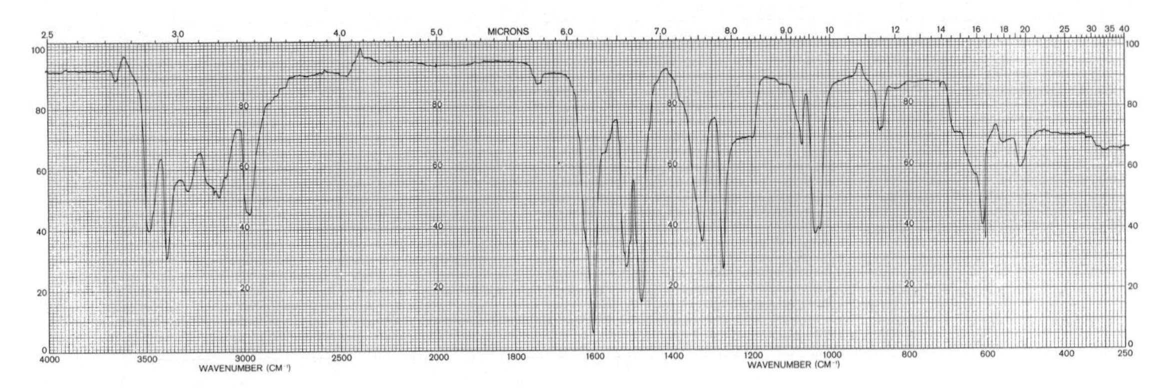

INFRARED SPECTRUM: 3.1% in chloroform

ULTRAVIOLET DATA: λ_{max}^{EtOH} 256 nm (ϵ 6,200)

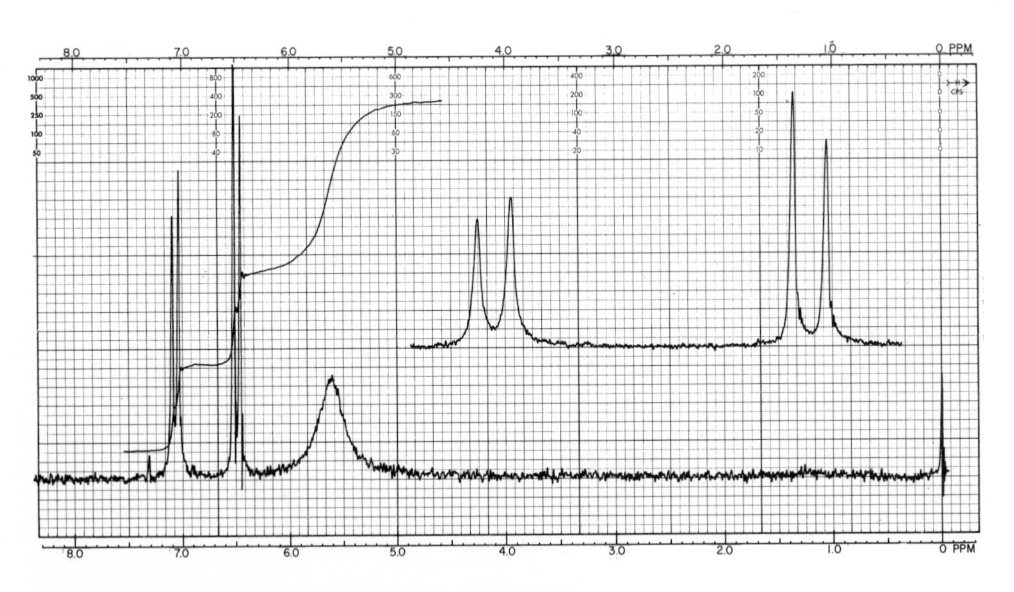

NUCLEAR MAGNETIC RESONANCE SPECTRUM:
15% in deuterochloroform, 500 Hz sweep width
Expansion: 100 Hz sweep width, 375 Hz sweep offset

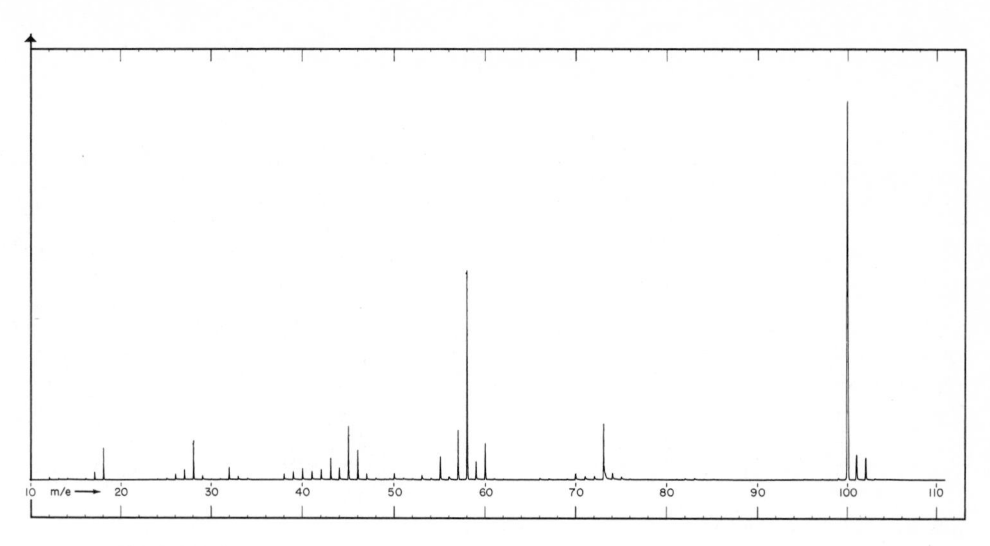

MASS SPECTRUM

MASS SPECTRAL DATA

m/e	relative intensity	m/e	relative intensity	m/e	relative intensity	m/e	relative intensity
17	0.5	39	1.3	53	1.3	71	1.0
18	2.2	40	1.9	54	0.5	72	0.9
25	0.2	41	1.7	55	4.7	73	15.0
26	1.0	42	1.9	56	1.0	74	1.8
27	2.0	43	4.7	57	10.4	75	0.8
28	8.1	44	2.2	58	74.0	77	0.3
29	0.9	45	10.4	59	4.2	81	0.4
30	0.2	46	5.7	60	7.8	82	0.4
31	0.3	47	1.4	61	0.3	99	0.5
32	2.1	48	0.4	62	0.2	100	100.0
33	0.8	50	1.3	66	0.4	101	6.9
34	0.4	51	0.3	67	0.4	102	6.4
38	1.2	52	0.2	70	1.9	103	0.3

COMPOUND 3-2

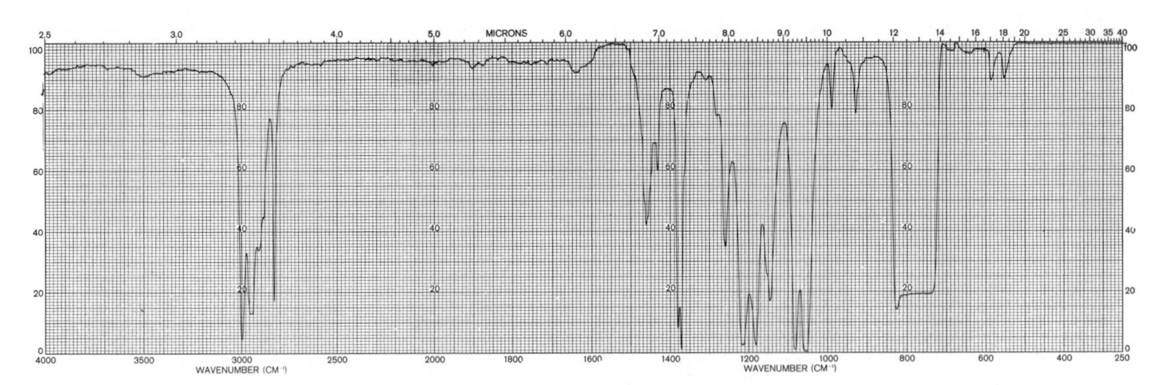

INFRARED SPECTRUM: 5.0% in carbon tetrachloride

ULTRAVIOLET DATA: no λ_{max} above 205 nm

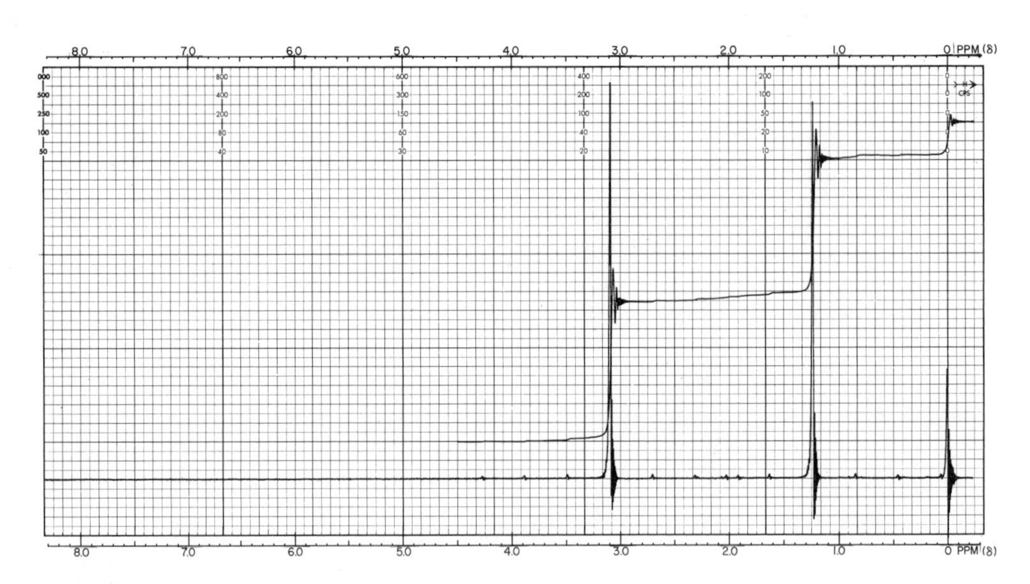

NUCLEAR MAGNETIC RESONANCE SPECTRUM:
15% in carbon tetrachloride, 500 Hz sweep width

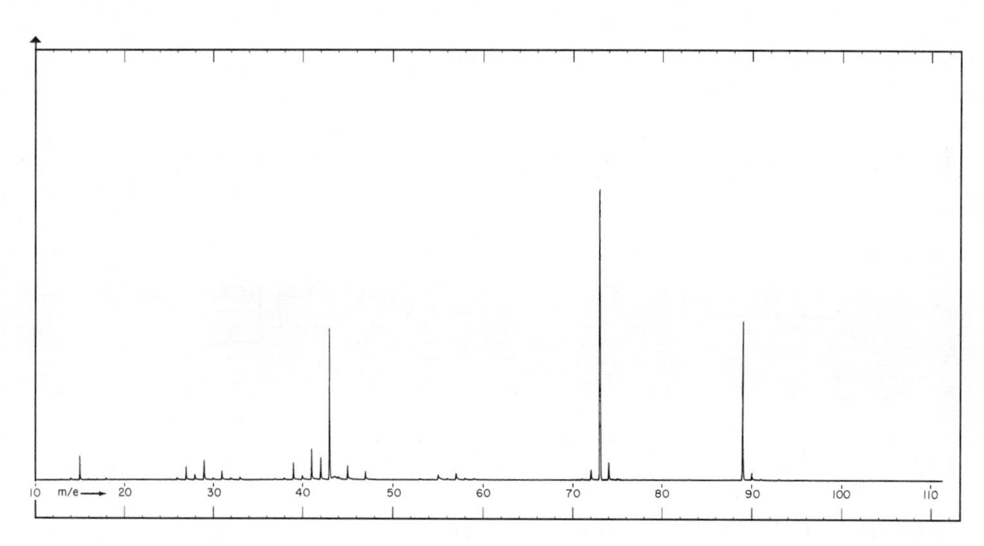

MASS SPECTRUM

MASS SPECTRAL DATA

m/e	relative intensity	m/e	relative intensity	m/e	relative intensity	m/e	relative intensity
14	0.9	33	1.4	45	7.7	70	0.3
15	8.5	37	0.6	47	4.8	71	0.8
16	0.8	38	1.2	53	0.6	72	6.1
26	1.2	39	8.5	55	3.2	73	100.0
27	6.1	40	2.4	56	1.2	74	10.1
28	3.2	41	15.2	57	4.4	75	0.8
29	10.5	42	12.3	58	1.1	89	55.0
30	1.3	43	56.0	59	1.0	90	2.9
31	4.3	44	2.0	69	0.5	91	0.8
32	1.2						

COMPOUND 3-3

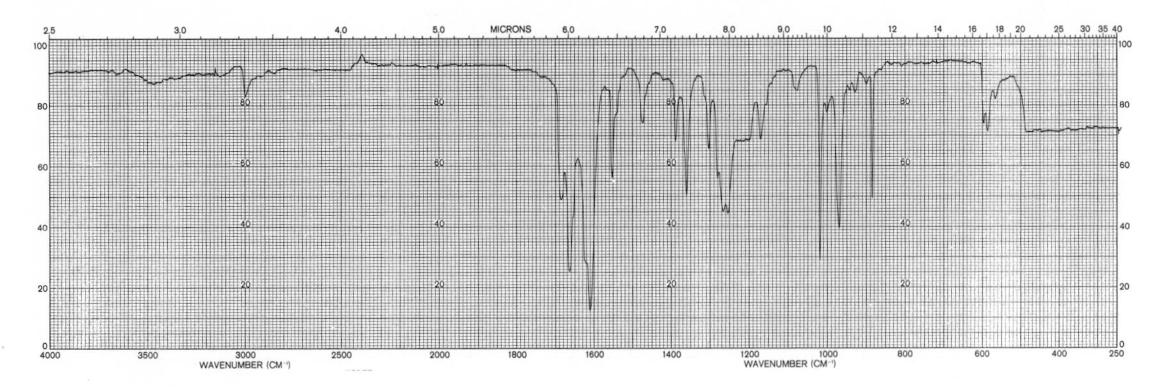

INFRARED SPECTRUM: 1.2% in chloroform

ULTRAVIOLET DATA: λ_{max}^{EtOH} 236 nm (ϵ 1,900); 316 nm (ϵ 23,000)

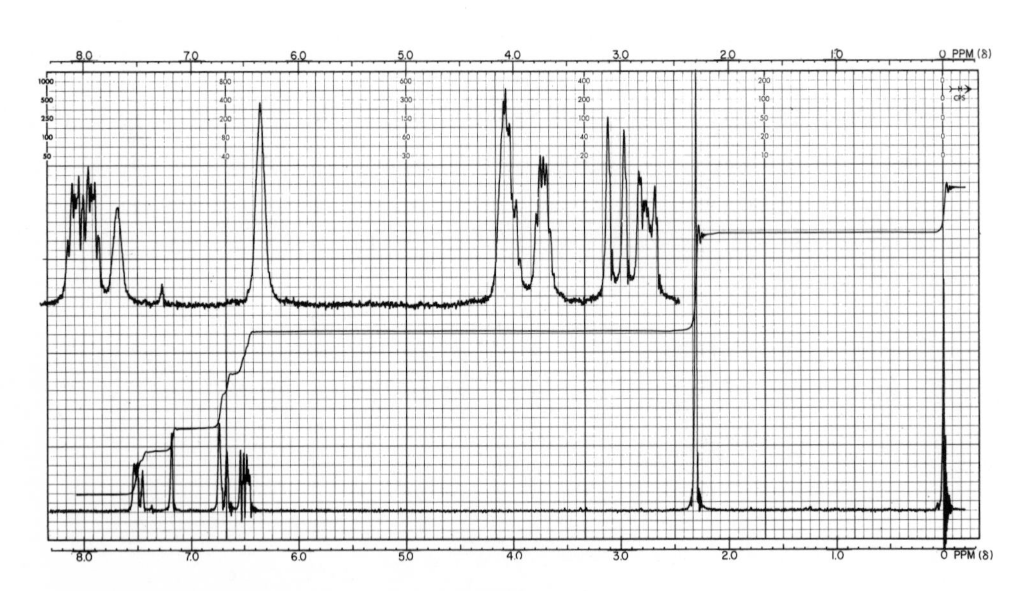

NUCLEAR MAGNETIC RESONANCE SPECTRUM:
15% in deuterochloroform, 500 Hz sweep width
Expansion: 100 Hz sweep width, 355 Hz sweep offset

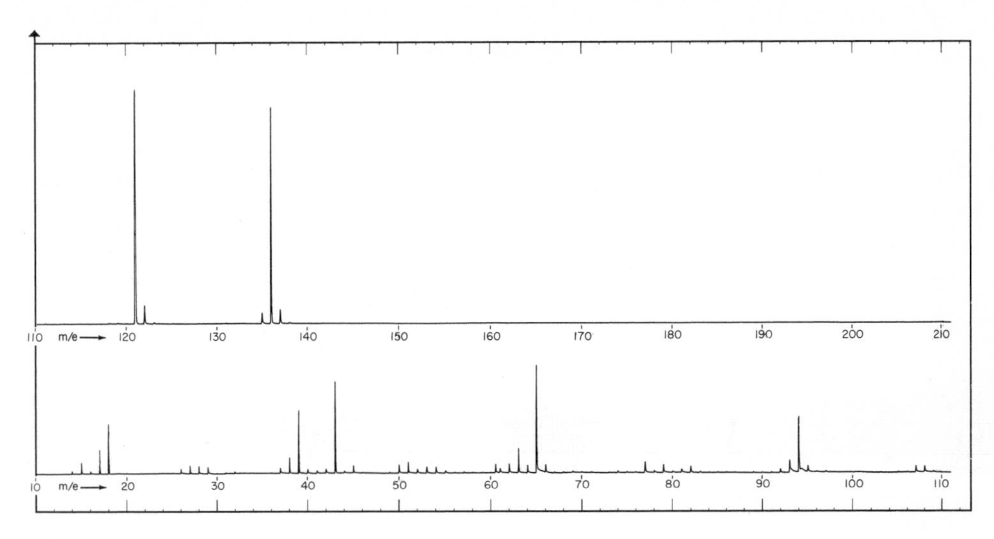

MASS SPECTRUM

MASS SPECTRAL DATA

m/e	*relative intensity*	*m/e*	*relative intensity*	*m/e*	*relative intensity*	*m/e*	*relative intensity*
14	0.8	45	5.5	69	1.0	94	41.0
15	3.4	49	0.8	71	0.7	95	5.3
16	0.4	50	4.3	73	0.4	96	1.0
17	3.1	51	6.4	74	1.2	97	0.9
18	13.8	52	2.1	75	0.7	105	0.7
25	0.4	53	4.6	76	0.4	106	0.5
26	2.0	54	3.7	77	6.9	107	5.3
27	3.2	55	1.9	78	1.4	108	4.7
28	3.0	56	0.4	79	5.9	109	0.7
29	3.4	57	0.8	80	1.0	111	0.5
31	0.7	59	0.4	81	2.4	118	0.6
32	0.6	60	0.5	82	4.6	119	0.7
32.5	0.4	60.5	4.0	83	0.8	120	0.3
37	2.4	61	2.7	85	0.4	121	100.0
38	7.3	62	5.6	86	0.5	122	14.9
39	29.0	63	15.0	87	0.4	123	1.4
40	2.3	64	5.5	89	0.8	134	0.4
41	1.7	65	65.0	90	0.7	135	8.2
42	2.5	66	6.6	91	1.0	136	98.0
43	44.0	67	1.1	92	3.0	137	11.8
44	1.5	68	1.0	93	9.1	138	1.2

COMPOUND 3-4

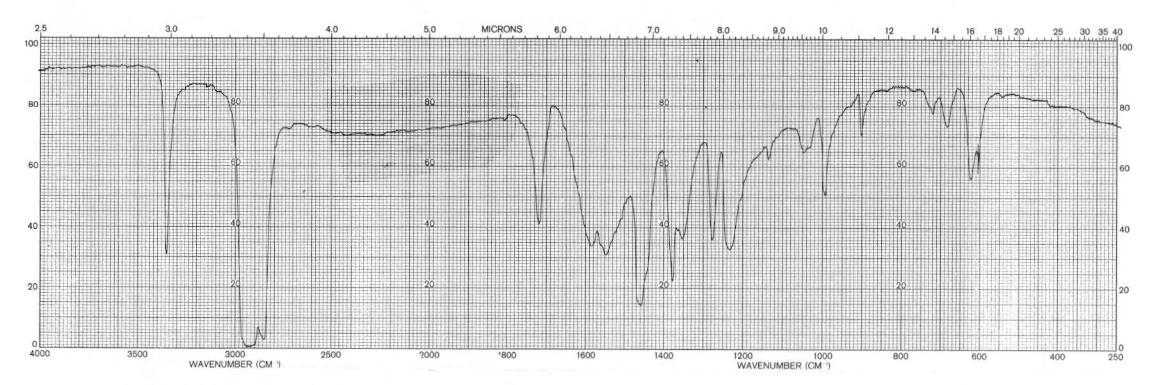

INFRARED SPECTRUM: Nujol mull

ULTRAVIOLET DATA: no λ_{max}^{EtOH} above 205 nm

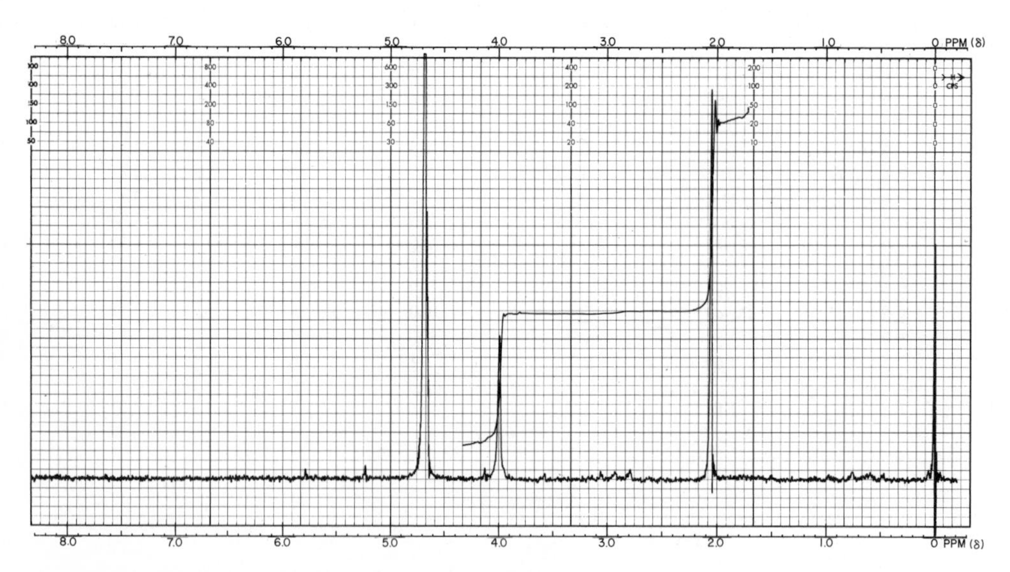

NUCLEAR MAGNETIC RESONANCE SPECTRUM:
15% in deuterium oxide, 60°, 500 Hz sweep width

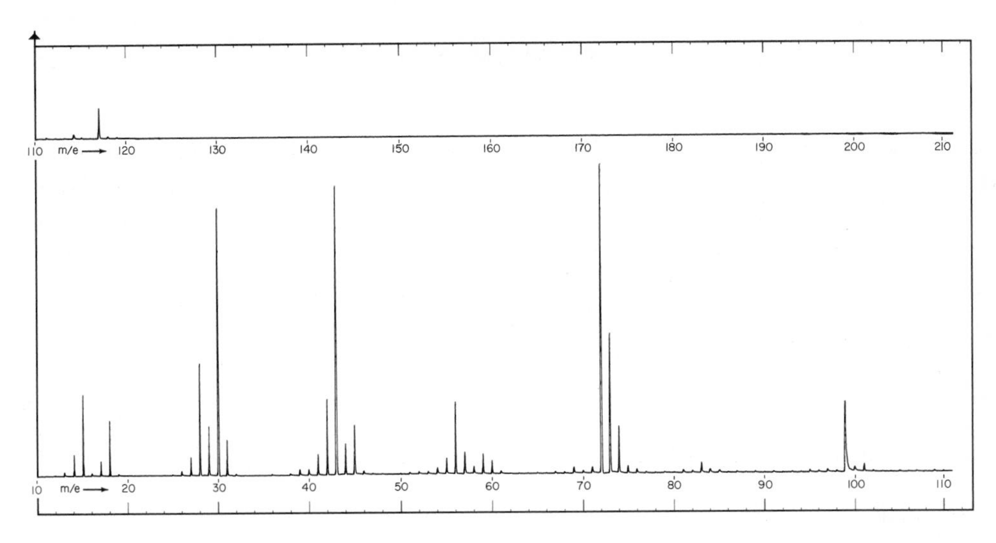

MASS SPECTRUM

MASS SPECTRAL DATA

m/e	*relative intensity*	*m/e*	*relative intensity*	*m/e*	*relative intensity*	*m/e*	*relative intensity*
13	0.9	39	1.5	57	6.0	81	0.5
14	4.6	40	2.0	58	2.2	82	0.5
15	19.0	41	4.9	59	4.2	83	2.1
16	0.5	42	21.3	60	2.5	84	1.3
17	1.3	43	100.0	61	0.7	85	0.6
18	5.4	44	8.8	67	0.5	95	0.5
25	0.4	45	14.0	68	0.4	96	0.5
26	1.3	46	1.2	69	1.2	97	0.7
27	4.1	47	0.4	70	0.7	98	0.5
28	32.0	50	0.3	71	1.7	99	22.9
29	13.2	51	0.5	72	98.0	100	2.0
30	100.0	52	0.7	73	46.0	101	2.1
31	10.4	53	0.5	74	15.1	102	0.8
32	0.6	54	1.8	75	2.0	117	9.6
36	0.4	55	3.2	76	1.3	118	0.8
38	0.7	56	20.6	79	0.2	119	0.4

COMPOUND 3-5

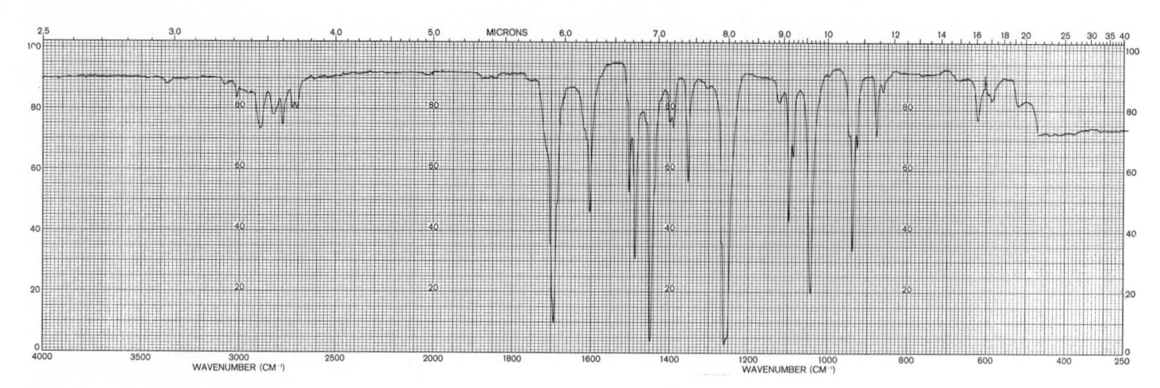

INFRARED SPECTRUM: 1.3% in carbon tetrachloride

ULTRAVIOLET DATA: λ_{max}^{EtOH} 275 nm (ϵ 6,300); 315 nm (ϵ 10,000).

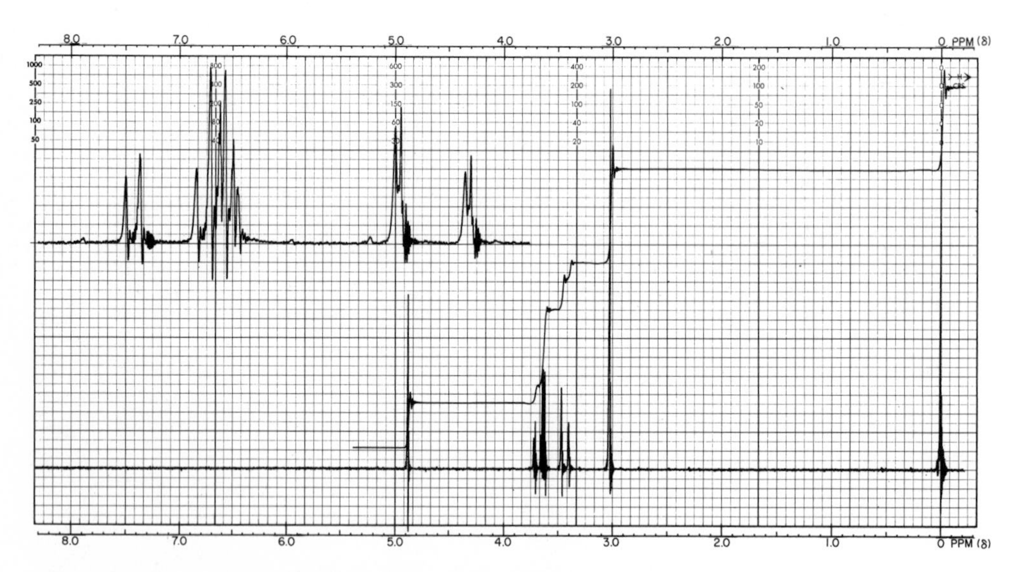

NUCLEAR MAGNETIC RESONANCE SPECTRUM:
15% in carbon tetrachloride, 1000 Hz sweep width
Expansion: 100 Hz sweep width, 355 Hz sweep offset

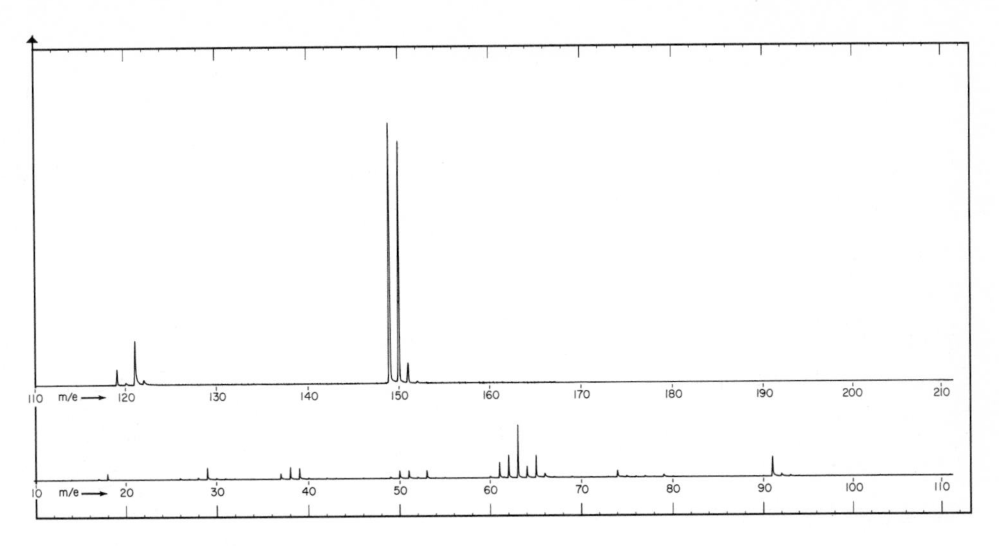

MASS SPECTRUM

MASS SPECTRAL DATA

m/e	*relative intensity*	*m/e*	*relative intensity*	*m/e*	*relative intensity*	*m/e*	*relative intensity*
16	0.7	40	0.6	58	0.5	77	0.7
17	6.0	41	0.5	60	0.6	79	1.0
18	2.3	42	0.8	61	6.5	91	7.4
26	1.0	43	1.4	62	7.7	92	1.2
27	1.0	44	0.5	63	19.6	93	0.5
28	5.2	45	0.5	64	4.3	119	6.7
29	6.0	46	0.4	65	9.3	120	1.1
30	0.9	49	0.7	66	1.7	121	16.3
31	1.2	50	3.1	67	0.3	122	2.1
32	0.9	51	3.1	69	0.3	149	100.0
36	0.5	52	0.5	74	2.6	150	87.0
37	2.3	53	3.5	75	0.6	151	7.7
38	4.8	55	0.6	76	0.4	152	0.8
39	4.2	57	0.5				

COMPOUND 3-6

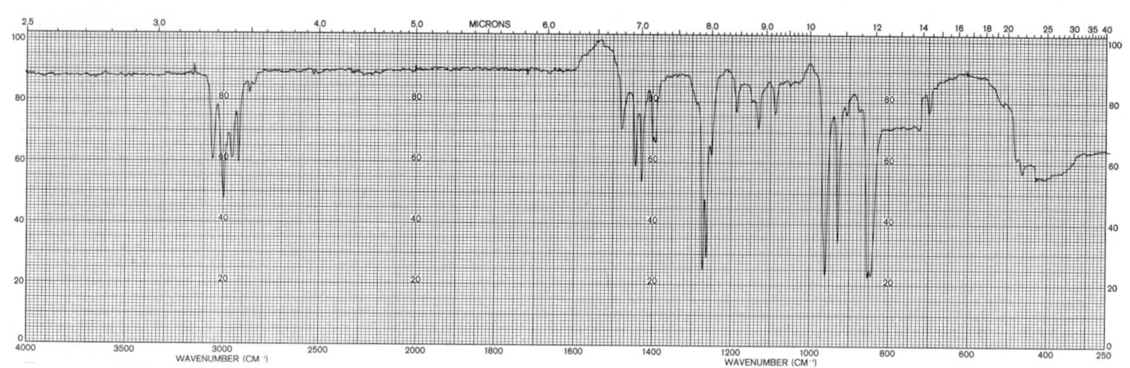

INFRARED SPECTRUM: 5.0% in carbon tetrachloride

ULTRAVIOLET DATA: no λ_{max} above 205 nm

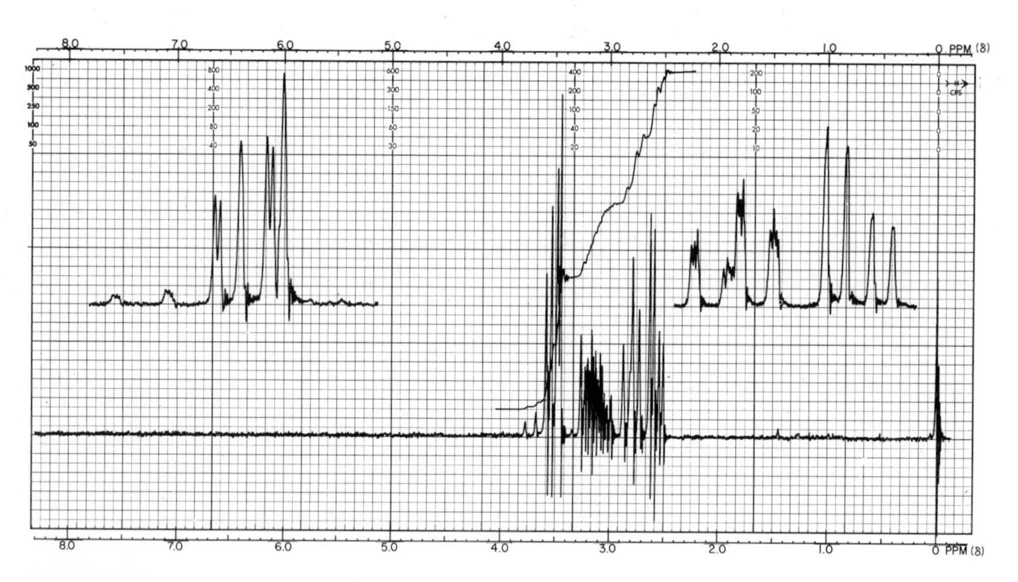

NUCLEAR MAGNETIC RESONANCE SPECTRUM:
15% in carbon tetrachloride, 500 Hz sweep width
Expansions: (left) 100 Hz sweep width, 135 Hz sweep offset
(right) 100 Hz sweep width, 146 Hz sweep offset

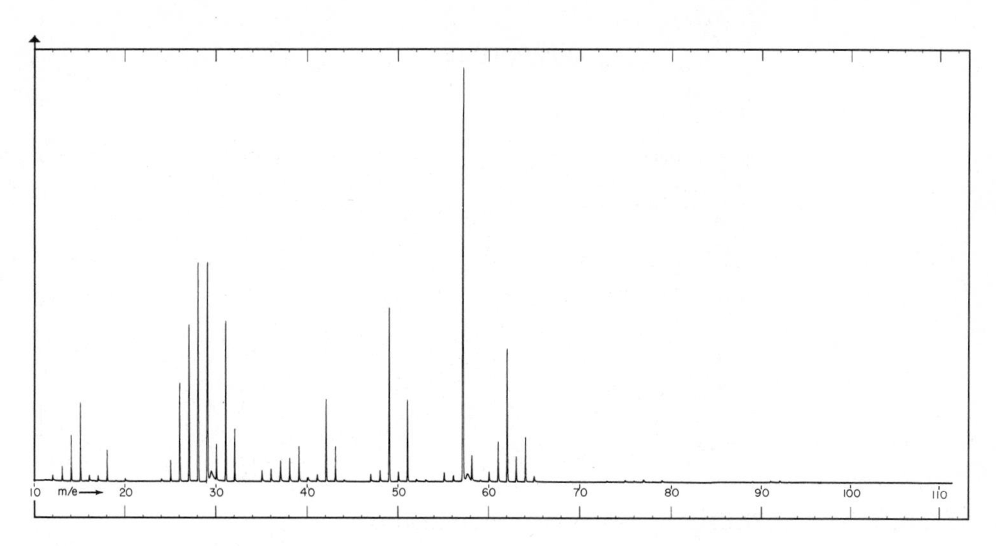

MASS SPECTRUM

MASS SPECTRAL DATA

m/e	*relative intensity*	*m/e*	*relative intensity*	*m/e*	*relative intensity*	*m/e*	*relative intensity*
12	0.8	31	45.0	49	36.0	64	6.5
13	1.9	32	7.6	50	1.4	65	0.8
14	8.1	35	1.9	51	11.8	69	0.1
15	26.0	36	2.7	52	0.4	73	0.2
16	0.8	37	2.7	53	0.3	75	0.2
17	1.4	38	3.2	55	1.3	76	0.1
18	5.5	39	5.2	56	1.0	77	0.4
24	0.3	40	0.6	57	100.0	78	0.1
25	3.2	41	1.2	58	3.6	79	0.2
26	21.8	42	11.8	59	0.3	81	0.1
27	92.0	43	5.1	60	1.5	89	0.1
28	58.0	44	0.3	61	5.3	91	0.2
29	53.0	47	1.3	62	21.2	92	0.2
30	5.3	48	1.9	63	3.6		

COMPOUND 3-7

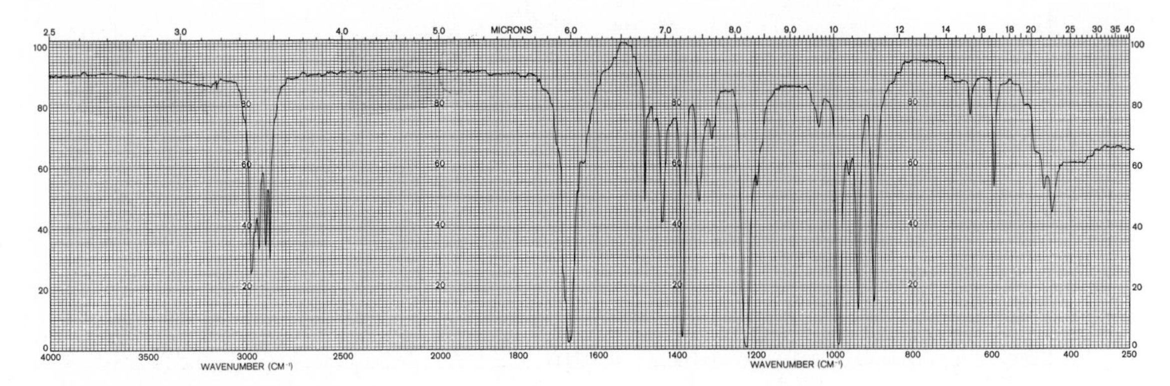

INFRARED SPECTRUM: 3.8% in carbon tetrachloride

ULTRAVIOLET DATA: no λ_{max}^{EtOH} above 205 nm

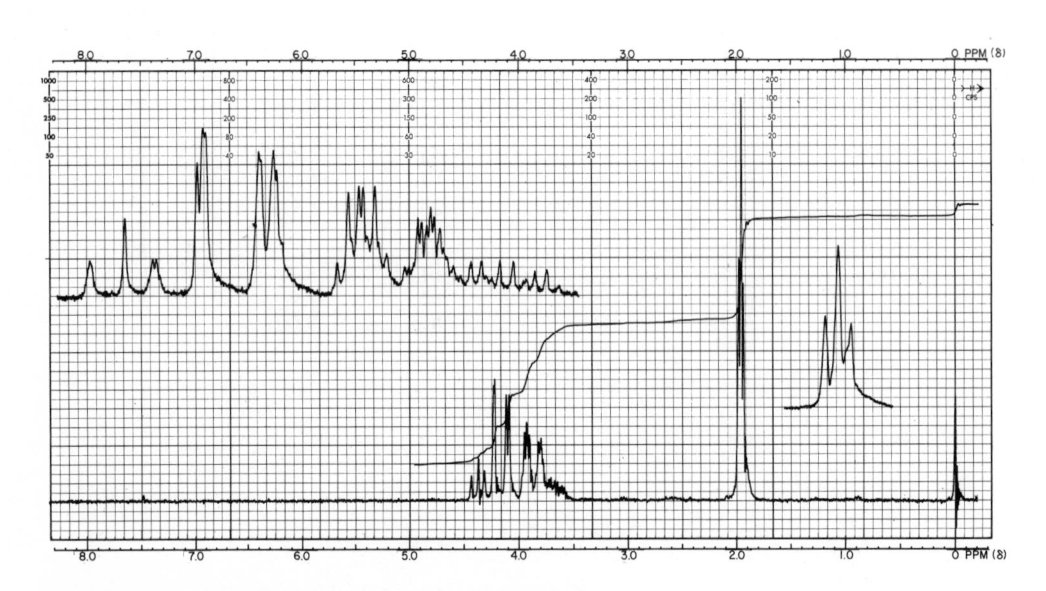

NUCLEAR MAGNETIC RESONANCE SPECTRUM:
15% in deuterochloroform, 500 Hz sweep width
Expansions: (left) 100 Hz sweep width, 170 Hz sweep offset
(right) 100 Hz sweep width, 105 Hz sweep offset

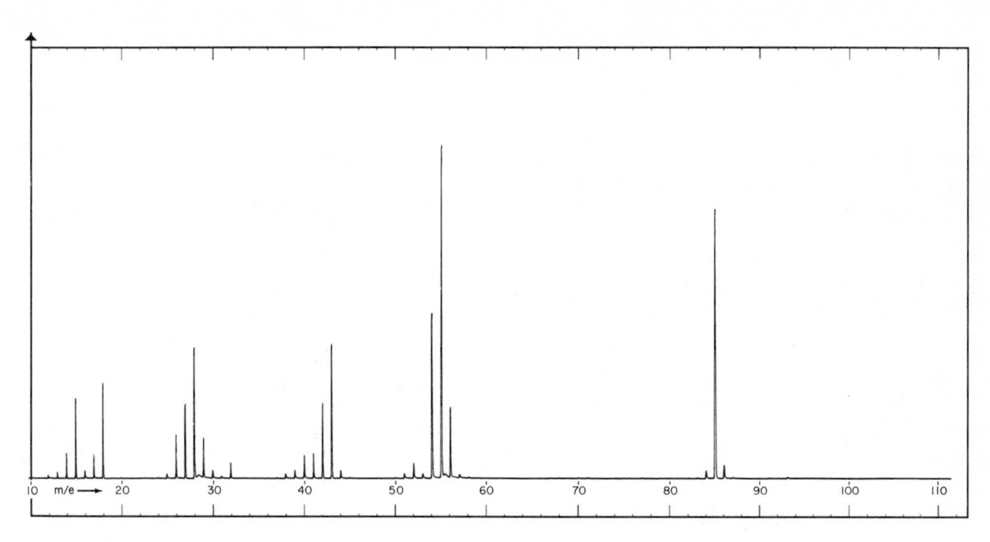

MASS SPECTRUM

MASS SPECTRAL DATA

m/e	*relative intensity*	*m/e*	*relative intensity*	*m/e*	*relative intensity*	*m/e*	*relative intensity*
12	0.4	26	7.1	39	1.2	54	29.0
13	0.7	27	13.4	40	3.0	55	100.0
14	4.8	28	30.0	41	3.7	56	10.8
15	30.0	29	6.0	42	10.7	57	0.6
16	1.0	30	1.1	43	24.0	84	1.1
17	3.5	31	0.4	44	1.3	85	44.0
18	20.0	32	2.2	51	0.7	86	2.2
25	0.8	38	0.7	53	0.8	87	0.1

COMPOUND 3-8

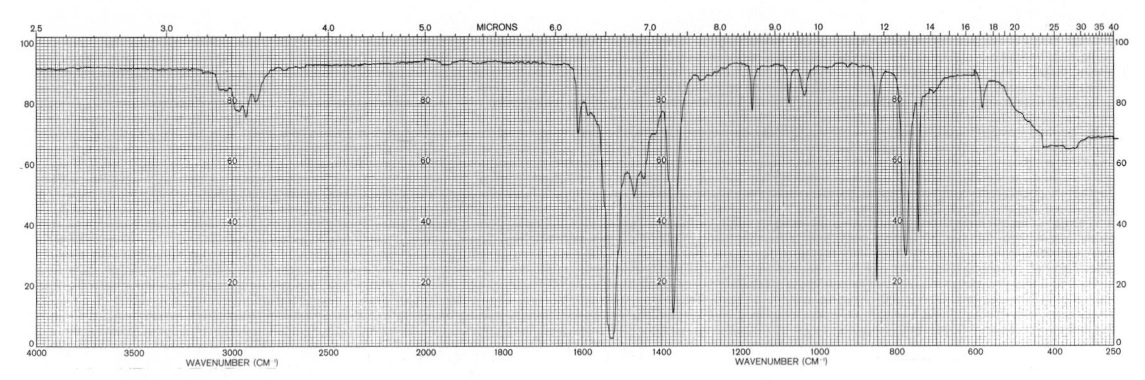

INFRARED SPECTRUM: Liquid film

ULTRAVIOLET DATA: λ_{max}^{EtOH} 208 nm (ϵ 12,500); 254(sh) nm (ϵ 1,000)

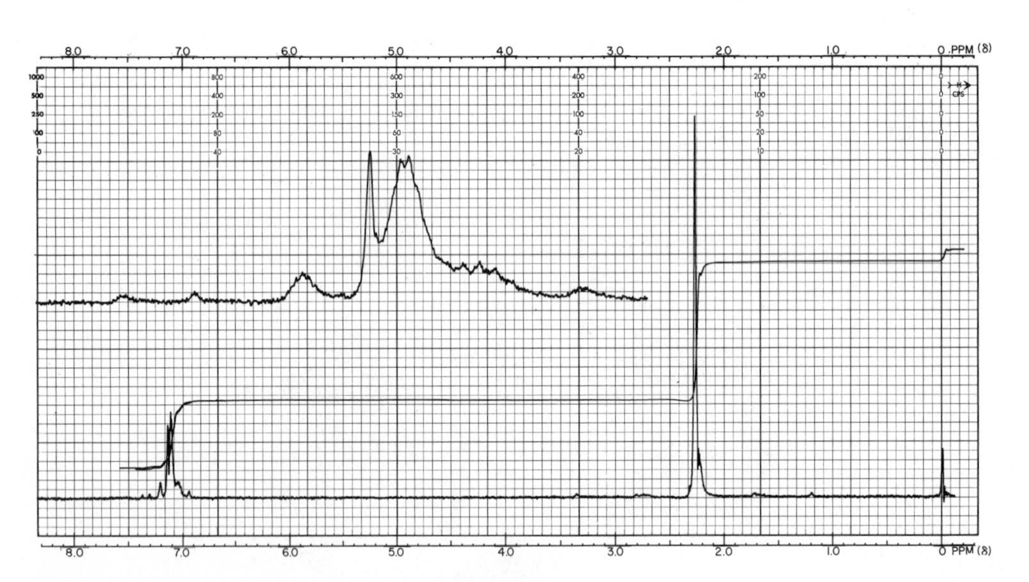

NUCLEAR MAGNETIC RESONANCE SPECTRUM:
15% in carbon tetrachloride, 500 Hz sweep width
Expansion: 50 Hz sweep width, 395 Hz sweep offset

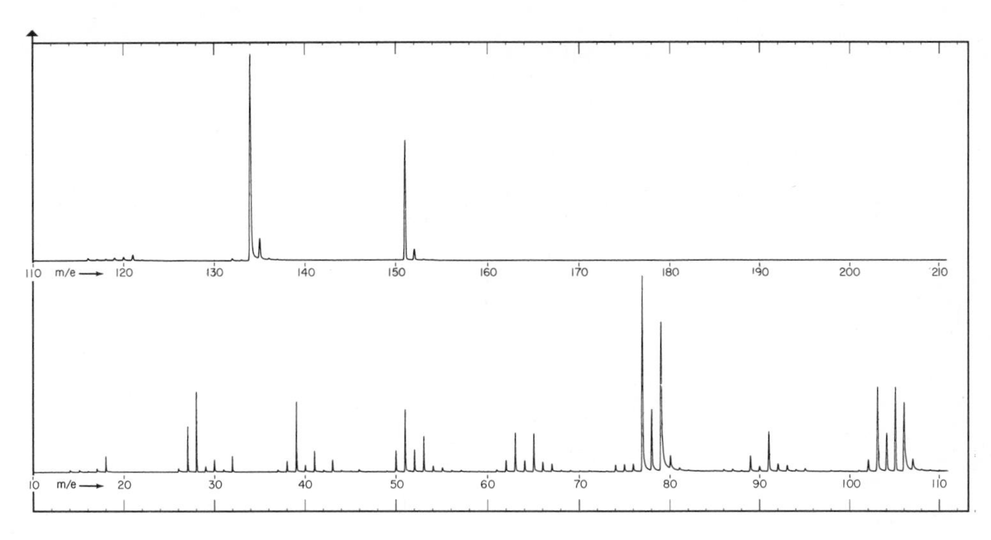

MASS SPECTRUM

MASS SPECTRAL DATA

m/e	*relative intensity*	*m/e*	*relative intensity*	*m/e*	*relative intensity*	*m/e*	*relative intensity*
17	0.4	53	9.2	80	7.1	106	31.0
18	2.7	54	1.7	81	1.8	107	5.9
26	1.3	55	1.3	82	0.8	108	1.3
27	14.8	56	0.6	85	0.7	110	1.0
28	12.3	57	0.5	86	1.2	116	1.5
29	2.0	61	1.0	87	1.5	117	1.0
30	4.9	62	4.6	88	0.5	118	1.1
31	1.0	63	15.4	89	6.7	119	1.5
32	1.9	64	4.7	90	2.5	120	2.1
37	0.9	65	14.4	91	16.4	121	3.3
38	3.2	66	4.0	92	3.6	122	0.8
39	24.7	67	3.0	93	2.7	132	1.3
40	2.6	68	0.7	94	1.1	133	0.7
41	5.6	69	0.7	95	1.2	134	100.0
42	0.7	74	3.0	98	0.6	135	10.7
43	3.7	75	3.2	101	0.8	136	1.5
44	0.6	76	3.3	102	5.5	137	0.8
46	1.1	77	82.0	103	35.0	151	59.0
50	7.7	78	25.2	104	16.1	152	5.3
51	21.7	79	62.0	105	35.0	153	0.8
52	8.1						

COMPOUND 3-9

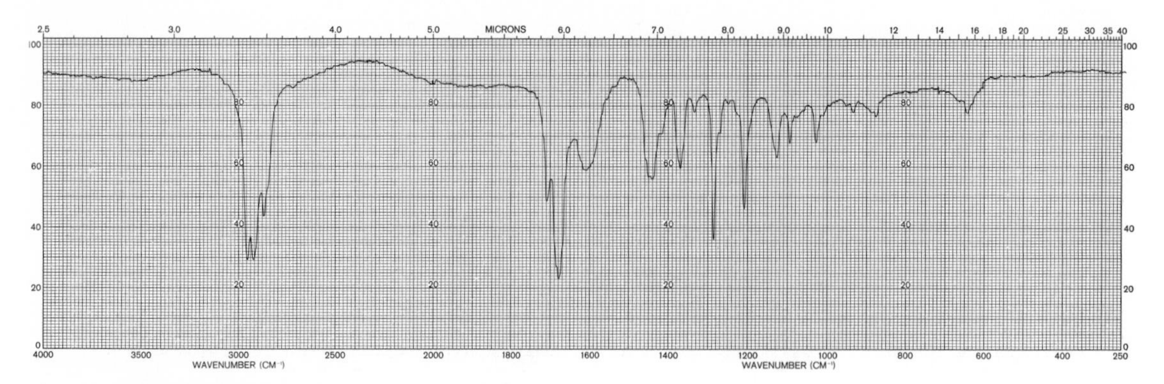

INFRARED SPECTRUM: Liquid film

ULTRAVIOLET DATA: λ_{max}^{EtOH} 253 nm (ϵ 8,150); 316 nm (ϵ 85)

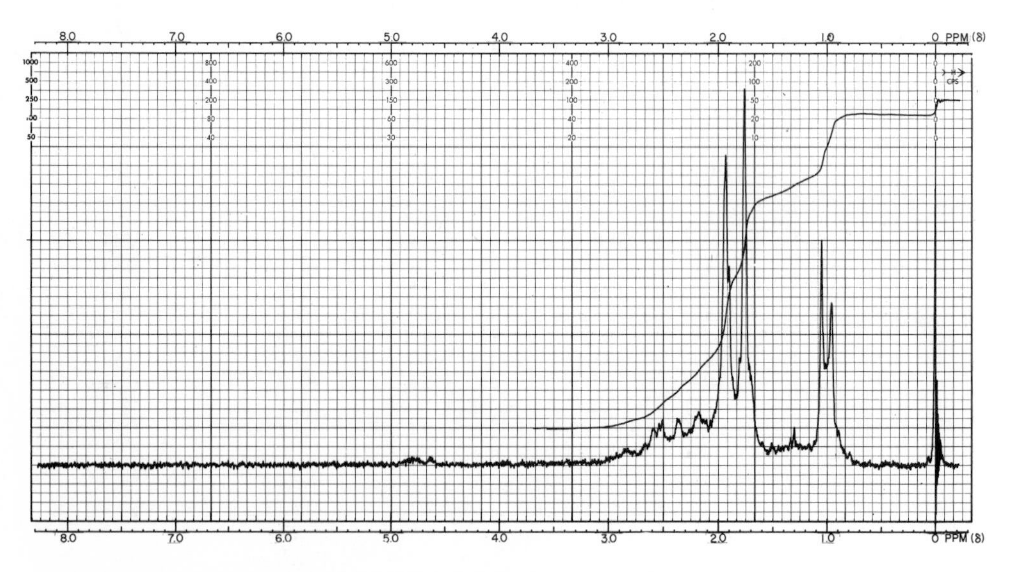

NUCLEAR MAGNETIC RESONANCE SPECTRUM:
15% in carbon tetrachloride, 500 Hz sweep width

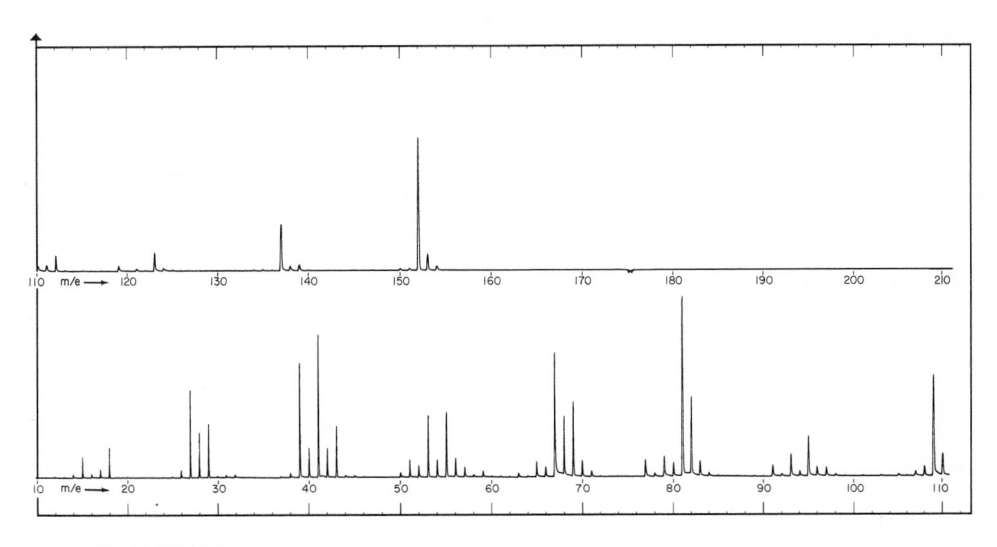

MASS SPECTRUM

MASS SPECTRAL DATA

m/e	*relative intensity*	*m/e*	*relative intensity*	*m/e*	*relative intensity*	*m/e*	*relative intensity*
14	2.6	54	9.5	80	7.9	112	11.6
15	10.1	55	34.0	81	100.0	113	1.1
16	2.0	56	10.9	82	42.0	115	0.8
17	3.2	57	6.8	83	9.5	117	0.7
18	13.6	58	1.5	84	2.8	119	4.3
25	0.9	59	3.7	85	0.8	120	0.8
26	5.2	60	0.6	86	0.6	121	1.9
27	47.0	61	0.4	91	7.0	122	0.8
28	28.0	62	0.6	92	1.9	123	12.6
29	30.0	63	2.7	93	13.8	124	2.4
30	0.9	64	0.9	94	3.9	125	1.4
31	1.4	65	9.5	95	21.0	126	0.6
32	1.7	66	5.7	96	6.2	133	0.7
37	0.7	67	74.0	97	5.9	134	1.0
38	2.8	68	34.0	98	1.6	135	1.4
39	58.0	69	42.0	99	0.5	136	0.9
40	15.2	70	9.2	101	0.5	137	31.0
41	87.0	71	4.0	103	0.5	138	3.5
42	15.5	72	0.7	104	0.5	139	4.3
43	28.0	73	0.3	105	1.6	140	0.9
44	1.4	74	0.3	106	0.6	150	1.5
45	1.1	75	0.4	107	3.5	151	1.6
50	2.7	76	0.3	108	6.2	152	92.0
51	10.2	77	9.6	109	59.0	153	11.9
52	7.1	78	2.4	110	13.5	154	3.4
53	34.0	79	12.4	111	4.6	155	0.5

COMPOUND 3-10

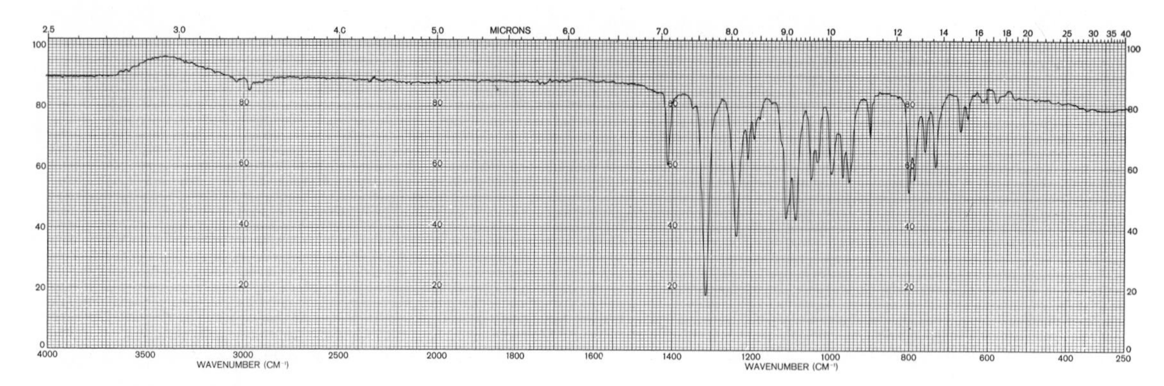

INFRARED SPECTRUM: Liquid film

ULTRAVIOLET DATA: λ_{max}^{EtOH} 243 nm (ϵ 505)

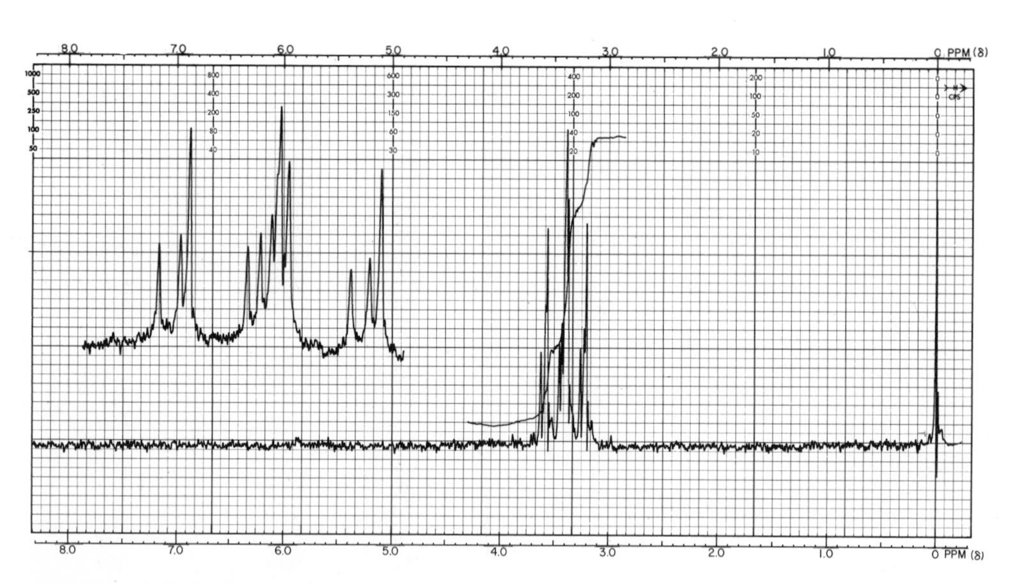

NUCLEAR MAGNETIC RESONANCE SPECTRUM:
15% in carbon tetrachloride, 500 Hz sweep width
Expansion: 100 Hz sweep width; 130 Hz sweep offset

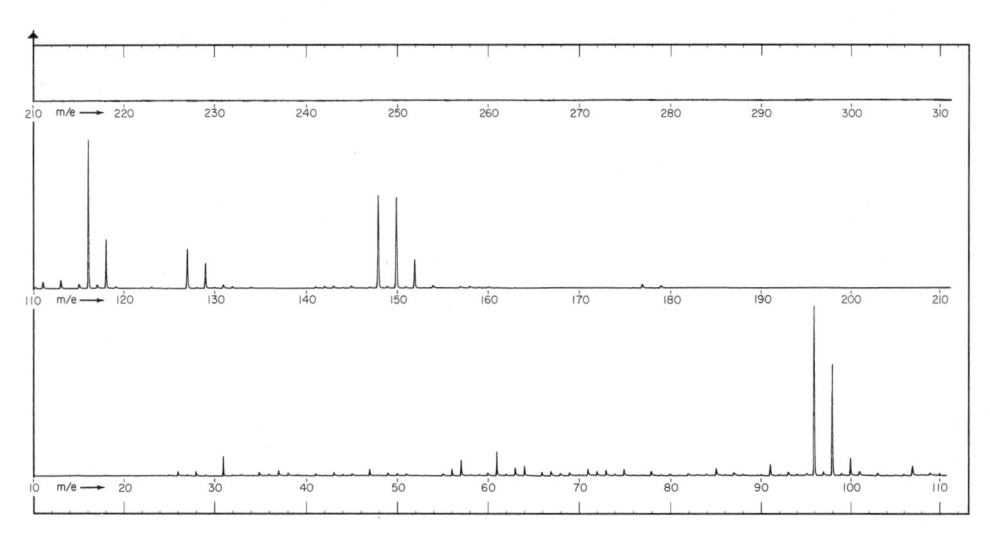

MASS SPECTRUM

MASS SPECTRAL DATA

m/e	relative intensity	m/e	relative intensity	m/e	relative intensity	m/e	relative intensity
25	0.5	64	5.9	96	100.0	131	3.2
26	1.8	65	0.3	97	2.9	132	1.7
27	0.7	66	3.0	98	70.0	134	1.1
28	2.0	67	3.0	99	1.7	141	1.4
29	0.6	68	2.1	100	11.0	142	2.0
31	8.7	69	2.2	101	3.2	143	2.3
32	0.6	70	0.8	103	2.2	144	0.9
33	0.7	71	3.9	104	0.3	145	2.3
35	1.9	72	2.9	105	0.6	147	1.0
36	1.2	73	3.8	106	1.1	148	74.0
37	2.6	74	1.1	107	7.6	149	1.6
38	1.6	75	4.2	108	0.5	150	71.0
39	0.7	76	0.3	109	2.2	151	1.6
40	0.3	77	0.4	110	1.3	152	22.3
41	1.3	78	3.3	111	4.9	153	0.7
43	2.1	79	0.6	112	0.4	154	2.5
44	0.9	80	1.4	113	6.4	157	1.4
45	1.1	81	0.5	114	0.3	158	1.8
47	3.5	82	1.7	115	3.4	159	1.1
48	0.4	83	0.7	116	100.0	160	1.2
49	2.0	84	1.2	117	3.1	161	0.4
50	1.4	85	4.8	118	35.0	163	0.6
51	1.3	86	0.5	119	1.7	165	0.4
53	0.5	87	2.3	121	0.3	176	0.6
55	1.7	88	0.8	122	0.8	177	3.7
56	4.2	89	0.3	123	1.3	178	0.6
57	9.3	90	0.7	124	0.3	179	2.5
58	0.6	91	7.9	125	0.6	180	0.3
59	0.9	92	1.0	127	29.0	181	0.4
60	1.9	93	3.0	128	1.2	212	0.6
61	13.6	94	1.1	129	18.7	214	0.6
62	1.2	95	1.7	130	0.8	216	0.2
63	4.5						

COMPOUND 3-11

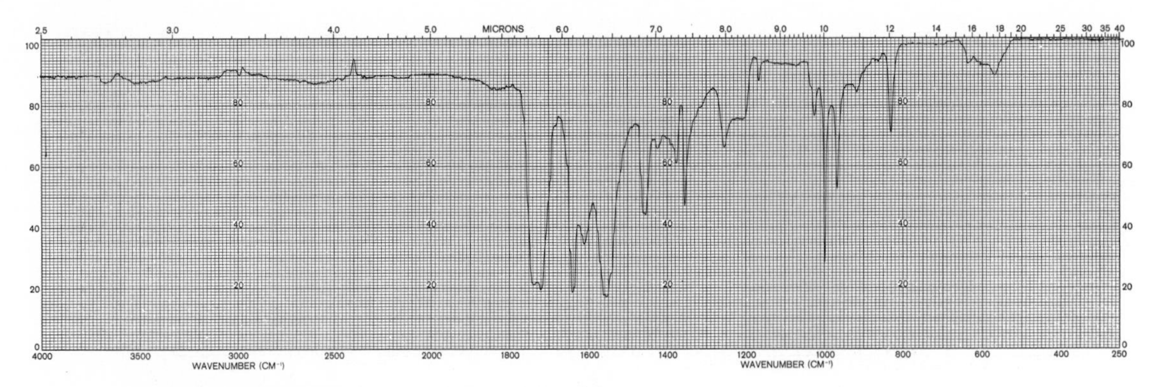

INFRARED SPECTRUM: 1.0% in chloroform

ULTRAVIOLET DATA: λ_{max}^{EtOH} 235 nm (ϵ 12,600); 310 nm (ϵ 10,000).

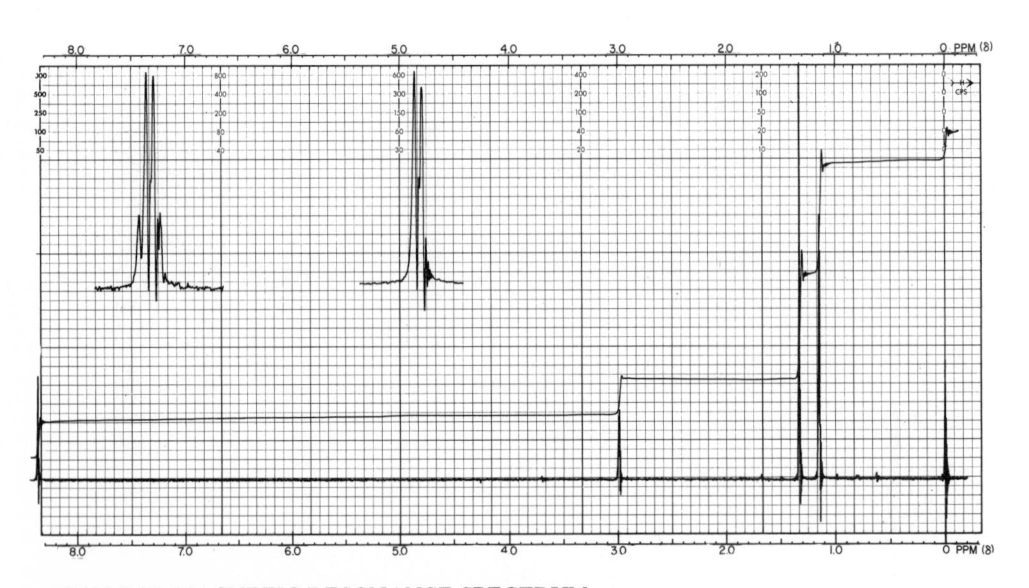

NUCLEAR MAGNETIC RESONANCE SPECTRUM:
15% in deuterochloroform, 1000 Hz sweep width
Expansions: (left) 100 Hz sweep width, 270 Hz sweep offset
(right) 100 Hz sweep width, 80 Hz sweep offset

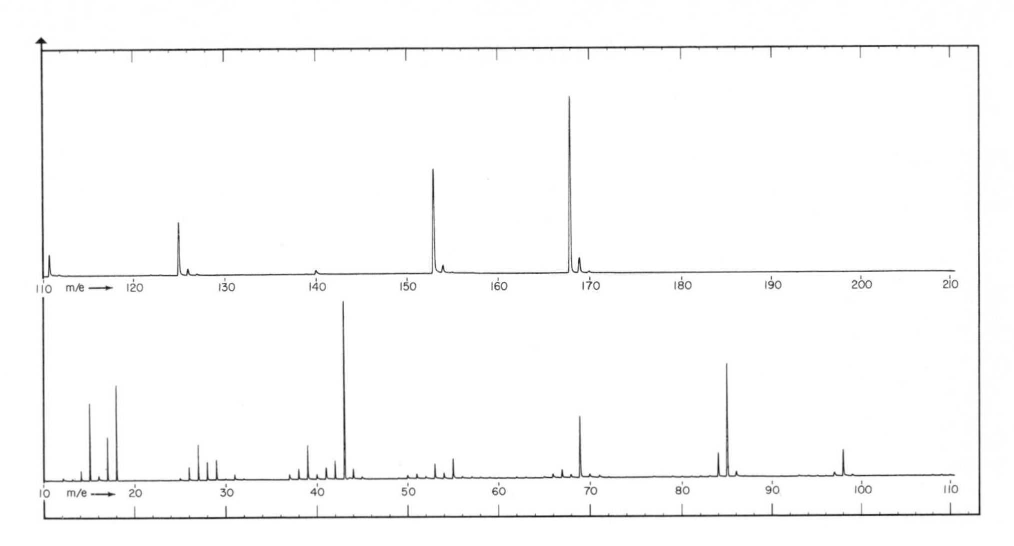

MASS SPECTRUM

MASS SPECTRAL DATA

m/e	*relative intensity*	*m/e*	*relative intensity*	*m/e*	*relative intensity*	*m/e*	*relative intensity*
12	1.2	39	10.9	59	0.5	97	1.4
13	0.8	40	2.2	60	0.3	98	9.9
14	3.6	41	4.1	61	0.4	99	0.7
15	25.0	42	6.3	62	0.5	111	6.7
16	0.9	43	100.0	63	0.5	112	0.6
17	6.0	44	3.3	65	0.7	125	15.6
18	26.1	45	1.0	66	1.4	126	2.4
25	1.4	49	0.3	67	2.8	127	0.9
26	4.8	50	1.4	68	1.4	135	1.5
27	11.2	51	2.0	69	18.2	153	25.0
28	5.4	52	1.2	70	1.4	154	2.7
29	7.0	53	5.5	71	1.2	155	0.5
30	0.7	54	2.5	84	7.8	168	45.0
31	1.9	55	6.8	85	28.0	169	5.1
37	1.7	56	0.7	86	2.0	170	0.8
38	3.7	58	0.6				

COMPOUND 3-12

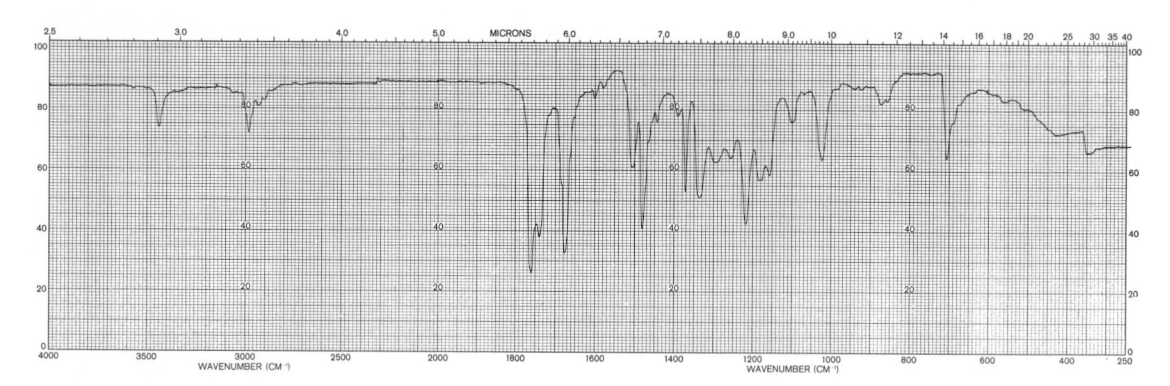

INFRARED SPECTRUM: 1.1% in carbon tetrachloride

ULTRAVIOLET DATA: λ_{max}^{EtOH} 227 nm (ϵ 11,800); 270(sh) nm (ϵ 640)

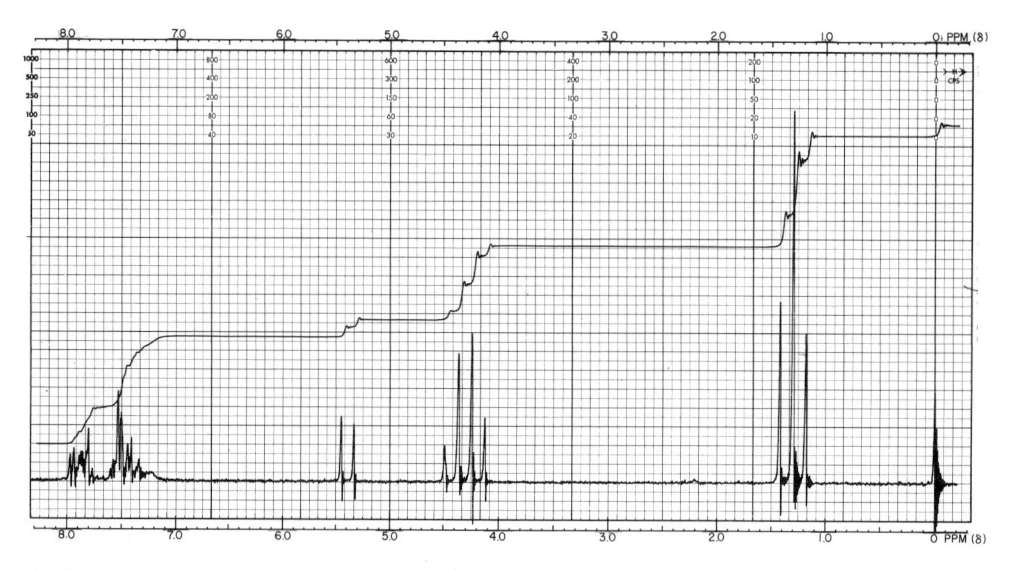

NUCLEAR MAGNETIC RESONANCE SPECTRUM:
15% in deuterochloroform, 500 Hz sweep width

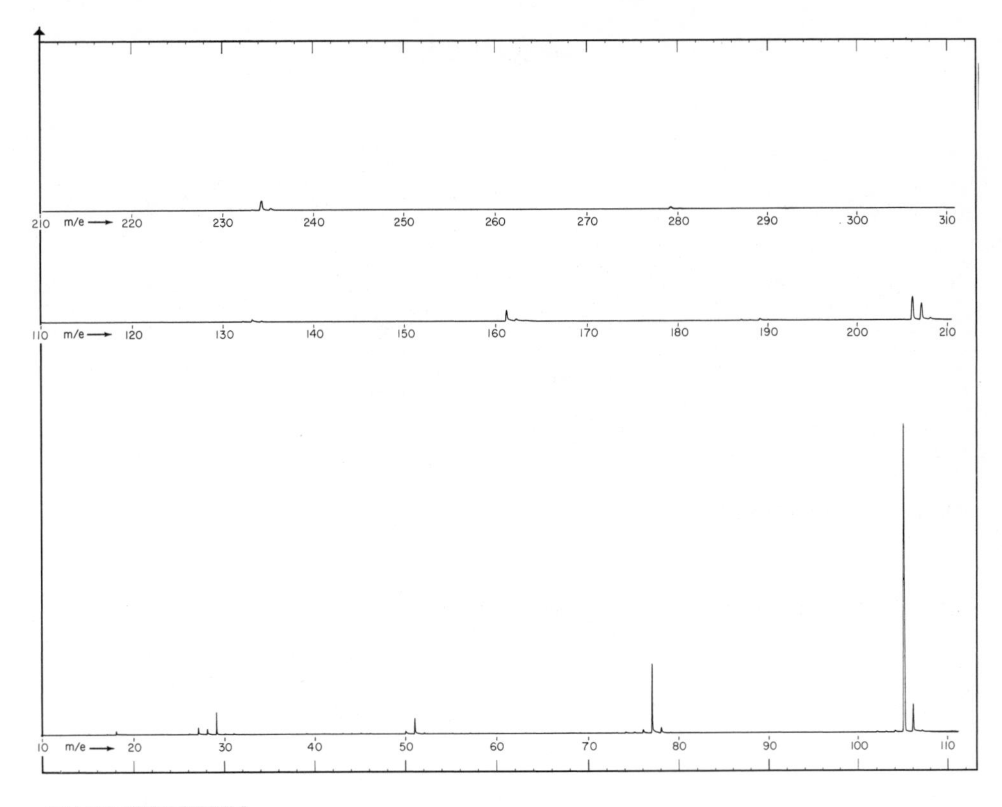

MASS SPECTRUM

MASS SPECTRAL DATA

m/e	*relative intensity*	*m/e*	*relative intensity*	*m/e*	*relative intensity*	*m/e*	*relative intensity*
18	0.7	50	0.2	105	100.0	189	2.3
26	0.5	51	8.7	106	16.0	190	0.7
27	4.1	52	0.8	107	1.2	206	19.0
28	3.7	74	0.8	132	0.7	207	15.0
29	11.9	75	0.8	133	2.6	208	2.3
30	0.6	76	2.5	134	1.1	233	0.7
31	1.0	77	41.0	161	10.6	234	6.8
39	0.5	78	3.6	162	2.5	235	1.6
43	0.5	102	0.8	163	0.5	236	0.5
44	0.5	103	0.7	187	0.8	279	1.3
45	0.7	104	1.2	188	0.6	280	0.4

COMPOUND 3-13

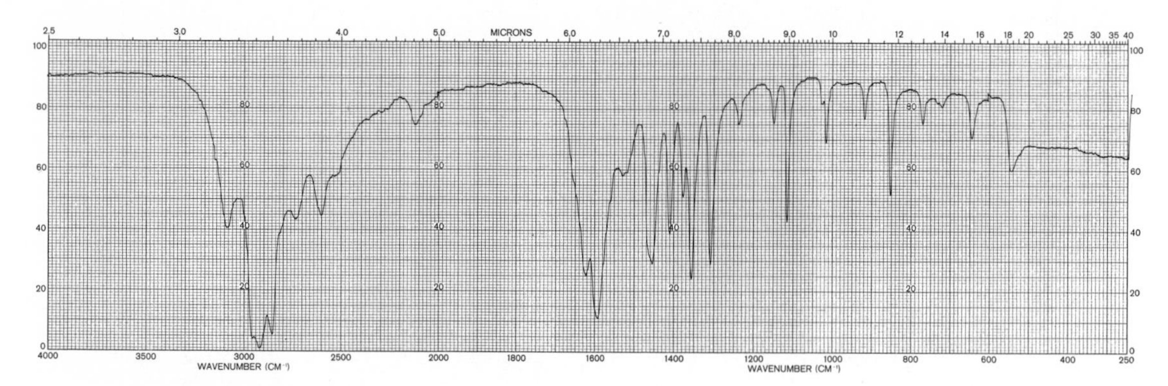

INFRARED SPECTRUM: Nujol mull

ULTRAVIOLET DATA: no λ_{max} above 205 nm

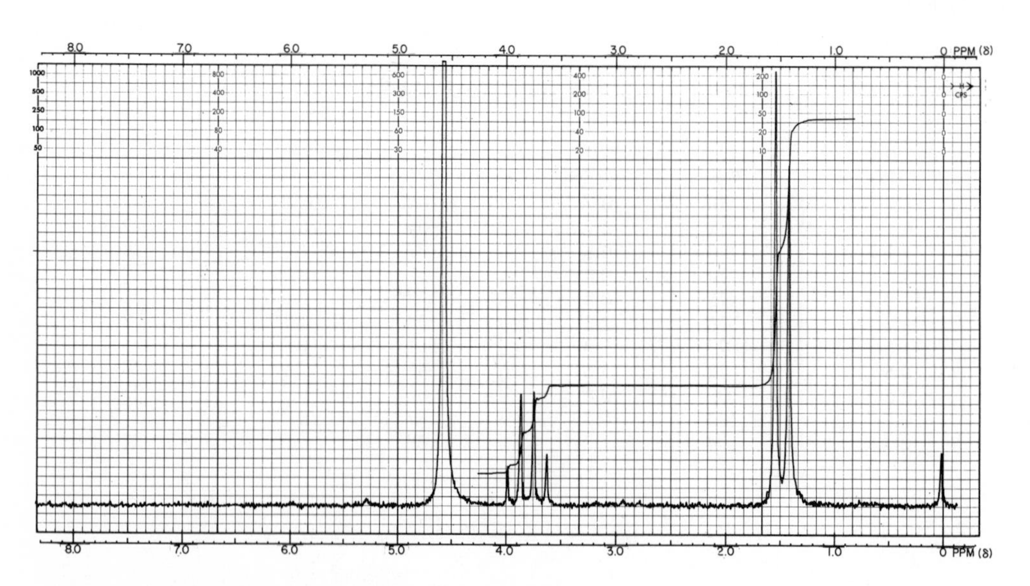

NUCLEAR MAGNETIC RESONANCE SPECTRUM:
15% in deuterium oxide, 60°, 500 Hz sweep width

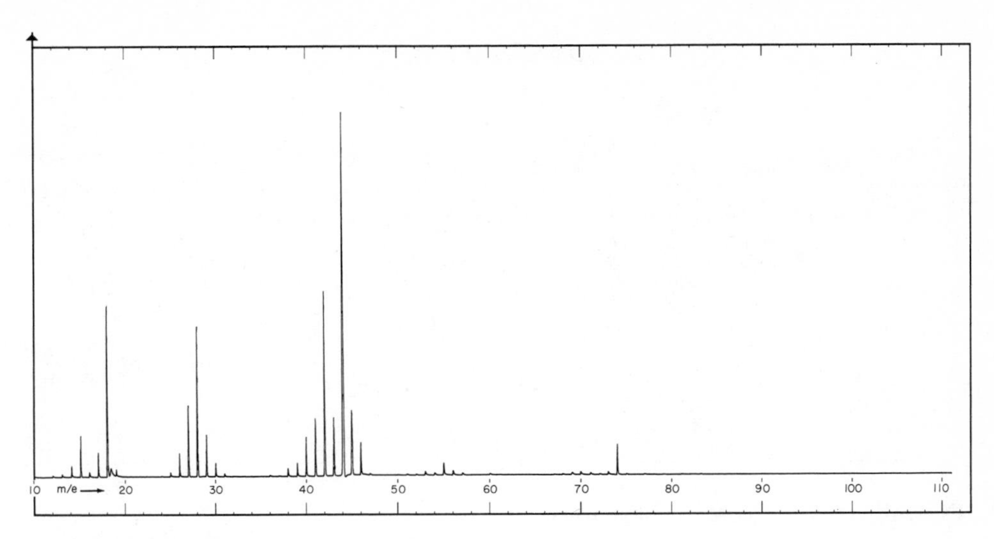

MASS SPECTRUM

MASS SPECTRAL DATA

m/e	*relative intensity*	*m/e*	*relative intensity*	*m/e*	*relative intensity*	*m/e*	*relative intensity*
14	0.5	27	6.6	40	3.3	46	3.1
15	3.3	28	20.9	41	5.7	53	0.4
16	0.5	29	4.2	42	18.2	54	0.4
17	1.8	30	1.2	43	5.9	55	1.3
18	16.9	38	0.7	44	100.0	56	0.5
19	0.7	39	1.1	45	8.6	74	3.1
26	1.8						

COMPOUND 3-14

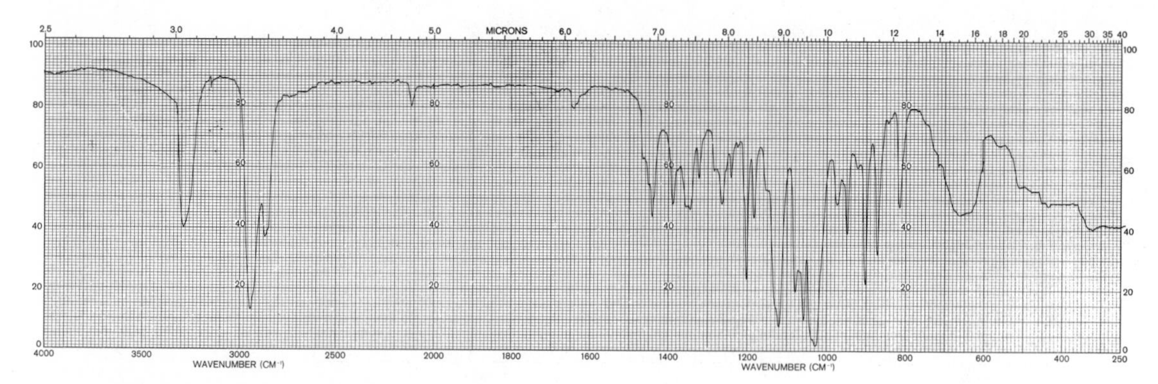

INFRARED SPECTRUM: Liquid film

ULTRAVIOLET DATA: λ_{max}^{EtOH} 242(sh) nm (ϵ 39)

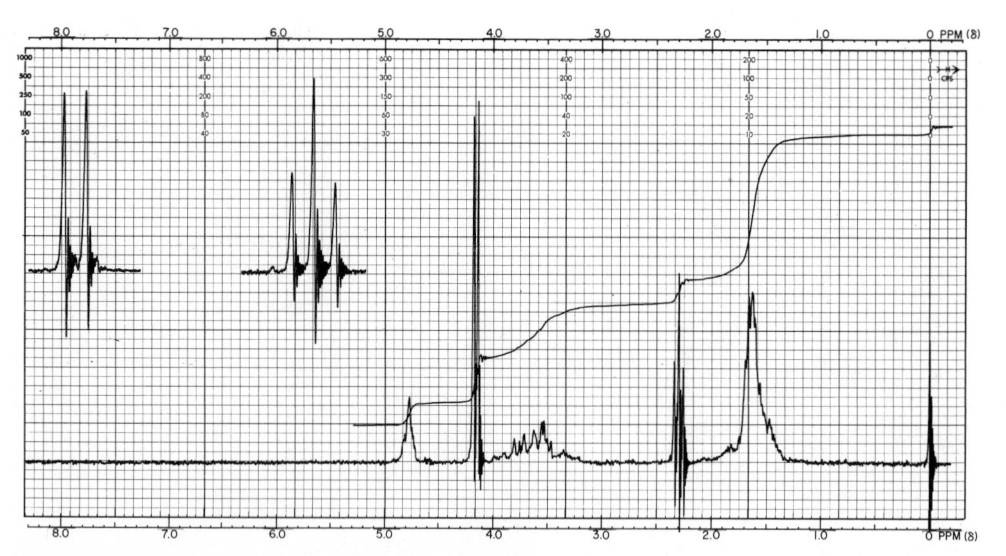

NUCLEAR MAGNETIC RESONANCE SPECTRUM:
15% in carbon tetrachloride, 500 Hz sweep width
Expansions: (left) 100 Hz sweep width, 155 Hz sweep offset
(right) 100 Hz sweep width, 70 Hz sweep offset

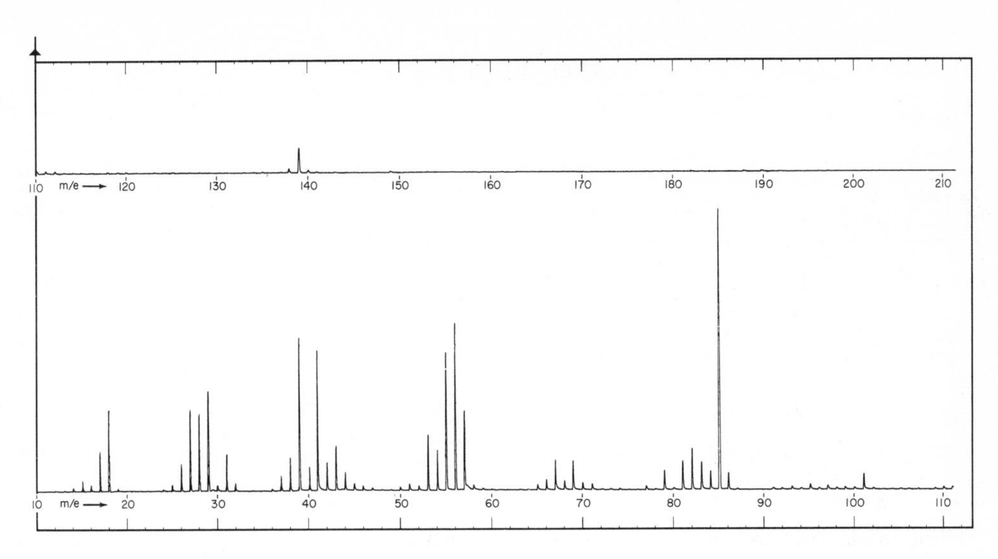

MASS SPECTRUM

MASS SPECTRAL DATA

m/e	*relative intensity*	*m/e*	*relative intensity*	*m/e*	*relative intensity*	*m/e*	*relative intensity*
14	0.6	42	8.5	68	3.1	95	2.0
15	1.9	43	13.6	69	9.1	96	0.7
16	0.8	44	5.0	70	2.8	97	2.0
17	5.5	45	1.9	71	2.2	98	0.8
18	20.5	46	0.7	73	0.5	99	1.2
25	1.4	47	0.9	74	0.5	100	0.8
26	5.6	50	1.3	77	1.3	101	6.2
27	22.3	51	1.9	79	6.1	102	0.8
28	18.3	52	1.9	81	9.5	109	0.6
29	29.0	53	17.2	82	13.2	110	1.2
30	1.8	54	11.1	83	9.5	111	1.1
31	8.3	55	45.0	84	6.5	112	1.2
32	1.5	56	49.0	85	100.0	113	0.4
36	0.8	57	25.0	86	6.3	118	0.6
37	3.9	58	1.8	87	0.6	119	0.6
38	7.7	59	0.8	91	0.9	138	1.1
39	58.0	65	2.2	92	0.6	139	10.6
40	6.1	66	3.5	93	1.1	140	1.3
41	46.0	67	8.7	94	0.8	141	0.5

COMPOUND 3-15

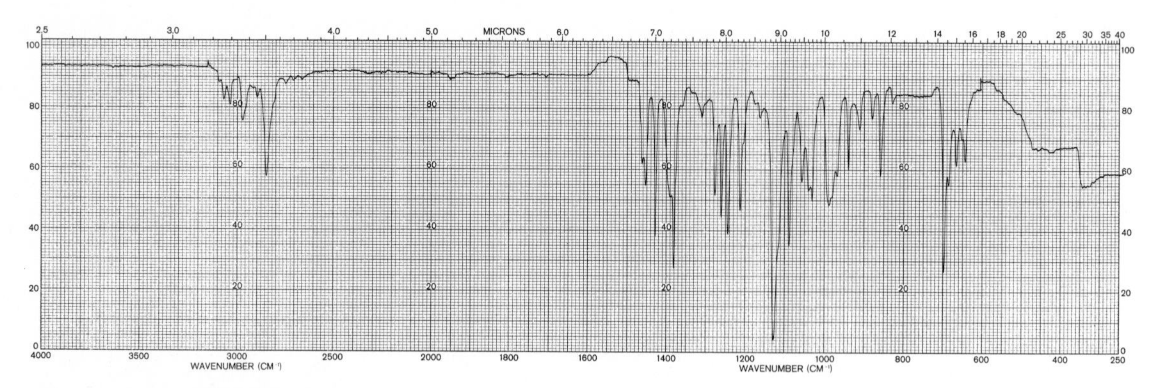

INFRARED SPECTRUM: 3.9% in carbon tetrachloride

ULTRAVIOLET DATA: λ_{max}^{EtOH} 205 nm (ϵ 6,000); 245 nm (ϵ 230); 250 nm (ϵ 320); 255 nm (ϵ 250); 259 nm (ϵ 175); 262 nm (ϵ 175); 266 nm (ϵ 100).

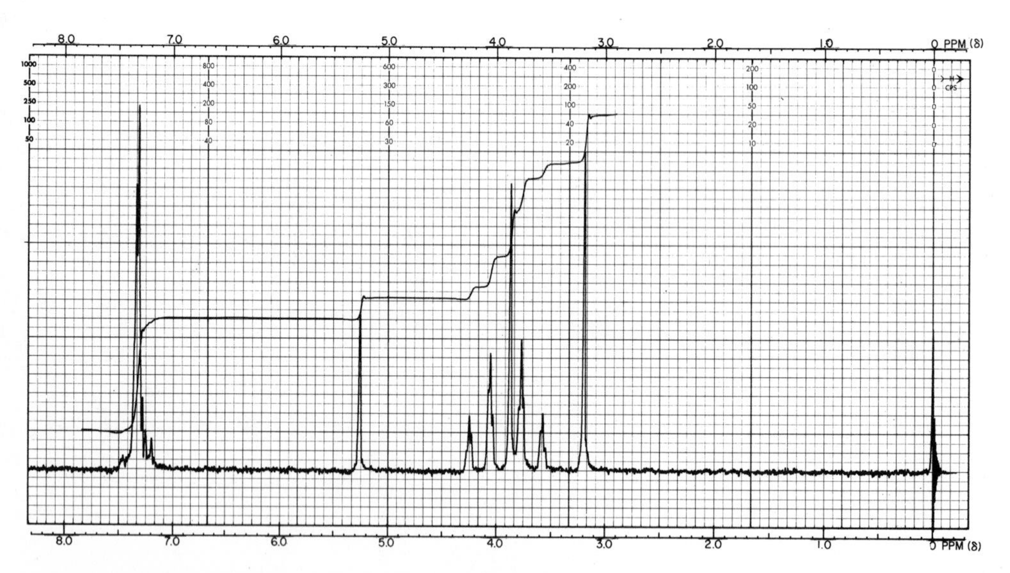

NUCLEAR MAGNETIC RESONANCE SPECTRUM:
15% in carbon tetrachloride, 500 Hz sweep width

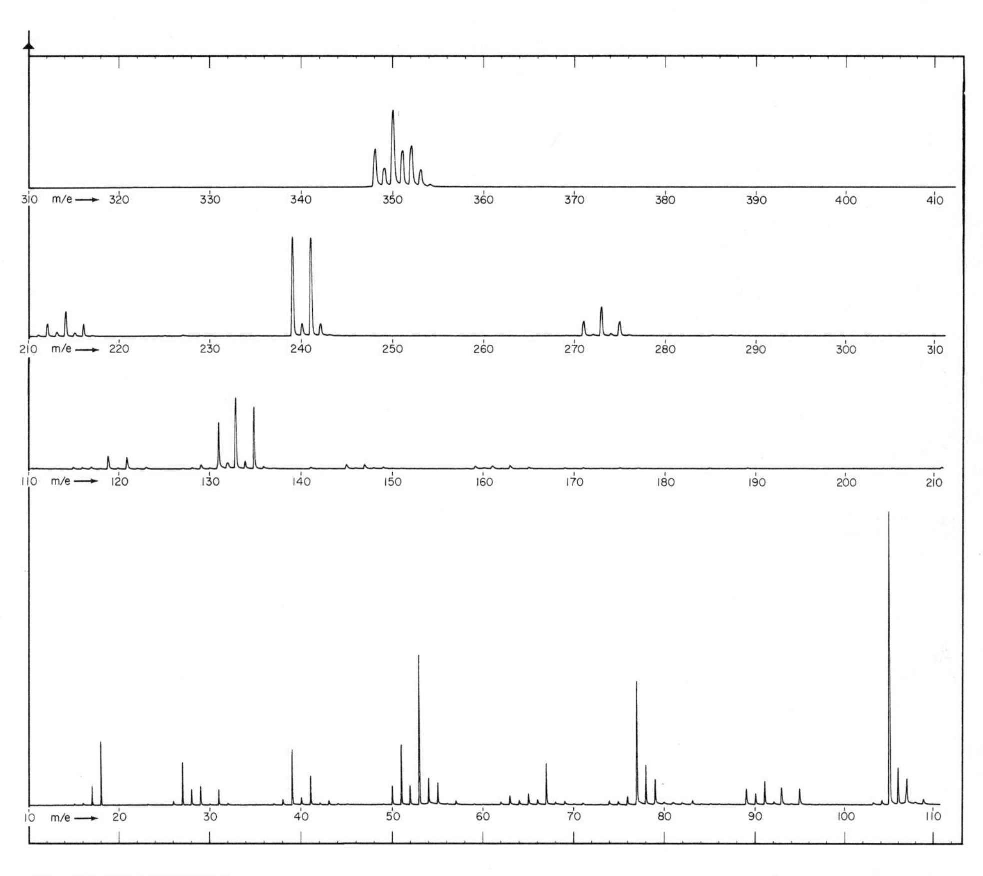

MASS SPECTRUM

MASS SPECTRAL DATA

m/e	relative intensity	m/e	relative intensity	m/e	relative intensity	m/e	relative intensity
17	0.7	68	0.8	119	4.2	169	0.7
18	2.2	69	0.5	120	0.5	171	0.5
26	0.8	74	0.8	121	3.9	211	0.6
27	8.6	75	1.2	122	0.4	212	2.2
28	3.2	76	2.7	123	0.5	213	1.2
29	4.4	77	35.9	128	0.4	214	4.5
30	0.3	78	11.9	129	1.6	215	1.1
31	2.0	79	8.0	130	0.6	216	2.3
38	1.3	80	0.9	131	16.4	217	0.4
39	13.9	81	0.6	132	2.5	239	34.3
40	1.8	82	0.8	133	21.6	240	4.7
41	5.1	83	1.1	134	2.8	241	34.3
42	0.4	89	4.9	135	20.5	242	4.5
43	0.9	90	3.8	136	0.9	243	0.6
50	4.5	91	7.2	137	0.3	271	5.3
51	14.8	92	1.0	141	0.7	272	0.8
52	4.9	93	4.5	145	1.8	273	10.9
53	40.0	95	4.5	146	0.4	274	1.1
54	7.6	103	0.7	147	1.6	275	5.3
55	4.9	104	1.8	148	0.4	276	0.6
56	0.6	105	100.0	149	0.6	348	40.8
57	0.9	106	12.1	159	1.1	349	20.5
62	0.8	107	9.4	160	0.5	350	84.0
63	2.6	108	1.3	161	1.3	351	38.8
64	1.5	109	2.2	162	0.2	352	46.0
65	3.0	116	0.7	163	1.2	353	19.3
66	1.4	117	0.7	165	0.4	354	3.0
67	11.1	118	0.8				

COMPOUND 3-16

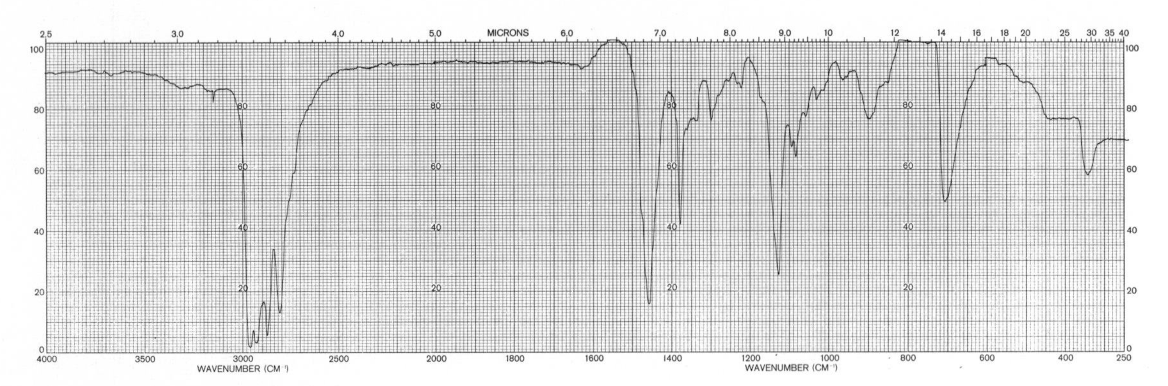

INFRARED SPECTRUM: 5.4% in carbon tetrachloride

ULTRAVIOLET DATA: no λ_{max} above 205 nm

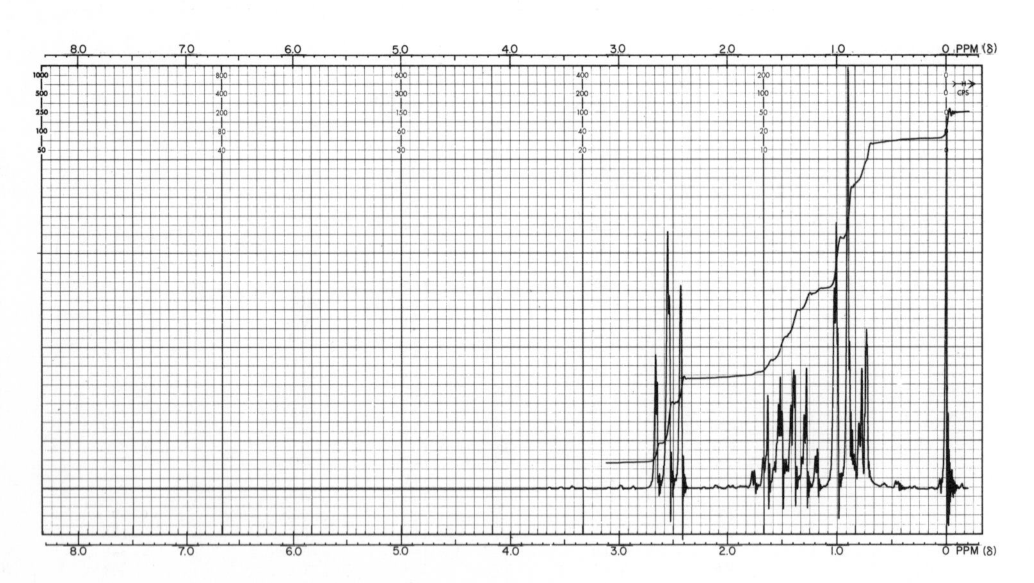

NUCLEAR MAGNETIC RESONANCE SPECTRUM:
Pure liquid, 500 Hz sweep width

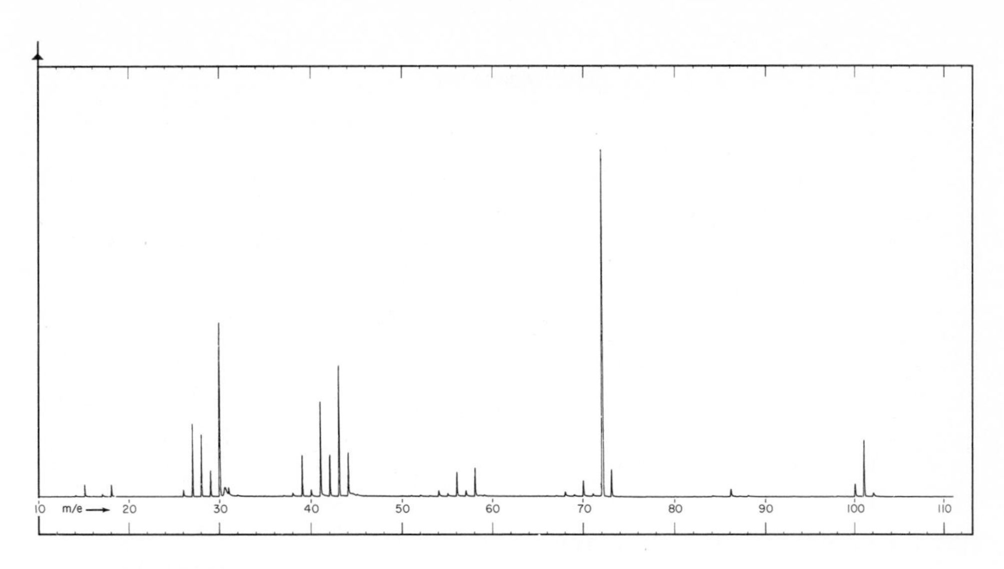

MASS SPECTRUM

MASS SPECTRAL DATA

m/e	relative intensity	m/e	relative intensity	m/e	relative intensity	m/e	relative intensity
15	1.1	31	1.4	54	0.9	72	100.0
17	0.2	38	0.6	55	0.4	73	4.9
18	1.2	39	5.9	56	3.7	86	1.5
26	0.9	40	1.0	57	1.1	88	0.5
27	10.1	41	13.7	58	4.2	100	2.2
28	8.4	42	5.6	68	1.0	101	10.4
29	3.7	43	20.4	70	2.7	102	0.8
30	40.0	44	5.8	71	0.6		

COMPOUND 3-17

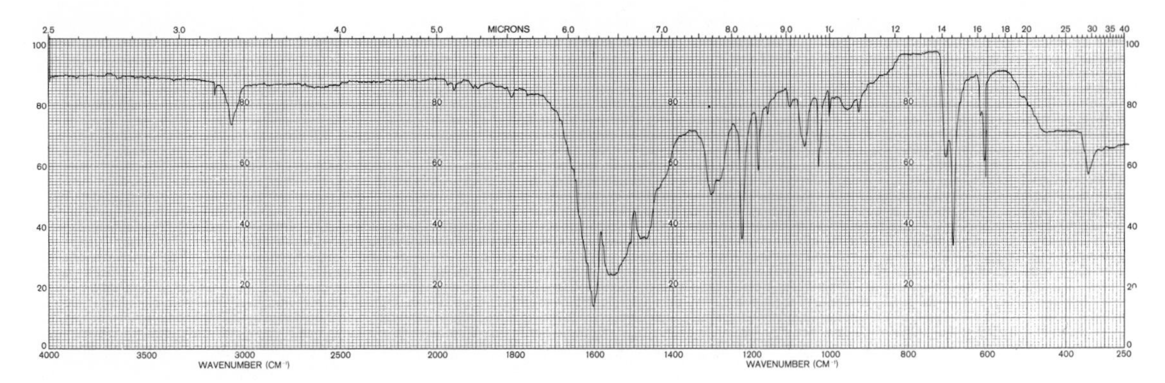

INFRARED SPECTRUM: 2.1% in carbon tetrachloride

ULTRAVIOLET DATA: λ_{max}^{EtOH} 254 nm (ϵ 10,000); 345 nm (ϵ 40,000).

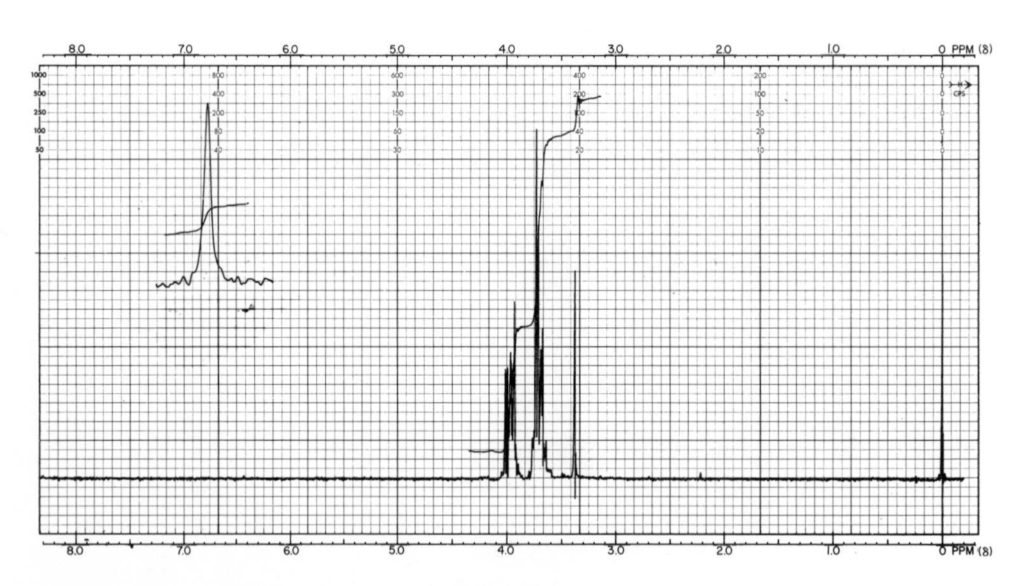

NUCLEAR MAGNETIC RESONANCE SPECTRUM:
15% in carbon tetrachloride, 1000 Hz sweep width
Insert: 1000 Hz sweep width, 200 Hz sweep offset, amplitude × 10

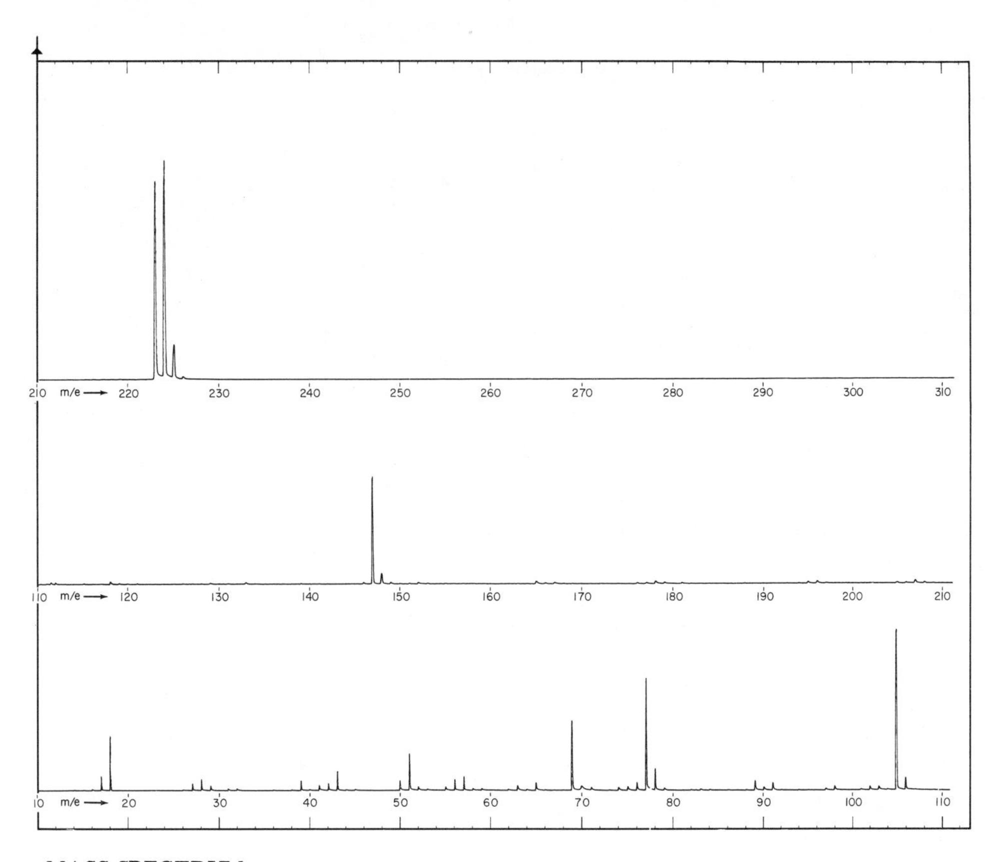

MASS SPECTRUM

MASS SPECTRAL DATA

m/e	*relative intensity*	*m/e*	*relative intensity*	*m/e*	*relative intensity*	*m/e*	*relative intensity*
17	2.2	56	1.3	91	3.1	165	1.6
18	9.6	57	1.7	97	1.1	166	0.6
27	1.3	63	1.7	98	2.1	167	0.9
28	2.9	64	0.6	101	0.6	176	0.8
29	0.9	65	2.8	102	2.0	177	0.7
31	0.7	69	24.1	103	2.0	178	1.5
32	0.6	70	2.3	104	0.8	179	0.7
36	0.6	71	1.0	105	62.0	181	0.8
38	0.6	73	0.5	106	5.0	195	1.3
39	2.6	74	1.6	107	0.2	196	1.4
41	6.1	75	1.6	111.5	1.2	205	0.8
42	4.5	76	3.0	112	1.0	206	0.6
43	5.1	77	38.0	118	1.5	207	1.7
45	0.7	78	7.2	133	0.8	208	0.9
50	3.2	79	1.0	146	1.0	223	90.0
51	10.3	83	0.9	147	46.0	224	100.0
52	1.5	86	1.0	148	4.8	225	16.7
53	0.7	89	4.0	149	0.9	226	1.6
55	0.7	90	1.7	152	1.0		

COMPOUND 3-18

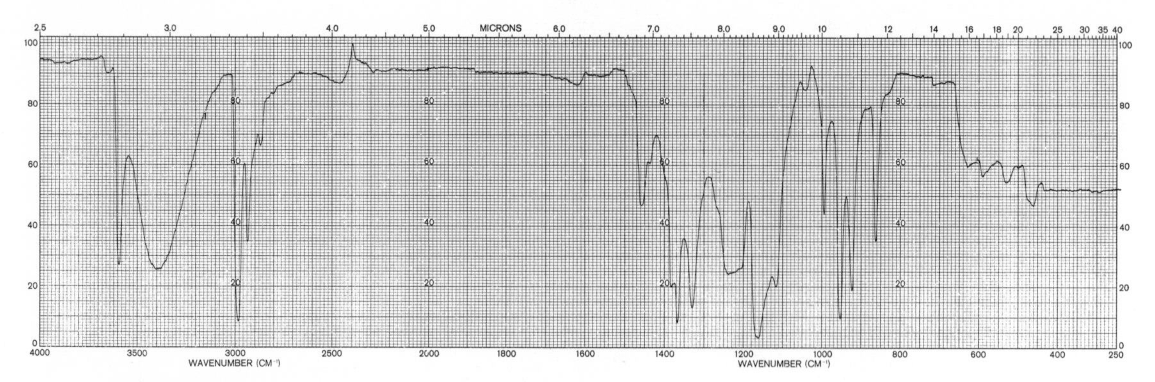

INFRARED SPECTRUM: 5.3% in chloroform

ULTRAVIOLET DATA: no λ_{max}^{EtOH} above 205 nm

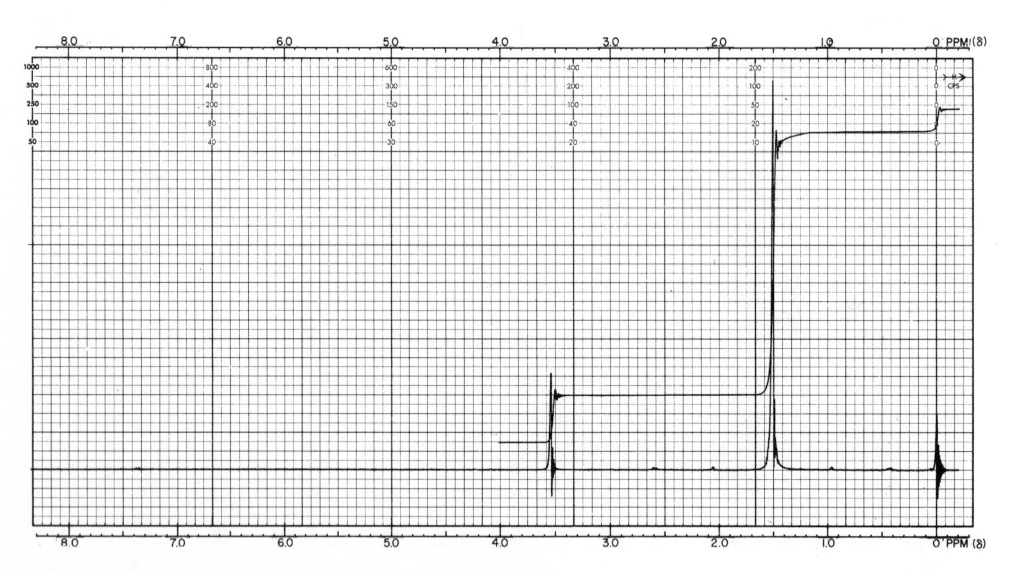

NUCLEAR MAGNETIC RESONANCE SPECTRUM:
15% in deuterochloroform, 500 Hz sweep width

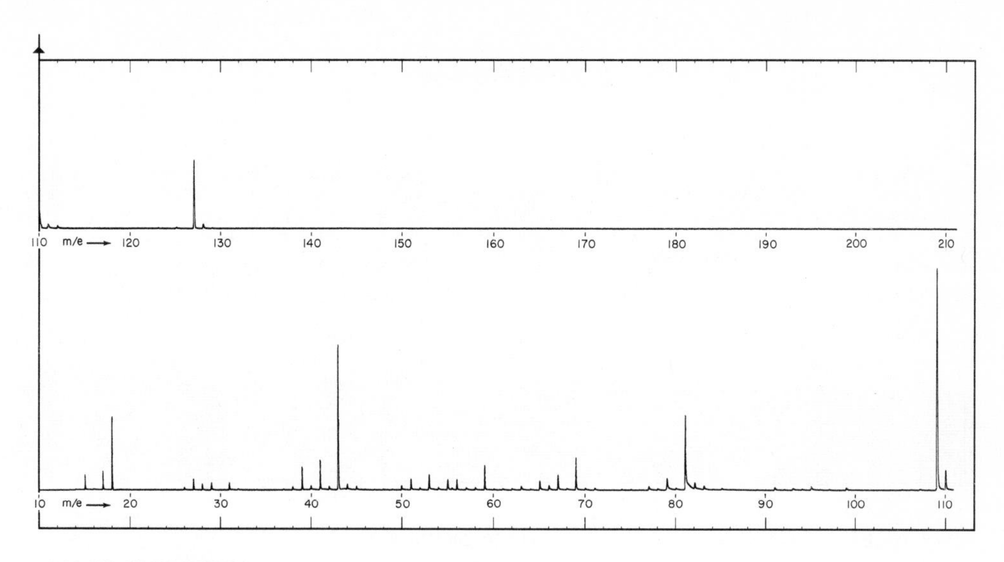

MASS SPECTRUM

MASS SPECTRAL DATA

m/e	*relative intensity*	*m/e*	*relative intensity*	*m/e*	*relative intensity*	*m/e*	*relative intensity*
15	2.7	51	3.3	68	0.7	93	0.8
17	2.3	52	0.9	69	11.7	94	0.4
18	9.7	53	5.6	70	0.8	95	1.7
26	0.7	54	1.0	71	1.0	96	0.3
27	2.8	55	3.6	74	0.4	97	0.4
28	1.8	55.5	0.9	75	0.5	99	1.3
29	2.1	56	3.6	77	1.7	107	0.5
31	1.8	57	0.9	78	0.6	109	100.0
38	0.7	58	1.0	79	4.9	110	9.7
39	6.9	59	7.7	80	1.0	111	2.6
40	1.3	61	0.5	81	29.0	112	1.6
41	8.1	62	0.8	82	3.5	123	0.6
42	1.4	63	1.8	83	1.9	125	0.9
43	51.0	64	0.4	84	0.6	127	32.0
44	2.2	65	3.4	85	0.8	128	2.8
45	1.2	66	2.1	91	1.4	129	0.2
50	1.7	67	5.6				

COMPOUND 3-19

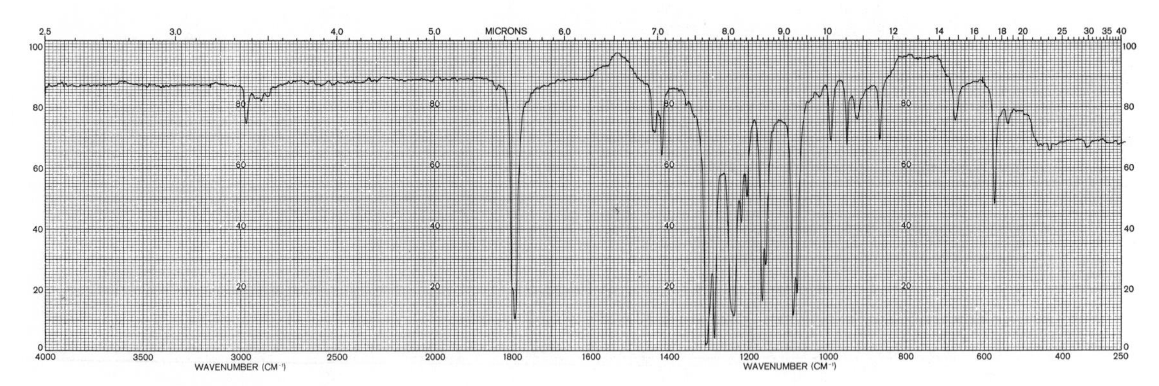

INFRARED SPECTRUM: 4.8% in carbon tetrachloride

ULTRAVIOLET DATA: no λ_{max}^{EtOH} above 205 nm

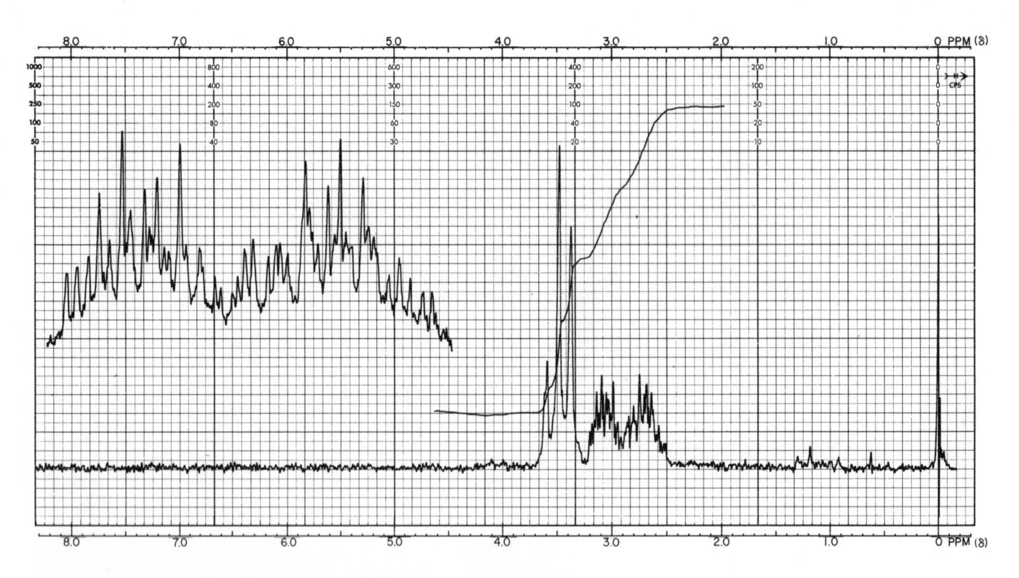

NUCLEAR MAGNETIC RESONANCE SPECTRUM:
15% in carbon tetrachloride, 500 Hz sweep width
Expansion: 100 Hz sweep width, 95 Hz sweep offset

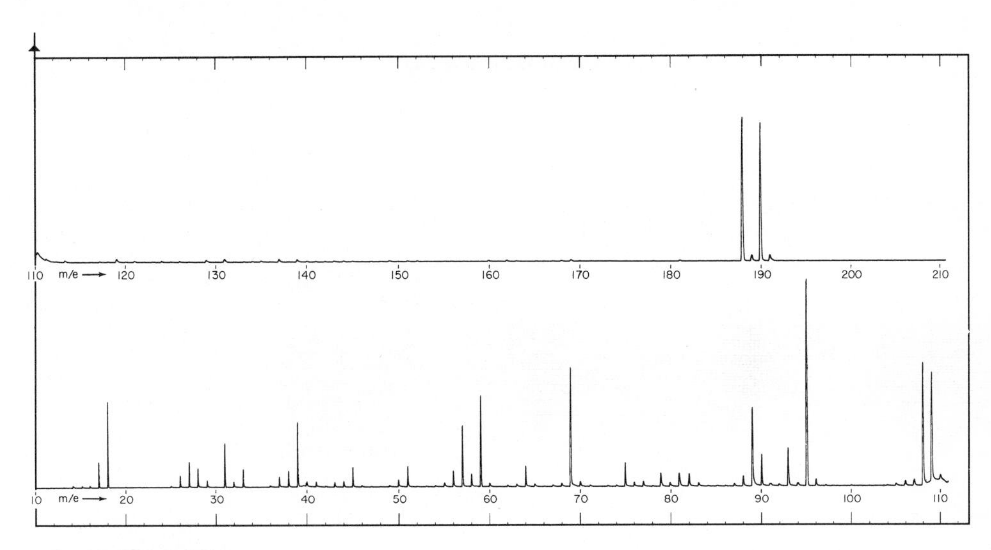

MASS SPECTRUM

MASS SPECTRAL DATA

m/e	*relative intensity*	*m/e*	*relative intensity*	*m/e*	*relative intensity*	*m/e*	*relative intensity*
14	0.7	46	0.7	79	5.5	110	5.1
17	4.7	49	0.7	80	2.0	111	1.5
18	18.6	50	2.7	81	6.0	113	1.6
25	0.7	51	6.8	82	5.6	117	0.9
26	3.7	55	1.5	83	1.7	119	1.9
27	8.5	56	6.0	87	1.1	121	1.1
28	6.4	57	23.5	88	4.5	124	0.8
29	1.8	58	4.8	89	34.0	126	1.1
31	14.6	59	34.0	90	13.4	129	1.2
32	1.8	60	1.5	91	1.4	131	2.0
33	5.6	63	0.7	92	1.2	137	1.4
36	1.0	64	7.9	93	16.1	139	1.7
37	3.4	65	1.1	94	1.8	140	0.7
38	5.3	67	0.7	95	100.0	160	0.8
39	24.4	68	1.7	96	3.5	162	0.8
40	2.5	69	46.0	105	1.5	169	0.8
41	1.9	70	2.0	106	2.9	188	74.0
42	0.7	75	9.7	107	3.9	189	3.7
43	1.6	76	1.6	108	57.0	190	73.0
44	2.1	77	2.2	109	51.0	191	3.4
45	6.3						

COMPOUND 3-20

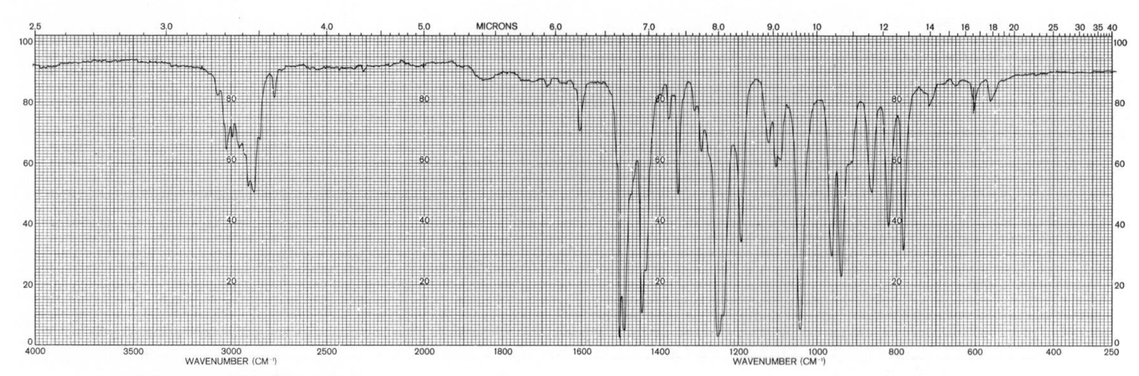

INFRARED SPECTRUM: Liquid film

ULTRAVIOLET DATA: λ_{max}^{Hexane} 264 nm (ϵ 15,500); 303 nm (ϵ 6,600).

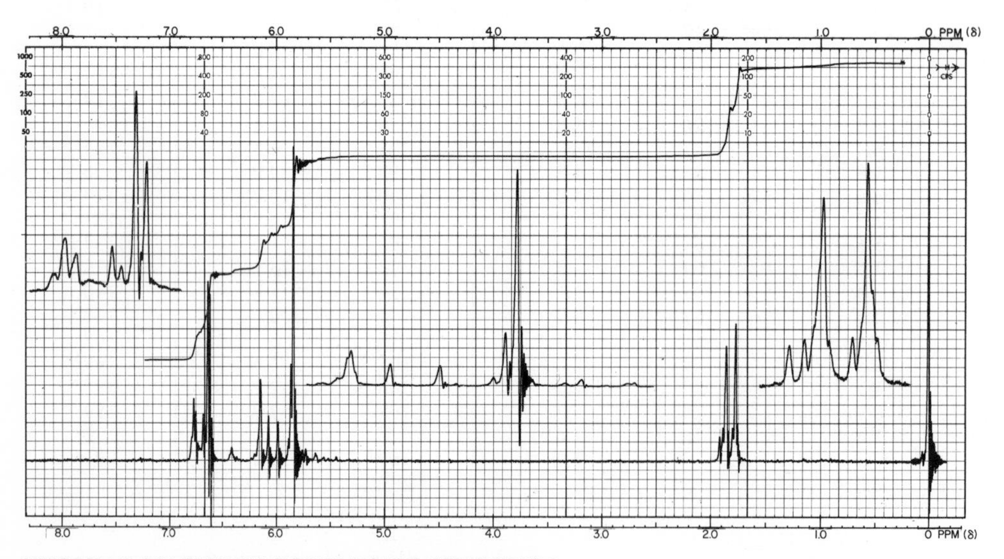

NUCLEAR MAGNETIC RESONANCE SPECTRUM:
15% in carbon tetrachloride, 500 Hz sweep width
Expansion: (left) 100 Hz sweep width, 310 Hz sweep offset
(center) 100 Hz sweep width, 305 Hz sweep offset
(right) 100 Hz sweep width, 100 Hz sweep offset

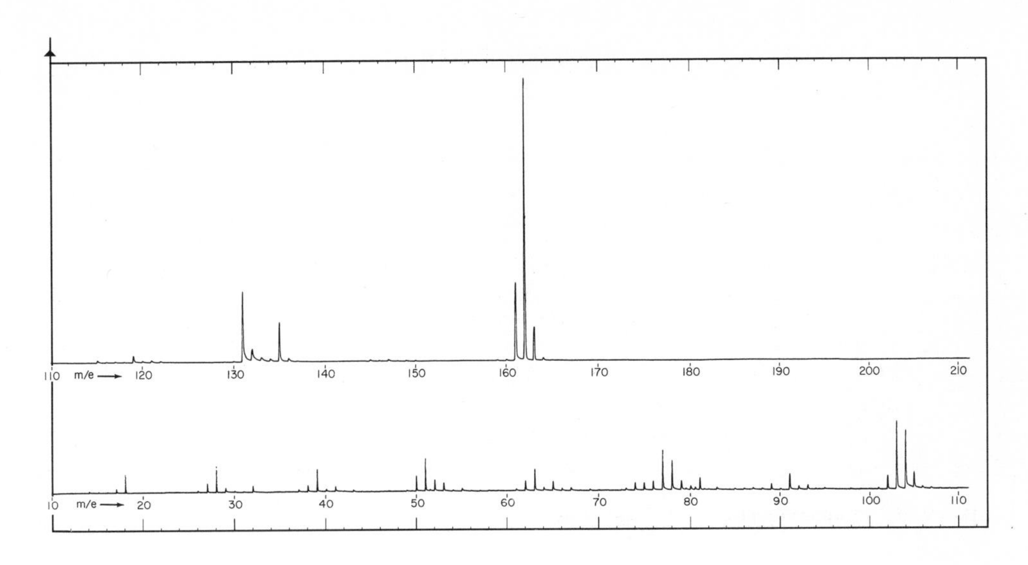

MASS SPECTRUM

MASS SPECTRAL DATA

m/e	relative intensity
17	0.6
18	2.0
27	1.4
28	1.5
29	1.1
37	0.4
38	1.1
39	4.3
40	0.6
41	1.2
43	0.6
49	0.5
50	3.6
51	7.1
52	2.6
53	2.3
54	0.5
55	0.7
61	0.7
62	2.2
63	5.0
64	1.1
65	2.3
66	0.9
67	0.8
73	0.7
74	1.7
75	2.0
76	2.6
77	10.7
78	7.7
79	2.5
80	1.2
80.5	1.0
81	3.4
87	0.7
88	0.6
89	1.9
90	0.6
91	4.3
92	1.3
93	1.3
95	0.7
101	0.7
102	3.1
103	20.0
104	17.8
105	5.3
106	0.9
107	0.7
115	1.1
116	0.4
117	0.7
118	0.4
119	2.3
120	0.7
121	0.9
122	0.7
130	0.4
131	21.0
132	4.1
133	1.5
134	1.2
135	12.9
136	1.3
145	0.6
147	1.0
149	0.6
159	0.6
160	0.6
161	25.0
162	100.0
163	11.8
164	1.0

COMPOUND 3-21

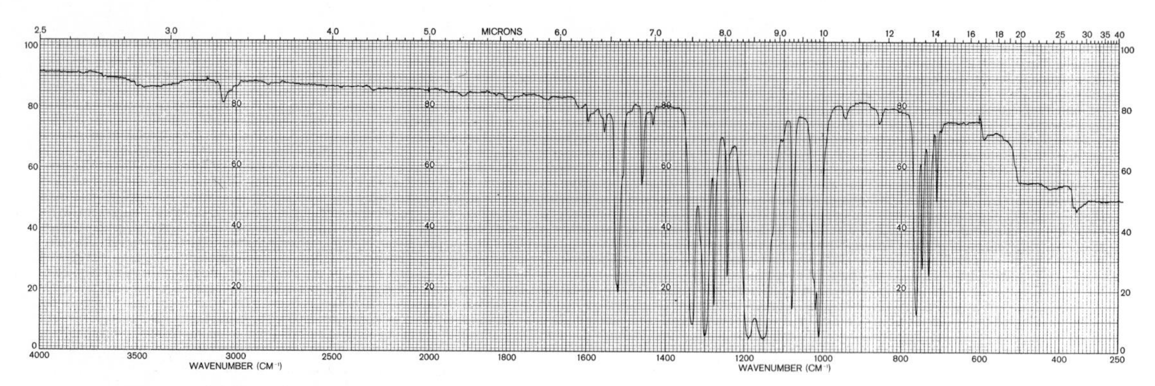

INFRARED SPECTRUM: Liquid film

ULTRAVIOLET DATA: λ_{max}^{EtOH} 218 nm (ϵ 12,600); 230 nm (ϵ 7,340); 261 nm (ϵ 5,820); 295 nm (ϵ 1,890)

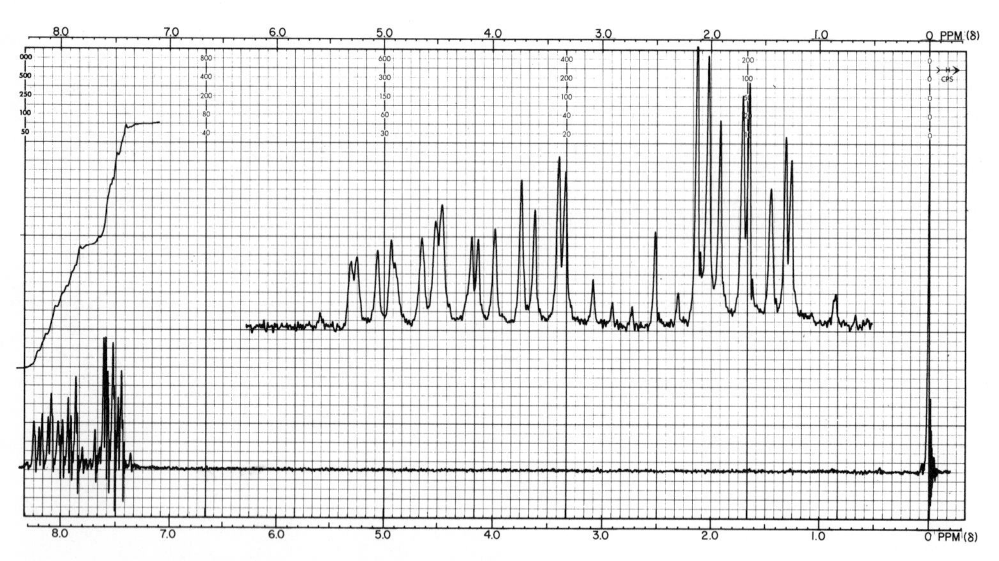

NUCLEAR MAGNETIC RESONANCE SPECTRUM:
15% in carbon tetrachloride, 500 Hz sweep width
Expansion: 100 Hz sweep width, 430 Hz sweep offset

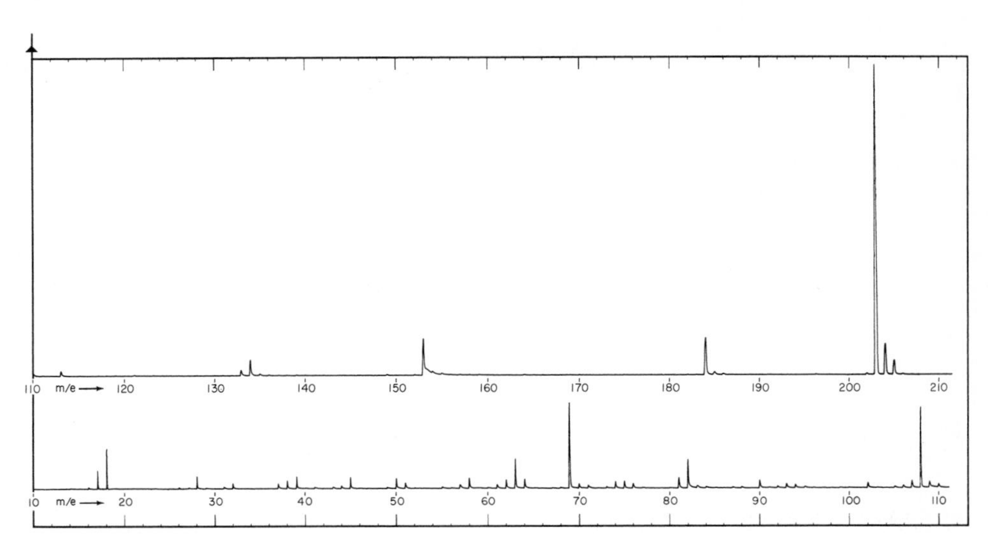

MASS SPECTRUM

MASS SPECTRAL DATA

m/e	*relative intensity*	*m/e*	*relative intensity*	*m/e*	*relative intensity*	*m/e*	*relative intensity*
17	1.2	62	2.3	90	2.4	134	5.1
18	4.2	63	7.6	93	1.7	135	0.8
28	2.1	64	2.2	94	1.3	136	0.2
37	1.2	69	22.8	102	1.6	153	10.9
38	1.7	70	1.3	105	0.5	154	1.1
39	2.9	71	0.7	106	1.1	155	0.5
44	0.8	73	0.8	107	2.3	184	11.6
45	2.6	74	1.8	108	24.2	185	1.0
50	2.7	75	2.1	109	2.3	186	0.5
51	1.0	76	1.5	110	1.0	203	100.0
57	1.3	81	2.9	113	1.9	204	10.1
58	2.4	82	7.5	133	2.1	205	5.0
61	1.2						

COMPOUND 3-22

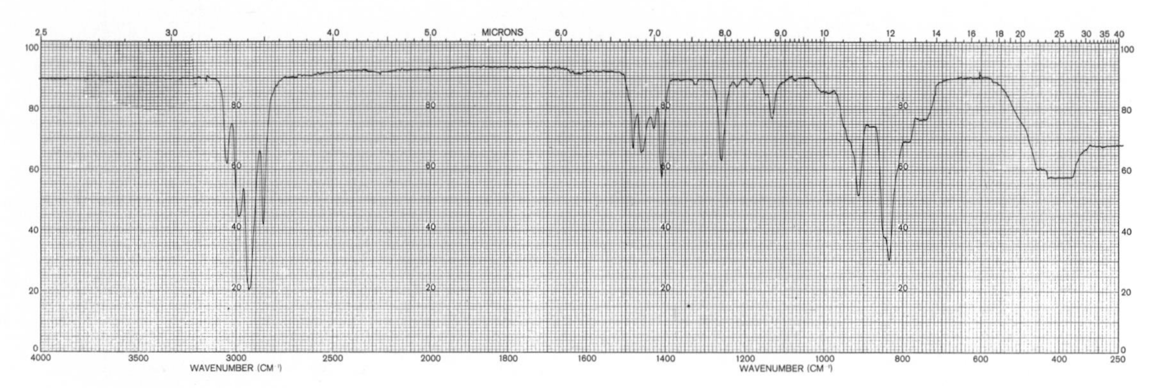

INFRARED SPECTRUM: Liquid film

ULTRAVIOLET DATA: no λ_{max}^{EtOH} above 205 nm

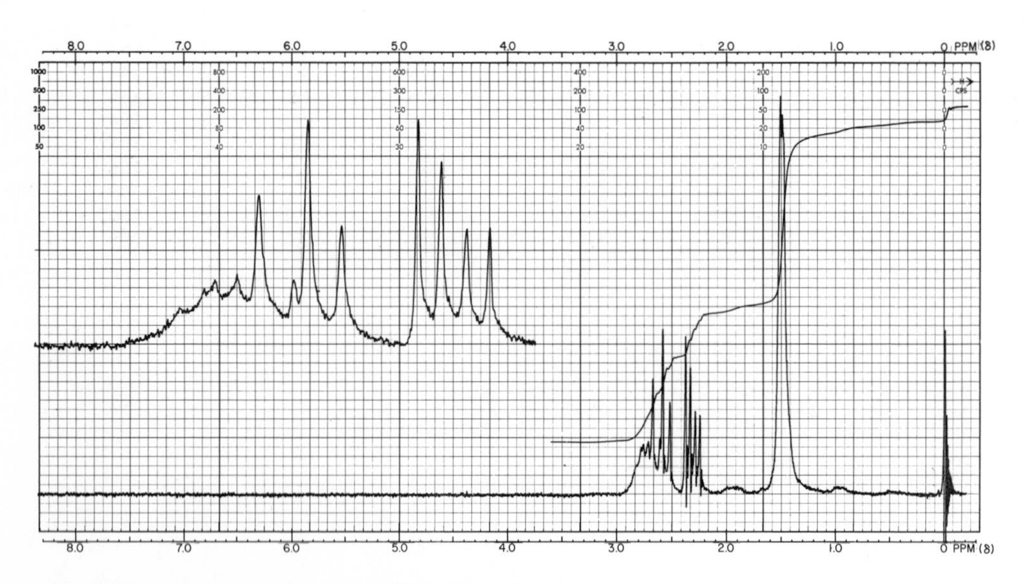

NUCLEAR MAGNETIC RESONANCE SPECTRUM:
15% in carbon tetrachloride, 500 Hz sweep width
Expansion: 100 Hz sweep width, 85 Hz sweep offset

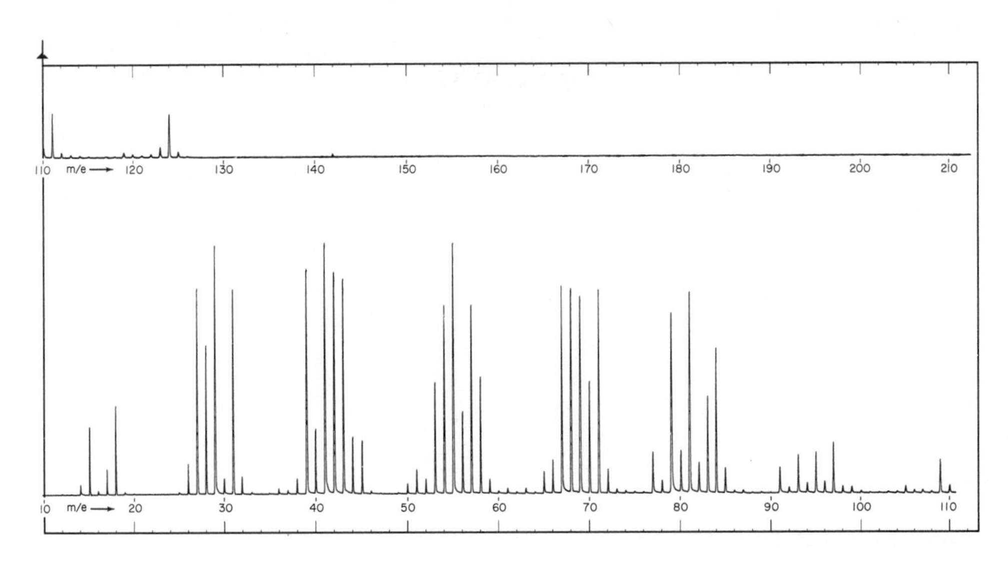

MASS SPECTRUM

MASS SPECTRAL DATA

m/e	*relative intensity*	*m/e*	*relative intensity*	*m/e*	*relative intensity*	*m/e*	*relative intensity*
14	2.4	49	0.5	74	1.0	101	0.6
15	24.6	50	3.5	75	0.9	102	0.4
16	1.5	51	7.9	76	0.6	103	0.8
17	5.0	52	5.0	77	15.0	104	0.7
18	21.4	53	39.0	78	5.0	105	2.7
19	0.7	54	61.0	79	69.0	106	1.4
25	0.7	55	91.0	80	16.2	107	2.4
26	10.1	56	27.0	81	81.0	108	1.0
27	93.0	57	65.0	82	12.4	109	17.7
28	46.0	58	40.0	83	36.0	110	2.8
29	94.0	59	4.6	84	57.0	111	11.0
30	6.8	60	1.1	85	10.3	112	1.5
31	81.0	61	2.0	86	1.4	113	0.9
32	2.4	62	0.8	87	1.3	114	0.8
33	0.6	63	2.2	89	0.5	115	0.5
36	1.1	64	0.6	91	10.1	117	0.4
37	1.0	65	7.9	92	2.3	119	0.9
38	4.7	66	12.0	93	14.5	120	1.1
39	88.0	67	82.0	94	4.1	121	0.7
40	21.1	68	75.0	95	16.5	122	0.9
41	100.0	69	71.0	96	4.7	123	2.6
42	82.0	70	40.0	97	18.9	124	15.0
43	93.0	71	80.0	98	3.1	125	2.0
44	18.2	72	7.9	99	3.1	142	1.3
45	15.2	73	1.6	100	1.3	143	0.4
46	1.0						

COMPOUND 3-23

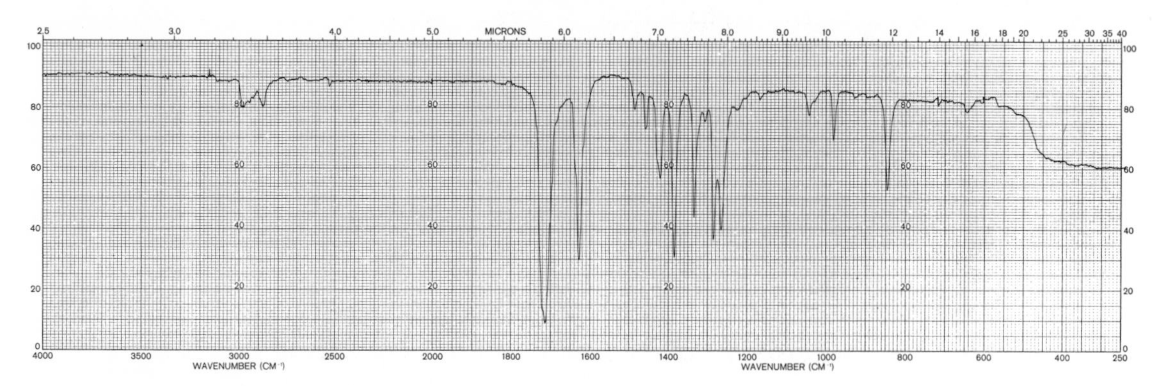

INFRARED SPECTRUM: 1.2% in carbon tetrachloride

ULTRAVIOLET DATA: λ_{max}^{EtOH} 240 nm (ϵ 12,500)

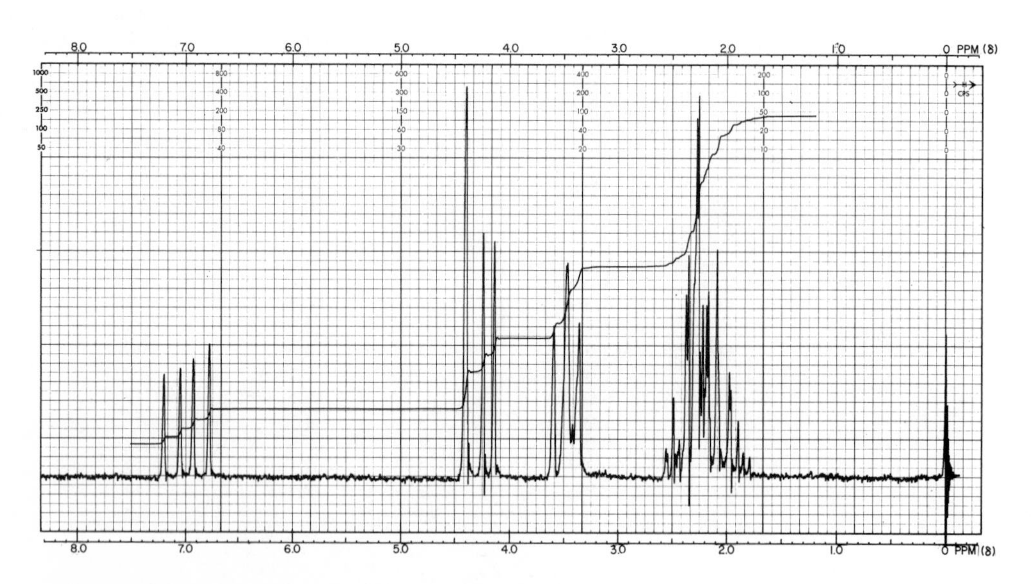

NUCLEAR MAGNETIC RESONANCE SPECTRUM:
15% in carbon tetrachloride, 500 Hz sweep width

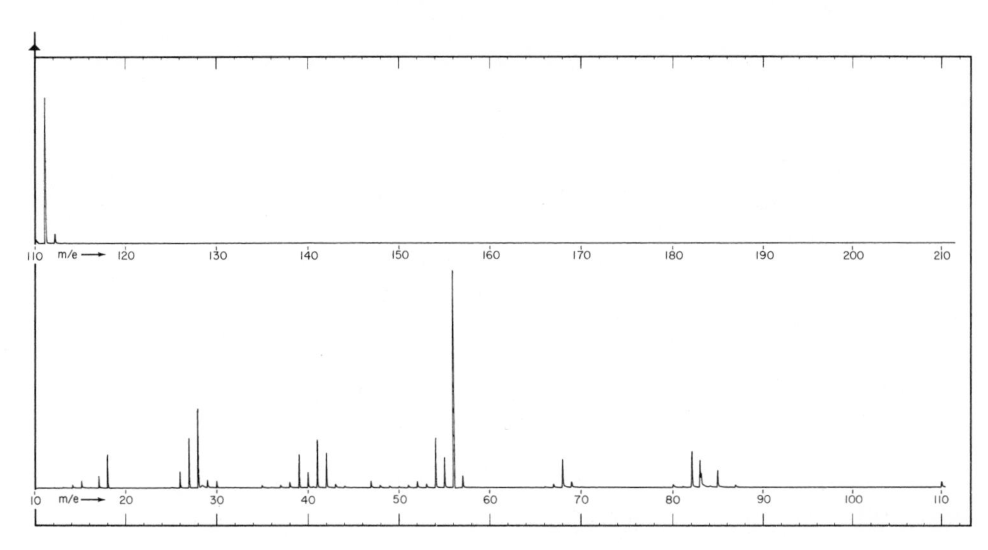

MASS SPECTRUM

MASS SPECTRAL DATA

m/e	*relative intensity*	*m/e*	*relative intensity*	*m/e*	*relative intensity*	*m/e*	*relative intensity*
14	0.8	37	0.9	51	0.7	70	0.4
15	1.4	38	1.9	52	2.5	80	1.4
16	0.2	39	10.9	53	1.6	82	13.7
17	2.3	40	4.9	54	17.4	83	9.4
18	6.4	41	15.2	55	11.4	84	0.8
25	0.3	42	11.8	56	100.0	85	6.1
26	4.6	43	1.5	57	4.4	87	1.0
27	15.2	44	0.6	58	0.3	110	2.6
28	25.6	47	1.9	66	0.3	111	70.0
29	2.6	48	0.7	67	1.6	112	5.1
30	1.9	49	0.7	68	10.0	113	0.4
35	0.8	50	0.5	69	2.4		

COMPOUND 3-24

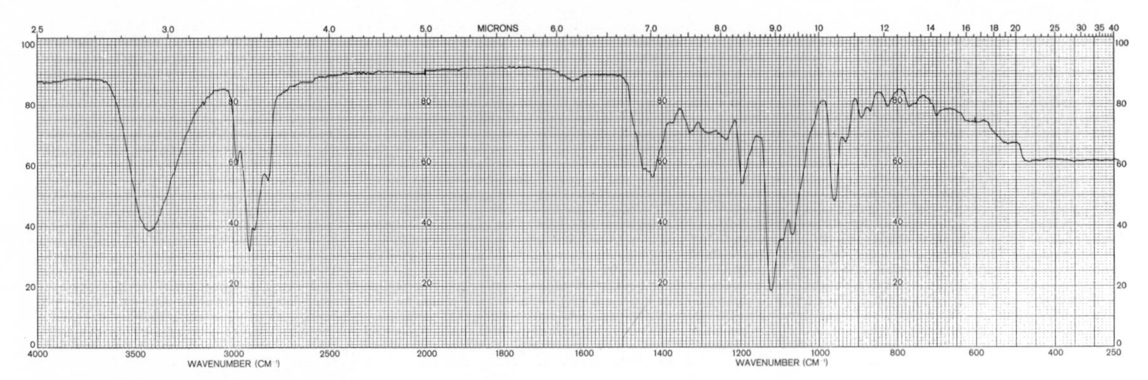

INFRARED SPECTRUM: Liquid film

ULTRAVIOLET DATA: no λ_{max} above 205 nm

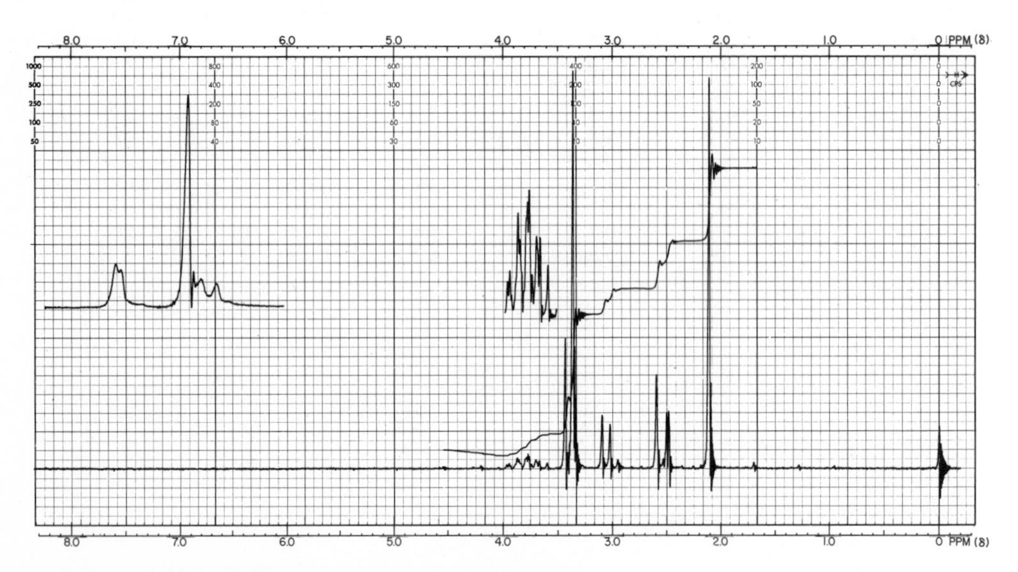

NUCLEAR MAGNETIC RESONANCE SPECTRUM:
15% in carbon tetrachloride, 500 Hz sweep width
Expansion: (left) 50 Hz sweep width, 160 Hz sweep offset
Insert: (right) amplitude × 10

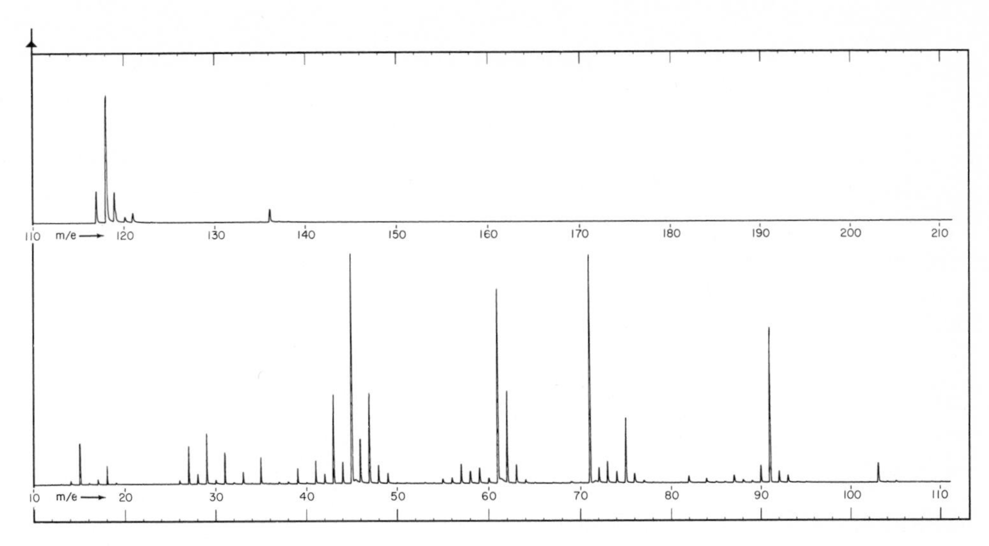

MASS SPECTRUM

MASS SPECTRAL DATA

m/e	relative intensity	m/e	relative intensity	m/e	relative intensity	m/e	relative intensity
14	0.9	40	0.5	61	78.0	87	3.2
15	12.3	41	6.7	62	32.5	88	1.7
17	1.3	42	3.3	63	7.5	89	1.0
18	4.4	43	29.9	64	1.7	90	6.8
19	0.4	44	6.8	69	0.7	91	65.0
26	1.3	45	96.0	70	0.3	92	5.0
27	10.3	46	15.6	71	100.0	93	3.3
28	3.8	47	29.9	72	6.2	103	8.2
29	17.1	48	6.0	73	8.0	104	0.7
30	1.5	49	4.1	74	4.6	105	0.9
31	8.9	50	0.4	75	24.7	117	12.1
32	0.7	51	0.3	76	3.7	118	55.0
33	3.0	53	0.5	77	1.2	119	12.5
34	0.5	55	1.7	78	0.3	120	2.8
35	6.8	56	2.3	79	0.4	121	4.1
36	0.3	57	7.3	82	2.6	136	7.0
37	0.9	58	5.3	84	1.7	137	0.6
38	0.7	59	6.5	85	0.5	138	0.4
39	5.0	60	2.4	86	0.6		

COMPOUND 3-25

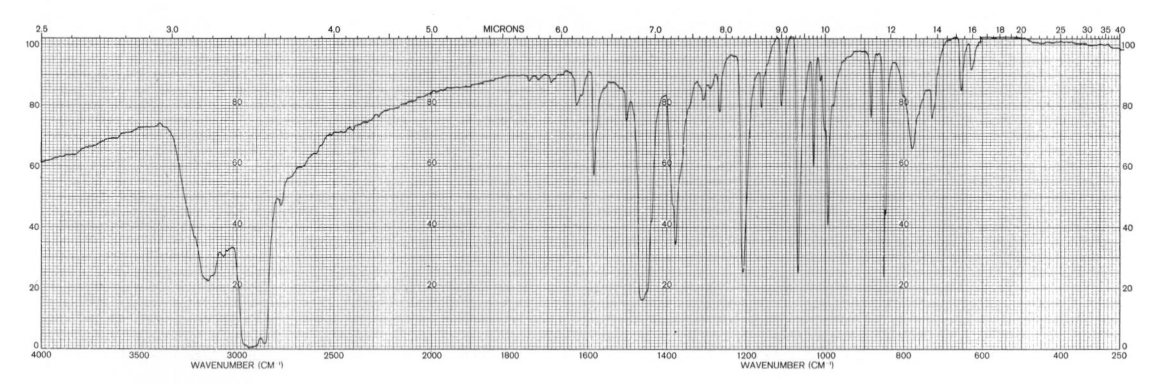

INFRARED SPECTRUM: Nujol mull

ULTRAVIOLET DATA: λ_{max}^{EtOH} 259 nm (ϵ 5,100); 264(sh) nm (ϵ 4,970) 283 nm (ϵ 4,520)

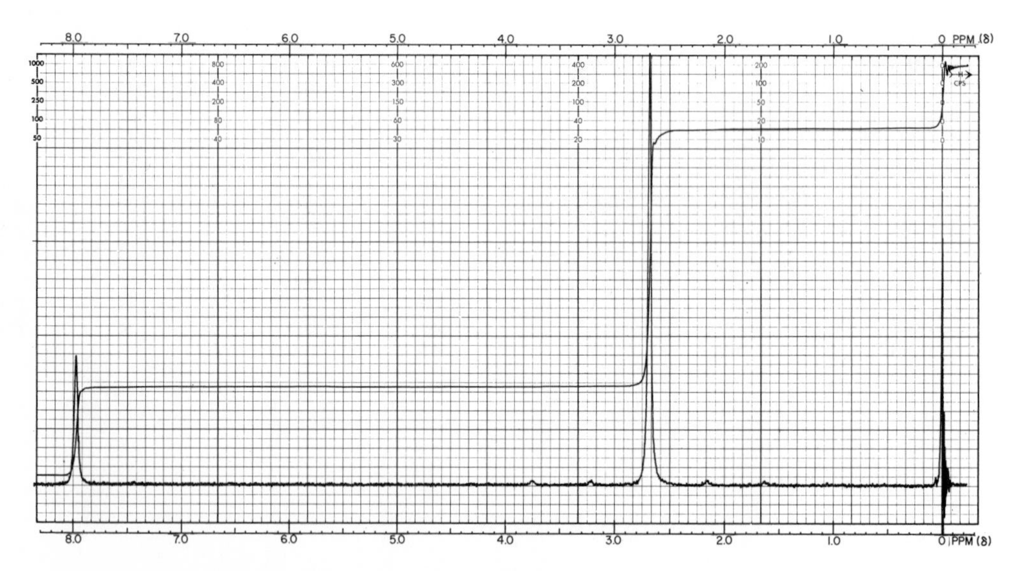

NUCLEAR MAGNETIC RESONANCE SPECTRUM:
15% in trifluoroacetic acid (solvent not shown), 500 Hz sweep width

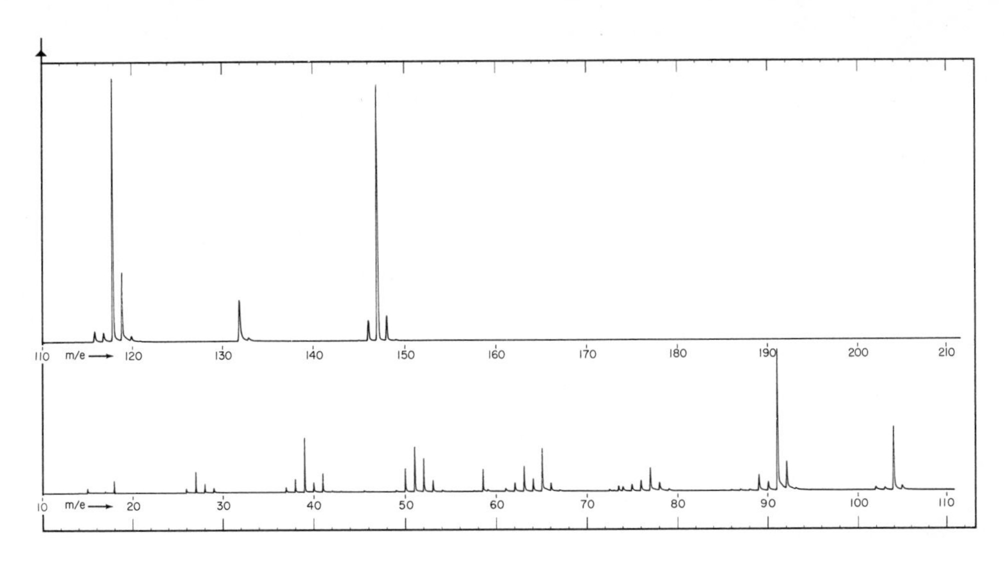

MASS SPECTRUM

MASS SPECTRAL DATA

m/e	relative intensity	m/e	relative intensity	m/e	relative intensity	m/e	relative intensity
15	1.8	52	11.5	73.5	2.2	103	1.3
18	1.1	53	4.1	74	1.9	104	21.0
26	1.7	54	1.0	75	2.6	105	1.9
27	7.3	55	0.4	76	3.8	116	4.8
28	2.3	57.5	0.6	77	7.6	117	4.2
29	1.8	58	0.6	78	3.2	118	100.0
37	1.9	58.5	7.6	79	1.1	119	28.0
38	4.7	59	1.2	86	0.6	120	2.6
39	19.7	61	1.5	87	0.8	121	0.6
40	3.2	62	3.4	88	0.8	132	17.6
41	6.4	63	8.7	89	5.3	133	1.9
42	0.7	64	4.5	90	3.1	145	0.5
43	0.5	65	14.7	91	46.0	146	7.9
45.5	0.9	66	3.2	92	9.5	147	98.0
49	1.0	67	0.9	93	1.1	148	10.3
50	8.2	72.5	1.0	102	1.4	149	0.7
51	15.5	73	0.7				

References to Answers

This section contains references in which the structural formulas for the problems in this book may be found. Where possible, references are made to handbooks, reference books, or chemical catalogs. The abbreviations for these are given below. Otherwise, reference is made to the original literature, sometimes with remarks pertinent to the preparation of the compound.

C refers to *Handbook of Chemistry and Physics*, 49th edition, 1968, Chemical Rubber Publishing Co., Cleveland, Ohio.

L refers to Lange's *Handbook of Chemistry*, 10th edition, 1967, Handbook Publishers, Inc., Sandusky, Ohio.

B refers to Beilstein.

E refers to Eastman Organic Chemicals List No. 44, 1966.

A refers to Aldrich Chemical Co., Inc., Catalog 14, 1969.

1-1 C, b647; **L**, 2406; **B**, **6**, 771; **E**, 274; **A**, 14,015-5.
1-2 C, b2510; **L**, 428; **B**, **1**, 136; **E**, 20; **A**, 12,409-5.
1-3 C, a568; **L**, 5131; **B**, **6**, 162; **E**, 3377.
1-4 C, p192; **L**, 4332; **B**, **1**, 486; **E**, 1828; **A**, 11,210-0.
1-5 C, e237; **L**, 2717; **B**, **5**, 598; **E**, 1207; **A**, B3370.
1-6 C, a925; **L**, 6180; **B**, **4**, 99; **E**, 616; **A**, 13,206-3.
1-7 C, e556; **L**, 969; **B**, **6**, 142; **E**, 1417; **A**, B7550.
1-8 C, b2926; **L**, 4357; **B**, **1**, 682; **E**, 3146.
1-9 C, a567; **L**, 5130; **B**, **6**, 161; **E**, 1900.
1-10 C, p1768; **L**, 148; **B**, **2**, 400; **E**, 5521.
1-11 C, b943; **L**, 5829; **B**, **5**, 431; **E**, 2295; **A**, T1959-3.
1-12 C, p267; **L**, 3846; **B**, **17**, 235; **E**, 6123; **A**, 14,440-2.
1-13 C, a583; **L**, 800; **B**, **9**, 441; **E**, 203; **A**, B1940.
1-14 C, b564; **L**, 2093; **B**, **6**, 844; **E**, 3044.
1-15 C, b2569; **L**, 1096; **B**, **1**, 123; **E**, 893.
1-16 C, t357; **L**, 793; **B**, **5**, 306; **E**, 910; **A**, B1790-5.
1-17 C, a1128; **L**, 2965; **B**, **12**, 159; **E**, 486; **A**, E1170.
1-18 C, s193; **L**, 2211; **B**, **2**, 609; **E**, 127; **A**, 11,240-2.
1-19 C, b2535; **L**, 1823; **B**, **1**, 120; **E**, 4283; **A**, 14,080-5.
1-20 **L**, 1163; **B**, **1**, 671; **E**, 652; **A**, I1550-5.
1-21 C, c92; **L**, 6392; **B**, **3**, 22; **E**, 939; **A**, U285-7.
1-22 **L**, 1059; **B**, **1**, 126; **E**, 613.
1-23 C, b1338; **L**, 4109; **B**, **14**, 422; **E**, 3403.
1-24 C, t749; **L**, 3309; **B**, **19**, 382; **E**, 3357; **A**, T8840.
1-25 C, f222; **L**, 3340; **B**, **18**, 272; **E**, 1366; **A**, F2050-5.
1-26 C, c339; **L**, 1501; **B**, **7**, 348; **E**, 77; **A**, C8068.
1-27 C, p1745; **L**, 208; **B**, **1**, 202; **E**, 559.
1-28 C, t202; **B**, **17**, 33; **E**, 6949; **A**, 10,847-2.
1-29 C, f119; **L**, 2173; **B**, **4**, 109; **E**, 1968.
1-30 C, b809; **L**, 476; **B**, **6**, 548; **E**, 2507.

1-31 **B, 1,** 444; **A,** 13681-6.

1-32 **B, 21, I,** 335; **A,** 11,189-9.

1-33 **C,** h289; **B, 1,** 259; **E,** 8778; **A,** D16,100.

1-34 **B, 6, II,** 559.

1-35 **C,** b679; **L,** 6467; **B, 6,** 492; **E,** 3527; **A,** 14,413-4.

1-36 **C,** p2051; **B, 22,** 46; **A,** 10,473-6.

1-37 **C,** b69; **L,** 1320; **B, 7,** 235; **E,** 563; **A,** 11,221-6.

1-38 **C,** p1159; **L,** 1825; **B, 1,** 127.

1-39 **C,** e439; **L,** 1056; **B, 6, II,** 410; **E,** 53.

1-40 **C,** p1384; **L,** 1443; **B, 2,** 249; **E,** 1409; **A,** 13,269-1.

1-41 **A,** 12,440-0.

1-42 **C,** s295; **L,** 1914; **B, 1,** 370; **E,** 310; **A,** B10,179.

1-43 **C,** b529; **E,** 7166; **A,** D4840.

1-44 **B, 5, I,** 35; **A,** 11,055-8.

1-45 **C,** p315; **L,** 408; **B, 1,** 385; **E,** 1355; **A,** P802.

2-1 **C,** b165; **L,** 4639; **B, 7,** 250; **E,** 182; **A,** N1084-5.

2-2 **C,** p167; **B, 2,** 684; **E,** 7427; **A,** D15,940.

2-3 **B, 10,** 764; **A,** B1350-0.

2-4 **C,** b2618; **L,** 2330; **B, 1,** 477; **E,** 5566; **A,** B8478.

2-5 The final product of the following sequence of reactions: 2-phenylcyclohexanone (1) sodium methoxide and benzene, (2) methyl iodide → intermediate I; (3) bromine in acetic acid → intermediate II; lithium bromide, lithium carbonate, and dimethylformamide → product. I thank Professor E. M. Burgess for this sample.

2-6 **C,** p1310; **L,** 5357; **B, 2,** 242; **E,** 1291; **A,** P5147.

2-7 **C,** a661; **L,** 1303; **B, 7,** 282; **E,** 832; **A,** C1968.

2-8 **C,** h391; **L,** 155; **B, 2,** 653; **E,** 5882; **A,** D7700.

2-9 **C,** a25; **L,** 10; **B, 1,** 608; **A,** A100.

2-10 **C,** e340; **L,** 2907; **B, 4,** 274; **E,** 1597; **A,** 11,016-7.

2-11 **C,** m12; **L,** 1510; **B, 2,** 768; **A,** C8260-4.

2-12 A reaction product of aziridine and ethyl chlorocarbonate: *Ann.*, **566**, 210 (1950).

2-13 **C,** c825; **B, 7, I,** 48; **A,** T1220-3.

2-14 **C,** c854; **B, 5,** 116; **A,** C10,920-7.

2-15 **C,** p461; **B, 1, I,** 235; **E,** 6723; **A,** 13,756-1.

2-16 **C,** t189; **B, 17,** 287; **E,** 6696; **A,** A2260.

2-17 **C,** t479; **L,** 2667; **B, 5,** 339; **E,** 103; **A,** 10,139-7.

2-18 A product isolated from the reaction mixture of *sym*-ethanetetracarboxylic acid, acetic anhydride, paraformaldehyde, and acetic acid: *J. Amer. Chem. Soc.*, **55**, 3684 (1933). I thank Dr. S. K. Gabriel for this sample.

2-19 **C,** m388; **L,** 4062; **B, 4, II,** 938; **E,** 3167; **A,** M885-1.

2-20 **C,** b1139; **L,** 905; **B, 11,** 57; **E,** 1681; **A,** 10,866-9.

2-21 **C,** p2006; **L,** 161; **B, 20,** 248; **E,** 6552; **A,** 11,005-1.

2-22 **C,** p153; **L,** 4947; **B, 1,** 164; **E,** 2396.

2-23 **B**, **6**, 183; **E**, 8793; **A**, F1320-7.

2-24 Compound II, *J. Amer. Chem. Soc.*, **73**, 3831 (1951). I thank Dr. K. C. Rice for this sample.

2-25 **B**, 2, **II**, 186; **A**, E5000.

3-1 **C**, t137; **L**, 377; **B**, **27**, 155; **E**, 5501; **A**, 12,312-9.

3-2 **B**, **1**, 648; **E**, 7846; **A**, D13,680.

3-3 **C**, b3057; **L**, 3319; **B**, **17**, 306; **E**, 2624.

3-4 **C**, g133; **L**, 90; **B**, **4**, 354; **E**, 5107; **A**, A1630.

3-5 **C**, b160; **L**, 5323; **B**, **19**, 115; **E**, 211; **A**, P4910-4.

3-6 **L**, 2879; **B**, **17**, 6; **E**, 507; **A**, E105.

3-7 **B**, **27**, 13.

3-8 **C**, b715; **L**, 4867; **B**, **5**, 378; **E**, 6147; **A**, N2835-3.

3-9 **C**, p1887; **L**, 5472; **B**, **7**, 81; **E**, 6739; **A**, P5570-8.

3-10 A reaction product of 1,1-dichloroethylene and chlorotrifluoroethylene with a trace of hydroquinone: *J. Amer. Chem. Soc.*, **81**, 2678 (1959).

3-11 **C**, h569; **L**, 1656; **B**, **17**, 559; **E**, 1624; **A**, D290.

3-12 **C**, m43; **B**, **9**, **II**, 185.

3-13 **C**, a754; **L**, 159; **B**, **4**, 387; **E**, 845; **A**, 13,522-4.

3-14 The product of the reaction of 2,3-dihydropyran and propargyl alcohol using a trace of *p*-toluenesulfonic acid: *Rec. Trav. Chim. Pays-Bas*, **86**(2), 129 (1967).

3-15 Compound III, *J. Org. Chem.*, **30**, 3308 (1965).

3-16 **C**, a893; **L**, 2772; **B**, **4**, 138; **E**, 1250; **A**, D21,745.

3-17 **C**, p1273; **L**, 1781; **B**, **7**, 769; **E**, 2197; **A**, D3345.

3-18 **C**, h587; **B**, **1**, 501.

3-19 Peninsular Chemresearch Catalog G5, p. 13, No. 01-10130-07.

3-20 **C**, i212; **L**, 5582; **B**, **19**, 35; **E**, 398.

3-21 **A**, 13,959-9.

3-22 **A**, 13,956-4.

3-23 **A**, V340-9.

3-24 An intermediate in scheme II: *J. Org. Chem.*, **32**, 2600 (1967). I thank Professor Jack Hine for this sample.

3-25 **A**, D15,000.